Boston Studies in the Philosophy and History of Science

Volume 350

The series *Boston Studies in the Philosophy and History of Science* was conceived in the broadest framework of interdisciplinary and international concerns. Natural scientists, mathematicians, social scientists and philosophers have contributed to the series, as have historians and sociologists of science, linguists, psychologists, physicians, and literary critics.

The series has been able to include works by authors from many other countries around the world.

The editors believe that the history and philosophy of science should itself be scientific, self-consciously critical, humane as well as rational, sceptical and undogmatic while also receptive to discussion of first principles. One of the aims of Boston Studies, therefore, is to develop collaboration among scientists, historians and philosophers.

Boston Studies in the Philosophy and History of Science looks into and reflects on interactions between epistemological and historical dimensions in an effort to understand the scientific enterprise from every viewpoint.

Senthil Babu D. • Roy Wagner
Editors

Normative Texts and Occupational Practices

Towards New Social Histories of Mathematics

Springer

Editors
Senthil Babu D.
Department of Social Sciences
French Institute of Pondicherry
Pondicherry, India

Roy Wagner
ETH Zurich
Zurich, Switzerland

ISSN 0068-0346 ISSN 2214-7942 (electronic)
Boston Studies in the Philosophy and History of Science
ISBN 978-3-032-28418-1 ISBN 978-3-032-28419-8 (eBook)
https://doi.org/10.1007/978-3-032-28419-8

The open access publication of this work was financed by the Swiss national Science Foundation (SNSF) Grant no. 10BP-2_238008.

This Springer imprint is published by the registered company Springer Nature Switzerland AG
The registered company address is: Gewerbestrasse 11, 6330 Cham, Switzerland

Acknowledgments We would like to thank everyone involved in the production of this book, including all contributors, anonymous reviewers, and participants of the associated workshops in Zurich and Pondicherry.

We would especially like to thank Alexandra Booz for her administrative support and Irida Altman for copy editing.

We further thank the team in the Social History of Mathematical Practices Program at the French Institute of Pondicherry, along with its librarians and staff in the administration, for their consistent support.

We thank the Series Editor Jurgen Renn and its Managing Editor, Lindy Divarci, for facilitating this publication. We also acknowledge the active cooperation of the publishing team at Springer in the making of this volume.

Some of the research for this book was supported by SNSF grant no. 182132: *History of Vernacular mathematics in medieval South India (Malayalam and Tamil, ninth–sixteenth centuries)*. https://data.snf.ch/grants/grant/182132.

Contents

Chapter 1
Introduction

Senthil Babu D. and Roy Wagner

The relationship between the categories "practice" and "knowledge" has always been value-laden. In reaction to a longstanding scholarly tradition of subordinating practice to knowledge, some have attempted to restore the credit due to practitioners for their role in knowledge making by embedding knowledge in their practice. But neither approach appears to offer a satisfactory analytic framework.

This volume, therefore, attempts to reframe this debate, at least in the context of measurement, quantification and calculation. Our strategy is to shift the discussion from the rather idealized practice-knowledge binary to somewhat more concrete objects of study: occupational practices and normative texts. We study the interaction of these two terms in various historical and geographical pre-modern and early modern contexts.

This introduction will start with a general discussion of practice vs. knowledge. We will then shift the discussion to that of occupational, practice-based knowledge vs. the knowledge asserted in normative texts that set standards for what's right and wrong and for how things should be done. Chiastically, we can view the discussion as one confronting the practices of expressing knowledge as text with knowledge asserted in occupational contexts. Building on this revised framing, the introduction will refocus the discussion on the context of mathematical practices, knowledge and texts. We will conclude the introduction by describing the contributions to this volume according to this framework and drawing some general conclusions.

Senthil Babu D. (✉)
Department of Social Sciences, French Institute of Pondicherry, Pondicherry, India
e-mail: senthil.babu@ifpindia.org

R. Wagner
ETH Zurich, Zurich, Switzerland
e-mail: roy.wagner@gess.ethz.ch

Senthil Babu D., R. Wagner (eds.), *Normative Texts and Occupational Practices*, Boston Studies in the Philosophy and History of Science 350,
https://doi.org/10.1007/978-3-032-28419-8_1

1.1 Occupational Practice vs. Knowledge

The challenge of recognizing the knowledge due to practitioners is twofold. On the one hand, we need to acknowledge that there is practical knowledge and that practitioners do not simply follow the knowledge created by someone else. Similarly, we should avoid the romantic myth that practitioners enjoy some unmediated access to nature or artifacts that is purely intuitive and thus alien to discursive knowledge. On the other hand, we should not assume that practical knowledge is the same as theoretical or scientific knowledge (to the extent that these categories make historiographical sense). Nor should we assume that the knowledge of practitioners is to be appreciated *only* because it has something in common with science or theory. In other words, the label "practical knowledge" should not be a means to project the (justified? unjustified?) prestige of science or theory onto practice, but a means to capture virtues that are often ignored when considering practitioners' work.

In western historiography, two contexts are almost invariably there, when the practice vs. knowledge question is raised. The first is the Aristotelian tradition (often mis-generalized as a universal Greek philosophical view), and the second is the historiographical debate on the relations between practitioners and the new scientific knowledge of European early modernity. This volume does not aim to focus on either of these reference points, but since these are, consciously or unconsciously, the reference points of many researchers and readers, it is better to spell them out rather than try to pretend them away. To do that, we will follow Pamela Long (2001), who explicitly links the two references.

Aristotle famously articulated the hierarchy of *techné* (which Long describes as material, technical, or craft production), *praxis* (action that requires judgment in contingent or uncertain circumstances, as in politics or war) and *episteme* (knowledge of unchanging things; see Long 2001, 2, 23, based on the *Nichomachean Ethics* 6.3–5, 1139b14–1141a8). This threefold distinction removes craftsmanship from knowledge twice, refusing to dignify it even with the status of judgment-guided action. Clearly, this was a scholarly-aristocratic view meant to elevate the philosopher above and beyond the political freeman, who was to be clearly distinguished from free practitioners who had to work for their living and those who were enslaved.

In Long's narrative (e.g., Long 2001, 250), the written expositions of craft practices in the late Middle Ages and early modernity transformed them into something akin to *praxis*. This legitimated a political valorization of crafts when used for displays of power and for large engineering projects. This new association of *techné-cum-praxis* with political authority paved the way to its integration into the emergent form of *episteme*, namely the new sciences.

This is Long's stance in a debate that is often anchored in Zilsel's thesis and Merton's work on the contribution of practitioners to the emergent early modern scientific practices and in the various objections to their claims (see Lefèvre's introduction to his 2021; for a more comprehensive, openly partisan, and therefore highly insightful, assessment, see Omodeo's 2019). The recent literature on this issue has reached a clear consensus that the practitioner/doer vs. scientist/thinker distinction is highly problematic and should be replaced by a continuum or multi-dimensional distinctions (e.g. Cormack et al. 2017, 2; Long 2011, 7; Roberts' and Schaffer's preface to Roberts et al. 2007).

There are, indeed, good reasons to reject this dichotomy. Think of a practitioner, such as a carpenter. We could imagine the carpenter as a rule follower, who learned the rules they must follow from mediators, such as written manuals and professional teachers, based on the work of knowledge makers who studied wood in a less production-oriented setting and systematized their empirical knowledge. But we could also imagine the carpenter as someone who, in their practical engagement, experiments systematically to discover new facts, comes up with new conceptual organizations of these facts (rather than simply collect them), poses and solves new professional challenges, and has to explain what they do to apprentices, colleagues and clients.

Even if we restrict our attention to the context of artisanal crafts, there is no one universal historical model of the division of labor between these forms of following, making, and transmitting rules. There is also no one conceptual system to distinguish following, making, and transmitting rules; whether a practitioner generates knowledge that is "systematic", "conceptually organized", or "new", whether they merely follow rules or also makes them up and transmit them, is often subject to debate. In a range that includes enslaved, caste-bound, hereditary, apprenticed, or schooled practitioners, trained by colleagues, family members, masters, or professors, and documented by colleagues, administrators, travelers, or intellectual authorities, in different cultural, economic, and political contexts, we are bound to come up with many different articulations.

This does not mean that there are no, or should be no, acknowledgment of a division of labor between practical knowledge and the knowledge embedded in normative accounts such as manuals, treatises, account books, or legal codes. Clearly, there are practitioners who do not teach, teachers who do not practice, accountants and administrators who organize knowledge without teaching or practicing it, and juridical or intellectual authorities who make rules that they neither teach nor follow (this complexity is attested, for example, in Lefèvre 2021). Still, the attribute "knowledge maker" need not be restricted to any one of these roles, even in those times, places, and circumstances, where they are kept more or less apart.

Moreover, sometimes conflicts may emerge between having to do something, having to describe or explain it, and having to capture it in a systematic framework of rules. Such conflicts may be due to keeping trade secrets, maintaining a unique personal style, discrepancies between the "language" of embodied motion and spoken or written language, and social hierarchies that distribute the authority to proclaim knowledge. But again, these conflicts do not mean that practitioners who do not describe, explain, or systematize have or produce no knowledge (Smith 2022 has a useful overview of the literature that attempts to articulate the knowledge of practitioners in Chap. 5 and adds her own take in Chap. 10).

Given these complexities, we would like to foreground the practitioner's knowledge by means of a historical study of practice. To us, this means foregrounding the work of the practitioner, allowing the site of work to become the focus of attention. The historical character of the work then becomes part of the study of *occupations*, that is, of the ecologies of knowledge made or transmitted at the site of work, and the social work-lives in such ecologies. And it is in this context of occupations that measurement and calculation will be studied in this volume.

With this approach in mind, a focus on the purpose of the practitioner in the context of occupations or practice becomes more prominent and challenges the historian's uncritical quest for "knowledge". Indeed, it is often practitioners' lack of orientation to some prevailing normative ideal of knowledge that makes their work perceived as less important, or as holding the potential to attain, but never actually properly attaining, that knowledge ideal. In such cases, the intermediaries who do the job of systematization take over the narrative of knowledge making, relegating practice to a fragment of the past, like a faded ruin. For us, histories of practice are hence a way to restore dignity to practitioners' work by grounding occupational orientations in their social context.

1.2 Practical Knowledge vs. Normative Text

Instead of attacking directly the practice/knowledge question, which was initially based on an untenable exclusive binary, we offer here a different articulation of the problem. We shift to asking about the relations between the knowledge embedded in occupational practices and that encoded in normative texts, such as manuals, treatises, accounts and legal codes. This rearticulates the problem, because, based on the literature cited above, we do not assume that practice is distinct from knowledge or that normative texts are necessarily associated with Aristotelian *epistemé*, early modern science, or even intellectual authority. Indeed, while the focus on the early modern practice-science nexus has led some strands of research into a conflation of textuality and science, Long's (2001) wider historical contextualization shows that the normative textual codification or representation of occupational practices need not be identified with any actor's category or analytic category that is more-or-less equivalent to "science" (many of our case studies confirm this view).

However, we also recognize that normative texts often gain a certain privilege in society, and that the processes by which such texts gather privilege in particular times and places is a matter of specific historical context. The privileged authority of the text may derive from adopting culturally privileged modes of textual expression, such as venerated grammatical, rhetorical, literary, or aesthetic codes, from embedding the practice in textualized ritual or theological settings, or from binding the practice with textual commitment to socially valued virtues. Sometimes such modes, embedding, and binding have nothing or little to do with the actual occupational practice and stem from attempts to construct scholarly privilege. In such circumstances, the latter may feed back into the practice and reshape it, potentially marking some practitioners (those who appear to adapt their practice to the demands of the texts) with the text's aura and prestige. In this context of the privilege of text, we should also note that the historian's focus on textual archives (as opposed to "archives" of practices) may contribute an additional, anachronistic privileging of normative texts at the expense of occupational practice.

In the same context, we note that we do not assume that practices and normative texts necessarily derive from the same community. The creators of the texts may be the practitioners themselves, but may also be people who interact with them or people who learn about their practice through intermediaries. The contexts of production and circulation of normative text may themselves support the privileges of texts with respect to practice; for example, the scholarly background of a writer who is not a practitioner may privilege their text over the practice, or the scholarly appearance of a text by a practitioner may mark this practitioner as an authority. It is likely that a text written by a practitioner and a text written by an outsider would include different strategies for codifying practices, but they are also likely to use (or at least try to use) some common strategies—those that the relevant culture recognizes as marks of authoritative texts.

Either way, we cannot assume that texts are or are not faithful expressions of practical knowledge. To tackle this question, it helps to have access to non-textual sources of information, and sometimes we are lucky enough to have anthropological or material evidence. All too often, however, all we have is texts, which we must consider as problematic witnesses, and interrogate them with suspicion.

Given the above complexity, the question of the resonance and tensions between practitioners' knowledge and the knowledge embedded in normative texts is precisely the focus of this volume.

1.3 Occupational Mathematical Practice vs. Mathematical Normative Text

We finally reach the specific focus of this collection: the relations between mathematical occupational practice and mathematical normative texts. Here too, however, the situation is very far from binary. Indeed, while we usually associate occupational practice with embodiment and mathematics with textual manipulations, the actual relations between practice, text, and mathematics are much more complex.

First, while some of the occupational mathematical practices are textual (such as calculating by writing numerals or consulting a written table), there are many non-textual occupational mathematical practices, involving tools to compare and manipulate quantities (such as using a compass to bisect a line or calculating on an abacus). So occupational mathematical practices include textual and non-textual components. Outside occupational contexts, such as learning general school mathematics or engaging in mathematical recreation, we also find a complex combination of textual and non-textual practices, which include reading texts, manipulating written symbols, using instruments, finger or mental arithmetic, and live discursive interactions. The question of the textual vs. non-textual is further complicated by verbal or mental repetitions of memorized written texts and by imagining the manipulation of written symbols without actually writing them down (these relations and tensions between practice and text have been a major axis of the "history of science, history of text" program, Chemla 2004).

Moving on with our problematization, the very identification of something as "occupational" or "mathematical", "practical" or "normative" is itself sometimes a bone of contention. One often distinguishes between "practical mathematics", which practitioners apply in their occupational practices, and "recreational mathematics", which may appear to be practical, but is actually never part of the practice. For example, while a land surveyor may often derive the area and the diagonal of a rectangular field from its length and width, problems that ask to derive the length and width from the area and diagonal, which appear in many treatises, are not actually encountered when surveying land. Such problems are thus often labeled "recreational". However, this distinction, which seems obvious at first, is also fraught with difficulties.

Indeed, we could imagine rare scenarios where such a problem may be used in practice (e.g. the surveyor finds the area and diagonal in an old document, but the previous page listing the side was lost). Are such rare, unlikely scenarios sufficient to make the problem "practical"? Moreover, as Høyrup (2007, 2008) argued, being able to solve such problems is often part of occupational prestige or identity building, thus contributing to one's authority as practitioner and one's occupational standing, especially if the practitioner is also a master-teacher or looking for patronage. Finally, even recreation has a practical value—it allows one to recharge, contemplate, and grow professionally. To make things worse, the very distinction between practical and non-practical mathematics is sometimes used to deny practitioners a claim to knowledge making, except when they supposedly step "outside" the scope of their occupation proper into the realms of recreation of theory (assuming they actually have access or incentive to become part of the prestigious scholarly networks in society).

The tension between occupational practice and normative texts in the contexts of measurement, calculation, and mathematics is significant because it may impinge on the authority of practitioners: the norms set in texts could limit the practitioner's freedom of design, procedure, or evaluation. The tendency to culturally identify textuality with mathematics and mathematics with universal and immutable truth often provides normative mathematical texts with a halo of prestige that the practitioner's "mere" experience-based judgment might not be able to compete with. Indeed, writing normative mathematical texts has been a means of asserting authority from the codification of measurements by Sumerian and Akkadian kings (Robson 2008, 115–123) to contemporary academic publications.

On the other hand, normative mathematical texts sometimes pretend to impose standards that are ignored in practice. Even what seems to be an obvious mathematical norm, such as the rule of three, may fail in practice. Indeed, if 1 banana cost 1 rupee, 100 bananas are likely to cost significantly less than 100 rupee, despite what we (and most agents recorded in this collection) learned in school or in their training. Measurement and calculation norms may thus acquire authority from repeating the prescriptions of fixed templates, which, in the context of mathematics, is often associated with immutable, universal truth. Ironically, when occupational practitioners embed their authority in careful repetitions of practice, they are often considered as mere followers of contingent rules, rather than holders of absolute knowledge. The tension between textual norms and occupational practices may thus lead to complex social struggles over reforming norms or over the right to interpret and implement them, ranging from the farcical to the tragic, as documented in a large array of contexts by Kula (1986), Scott (1998), and Lugli (2019).

Complementing the above tensions between mathematical practices and normative texts, recall that the working world of the practitioner is imbued with non-textual power and authority as well. This is particularly salient in studies of the social history of mathematics, which situate mathematical practices in revenue accounting, administration, surveying, school-teaching, or even the work of scholar-practitioners as they relate to the people around them. These contexts link measuring and calculation to the determination and authentication of value as occupational practices. Measure and value may then be objectified mathematically, as if they were transparent and subject to objective verification (for example, the determination of rent and revenue in land administration). This determination of value is delegated to the professional and demands compliance and rule following by all members of society. Mathematical activities are thus immersed in the social and economic lives of practitioners and of the people whose life and labor they routinely evaluate. Such practices of evaluation sometimes have to be reconstructed from available texts along with contingent historical evidence, an exercise that helps us move beyond texts and dig deeper into the realms of practice.

This leads us to consider the locus of production of normative mathematical texts. While we would expect normative mathematical texts to emerge from within occupational contexts, there are also mathematical texts that may be (or claim to be) established independently of occupational contexts, and yet impose their authority on the latter. This authority may take the form of determining what mathematical claims are true or false, how things should be measured or calculated, and how to evaluate labor or its products. To make things even more complex, we note that it is sometimes difficult to even pose the question of the independence of text from occupational practice. Indeed, a textbook written by a practitioner may be derived from a textual tradition that has little to do with the author's occupational practice as a crafts-person, but may be part of their occupational practice as master-teacher of the occupation. In some cases, such texts are written or co-created by practitioners who may not be acknowledged as engaging in normative endeavors such as teaching, accounting or administration.

Moreover, normative mathematical texts that are embedded in actual occupational practices sometimes either dissimulate this origin or take on a life of their own. Reframing knowledge that emerged from occupational practice in ways that hide its original context (a process often referred to as "abstraction") may discredit practitioners as generators of knowledge. In this context, it helps to bear in mind that the mathematical practitioner is not necessarily creating an abstract blueprint of the world they work and live in, but is rather measuring and making things along with rules that they follow and transmit. Mathematical authors sometimes follow up on these activities by making a conscious effort to distance themselves from occupational contexts in order to assert their own status and autonomy as intellectuals (for example, in the case of the emergence of classical Greek mathematics as a reaction against practical mathematics, as told by Netz (1999) and Asper (2008). Once recognized as authoritative, a mathematical text that has its roots in some occupational practices may become a canon that must be mastered by a wide community of learners, even learners who will have nothing to do with those occupational contexts (as in contemporary schools). Again, this allows the text to attain an authority that the practitioners themselves do not.

Since the range of possibilities for relating occupational practice, mathematics, and texts is so rich, it is very difficult to make wide-ranging generalizations. The purpose of this collection is, therefore, to document this range based on historical evidence. As the papers show, even this task is extremely difficult, as often the evidence is quite thin. So we now turn to describing the various contributions to this volume.

1.4 Overview of the Contributions to the Collection

In her paper, *Slavery, School Texts and Roman Accountants*, Melissa Kutner discusses the possibility of unpacking the mathematics of the enslaved people in Greco-Roman times with respect to the practice of accounting, the variety of tasks that this practice demanded from its practitioners, and the possible training they received in educational settings. She discusses educational texts, such as *P.Math*—a papyrus codex from the fourth century CE that includes a collection of documents, model contracts, metrological tables, and mathematical problems. While this organization of the codex points to a structure of lessons, Kutner demonstrates how such school texts may not anticipate the entire range of tasks that the professional accountant tackled on the job.

While learning on the job characterized the training of accountants and managers, the educational texts also carry problems that do not necessarily correspond to the job of the practitioner. This points to a culture of professional identity formation, where learning at school diverges from the practical necessities of everyday work. This is pertinent mostly to Roman Egypt, which had more free born accountants and managers; however, Kutner's analysis also extends to the enslaved accounting practitioners of the Roman West, who were trained at the master's household. Kutner discusses in detail how this should not lead us to an overly-romanticized image of the enslaved-accountant, but rather remind us how education and hierarchies were intertwined in the practice of accounting in this region. Accountants here worked within contexts of pervasive exploitation, but also received a narrower mathematical education that seems to have lacked challenging geometrical problems, with perhaps important consequences for their professional identity formation.

Kutner's essay emphasizes how school-practices and work-practices in accounting do not follow an exact correspondence. The pedagogical organization and problem posing in the texts clearly suggest that the texts played a normative role, but many of the skills that accountants and managers practiced were actually acquired on the job, rather than from these texts. Together, the school and the workplace played a role in constituting a culture of problem solving for Greco-Roman accountants.

In her paper on the *Metrology of Volume in Greco-Roman Egypt*, Julia Lougovaya draws our attention to another set of sources from the Greco-Roman context to illustrate how normative texts and contexts of occupational practices do not always coincide. She discusses the practice of calculating volumes of solids made of materials such as stone, wood and earth, turning linear units into volumetric units. The first set of normative texts that she draws upon comes from the Greco-Roman corpus of metrological and mensurational mathematical literature, such as the treatises collected under the name of Heron of Alexandria, which measure geometric solids while illustrating computational procedures. The second set of texts that she uses are Egyptian papyrus manuscripts, likely produced in occupational contexts. They include instructions given to practitioners on how to measure objects, perform metrological conversions, and prepare accounts, receipts, and contracts. These texts also include problems on the mensuration of solids.

Lougovaya provides us with a survey of practical metrology using the available information from these texts, as well as some inscriptional evidence, to emphasize the practical contexts in which objects made of stone, earth, and wood were measured and quantified. She shows that the practical conditions of working with actual materials, along with the occupational contexts in which practitioners directly engaged with these materials, affected their measurement and quantification. The volume computations of these objects were not the same for any specimen of a given geometric solid, but depended on the specific materials. Lougovaya correlates the practical circumstances of calculation, the particular units of linear and volume measures, the corresponding computational procedures for each kind of material, and the typical objects made of that material.

Using papyrological evidence, Lougovaya effectively points to how mensurational problems in Greco-Roman Egypt were not always exercises in pure solid geometry following abstract mathematical procedures. They were more like pedagogic or practical instructions for those who had to measure and quantify in particular circumstances, such as at the construction site, while excavating earth, or at the stone-mason's workshop.

Moving on to early modernity, Thomas Morel and Raffaele Danna, in their essay *Texts and practices: situating mathematics in practical arithmetic manuals. Evidence from German and Italian Cases (fourteenth–sixteenth c.)*, draw upon the traditions of practical arithmetic in early modern Italian and German vernacular mathematics and discuss the practice of reckoning in the contexts of Italian currency exchange and German mining and minting. The normative texts that are discussed in this comparative paper are the manuals of fifteenth–sixteenth century Italian *abaco* and the German *Rechenmeister* traditions, where practical arithmetic crystallized, as the authors put it, problem clusters drawn from past experience and the necessity to confront emergent needs.

The authors emphasize how it is important to situate mathematical problems in their contemporary socioeconomic dynamics and educational practices to ascertain the social conditions and the professional context in which normative texts were created and used. In this context, one can see that these texts were not mere repositories of static knowledge, but "distillates of experiences", which reflect how mathematics responded to practitioners' needs. They offered the practitioner a toolkit and developed a pedagogic culture for forming reckoning curricula in the contexts of the commercial revolution of fifteenth century Italy and the Saxon mining boom. It emerges that the technical and economic problems, along with the lack of trained officials and reckoners, fueled the circulation of people and knowledge. The authors also point out how such productive periods for practical arithmetic came about in the context of "loose" forms of institutionalization.

The essay presents in more detail two distinct cases: Dino from Pisa's practical arithmetic and the work of Adam Ries, a prominent teacher and mining official. Both clearly illustrate the nature and circulation of practical arithmetic in the two different settings that Morel and Danna compare. In particular, they analyze the problems pertaining to alloying and assaying in coinage and the arithmetic techniques that developed in these two different contexts. They argue that the mathematical attitude was not deductive, but analogical, focusing on similarities between problems.

In another early modern European case, Michael Friedman's *Winding and Warping the Yarn Arithmetically in Early-Modern Europe: Leibniz, Krüger and Practices of Measuring Threads During the seventeenth century* concerns the practice of spinning, winding, and threading yarn using various instruments in the early modern textile cottage industry. The normative text is a collection of calculation rules and tables relating measurements of raw materials to processed products, collected or edited in seventeenth century Germany by Joachim Jungius, Gottfried Wilhelm Leibniz, and Daniel Krüger.

The paper investigates the relations between yarn processing practices and the documented calculations in a context that involves various systems of measurements, different instruments, and several legal edicts on measurement and remuneration. It raises questions such as: Were the calculations formulated by practitioners or by the scholars who documented them? What was the industrial, commercial, and political context that gave rise to this attempt at arithmetical codification? Was this codification ever effectively used by people actually involved in yarn processing? What was the ideological force of codifying the calculations and exporting the work of calculation (or estimation) to paper-tools in the form of tables, especially in a context of varying and unstable standards of measure, which undermined the usability of these tables? And what kind of hierarchies of knowledge do the normative texts generate?

In a fascinating twist of the thread of the argument, Friedman weaves together the above questions concerning the normative force of numerate texts with social-normative judgments on the yarn processing industry. Following descriptions of this industry ranging from ideal feminine occupation to a locus of sexual debauchery, the normative power of text in relation to practice attains additional gender and class dimensions that are overdetermined by the context of arithmetical rationalization.

Moving on to eighteenth century Italy, Maria Giulia Lugaresi, in her paper *Between Theory and Practice: Teodoro Bonati and the teaching of hydrostatics at the University of Ferrara*, brings us to the interesting dynamics of theory and practice in the normative curricular settings of teaching university mathematics. The practice under consideration is solving problems of water flow in the territory of Ferrara and in other Italian provinces. Here, the normative texts are books that were part of the University curriculum, such as Euclid's *Elements*, Sacrobosco's *Sphaera*, the *Theorica planetarum,* and guidebooks such as *L'idea del perfetto giudice d'argine*. The lecture hall at the University became the site for the professional training of practitioners, who were expected to solve real problems in the canals and rivers of the region.

Lugaresi explains in detail the late eighteenth century context, which saw a particular Italian tradition of hydrology studies grounded in practical problem solving related to water flow. Teodoro Bonati from Ferrara was trained in this environment, which saw persistent initiatives and experiments in the managing of riverbanks, flood-control, and land-surveying for water management. Trained in this milieu, Bonati conducted studies and experiments, for example, on the confluence of rivers. He was in scientific exchange with several of his contemporary peers in Italy and was known also for the instruments he created to measure water flow.

The science of hydrodynamics combined a nascent, limited, normative-textual knowledge together with newly emergent practices that shaped two relevant figures: the mathematician as a technical consultant and the practitioner who deployed the solutions in practice. The mathematics teaching at the eighteenth century university became a site for professionalizing officials and technicians who together produced the ideal of the "hydrostatic mathematician". Lugaresi also points out that this combination of the professor in the university and the technician outside of it, while not actually solving all the problems of canal and river water flow, did contribute to the ethos for a particular way of doing science.

In Arun Ashokan's paper, *Sawing Measure: Thingking Wood Problems in Malayalam Kaṇakkatikāram and other Indian Mathematical Treatises*, the practice under study is that of wood workers who turned timber into standardized pieces of wood by cutting, measuring, and evaluating them. The normative textual corpus is a family of treatises of elementary mathematics written in Malayalam, the local language of Kerala in southwest India, between the fifteenth and nineteenth centuries. These treatises describe numbers, measurement scales, and arithmetic operations, and go on to pose and solve problems in a variety of areas, such as land measurement, gold mixing, and woodcutting. This paper analyzes some of the woodcutting problems in these treatises.

Theoretically, the paper combines a social-history framework that foregrounds the role of practitioners in the formation of mathematical knowledge and a new-materialist framework that highlights the role of nonhuman materiality in forming knowledge—a material agency meant to be captured by the neologism *thingking* in the title. After highlighting some of the unique features of wood problems in this corpus compared to problems in the Sanskrit and Tamil literature, the paper follows various networks that contribute to the formation of practical and textual aspects of wood-related mathematical knowledge. These networks include the educational setting where this knowledge is processed, recited and written, the network of production where humans, tools, and trees interact to produce physical and codified shapes, the network of wood circulation where timber is transported for trade and construction, and the ecological network where environment, trees, and humans interact to maintain reproduction.

Following these networks allows Ashokan to highlight the various tensions and relations between woodwork practices and mathematical word problems while expanding our focus beyond human-centered perspectives on labor and mathematics. The relations explored in this paper are thus not only those between practice and normative text, but also between humans and nonhumans in practical and textual contexts.

In their essay, *Evaluating Prestige Commodities in South India: Calculating with Pearls*, Muthu Prakash and Sengupta discuss the practice of calculating the value of pearls in south Indian pearl fisheries and trade in the colonial period. The normative text that they study is a Tamil text, reflecting pre-colonial sources, called the *Muttukkanakku*, or Pearl Computations, which details the stages of calculation of pearl value starting with the physical apparatus of the pearl colander, followed by grading, sorting, weighing, and the calculation of the final value called the *cevvu*, which factored into the price of the pearls.

The essay provides the historical context of pearl fisheries in the Gulf of Mannar in south India, which witnessed a new set of conflicts in the world of maritime trade. This presumably gave rise to transparent protocols and standards, which the *Muttukkanakku* organizes in the form of Tamil verses. In such markets of speculation, given the uncertainties of pearl harvest, the apparently transparent computational protocol served to conceal the debt bondage that defined the labor of the divers of the *Parava* community of fishers. Hence, the logic of the textual organization of the *Muttukkanakku* does not reflect a mathematically self-evident manner of fixing the measure and value of pearls, but demands from the historian of mathematics attention to practices of valuation in relation to the social lives and labor of the divers, about which the normative text would not speak.

Inside the textual apparatus, however, one is provided with a foray into the world of Tamil measures and the memory-based learning of Tamil numbers that served the practice of pearl evaluation. The text operated like a ready-reckoner in which the arithmetic procedure for different units of weight were laid out, pointing to the significance of measuring and calculation in an otherwise long chain of practices yielding the value of pearls, linking the seabed to the markets of the empire.

In Michel Coulon's *Magic Squares, Mathematical Tricks and Angels in Islamicate Occult Sciences: Perspectives from the Būnian Corpus*, the practice is that of producing magic squares for occult artifacts, such as Talismans. The normative texts are a corpus of treatises dealing with the creation of magic squares and their properties, which is attributed (sometimes erroneously) to the medieval Sufi master al-Būni.

The treatises expose various methods of constructing magic squares. We see in the presentations the tension inherent in attempts to formulate rules in a robust yet easy to follow manner, so that practitioners can memorize and execute them with little effort. In contrast to other Arabic mathematical texts, we see here a focus on simplicity of construction rather than mathematical systematization or justification. The texts also show the limitations of this attempt. They often rely on memorization of verses that encode the numbers to be placed in the magic squares and sometimes fail due to errors in rule formulation or in the transmission of the texts.

Another tension is that between the mathematical constraints and the need to adapt the squares to their purported occult functions, where mathematical regularities are no longer the primary consideration. In this context, calculations are designed to transform magic squares into occult names or expressions and vice versa. Here too, the normative texts provide an elaborate system, but the magic squares actually used in surviving artifacts tend to follow a few known templates rather than build on this elaborate edifice. In many cases, "magic squares" codify some occult names or phrases, but do not have the mathematical properties of a magic square. We see here a practical shift from a system of mathematical regularities to a system of mathematical-textual encoding.

The coherence of the corpus poses a further challenge. The attribution to al-Būni projects an image of consistency on the corpus, but the corpus is, in fact, a collection of rules from various different sources and authors. We see here how the normative text is meant to generate the appearance of stronger authoritative coherence beyond the actual, not always coherent practices. These attempts to combine mathematical constraints, systematization of the occult, practical constraints, and authoritative coherence generate various kinds of tensions, and may have been somewhat dissociated from the simple prevalent practice of copying common talisman templates.

In Alexis Trouillot's *Two Sciences, Two Methods for Handling Fractions: Arithmetic and Inheritance Shares in Nineteenth-Century West Africa*, the practice is that of dividing inheritance between heirs. The normative texts are a collection of juridical-religious Islamic texts, general calculations manuals, and inheritance calculation manuals from nineteenth century Muslim West Africa. The problem of dividing inheritance is famous in Islamic law for the subtle rules that mediate all sorts of confusing priority disputes between potential heirs.

While Islamicate general arithmetic manuals included a robust arithmetical system of fraction calculations, when it came to the inheritance calculation manuals surveyed in this paper, the strategy was to bypass fractions by rescaling the numbers in the problems so as to yield integer solutions. One would first figure out the least common denominator of all fractions involved and then use this as an *ersatz* for the total inheritance to be divided. This procedure and the rules for deriving the relevant numbers were designed to minimize fraction calculations. However, when these rules were to be implemented in practice, one had to bring back the integral results obtained to the actual property to be divided. In reviewing actual records of inheritance, one sees that inheritance divisions do not quite fit the prescribed calculations.

This highlights the tensions between the elegant integer-oriented system developed in the context of inheritance calculation manuals and the actual practices of dividing inheritance.

In the same cultural context, general arithmetical manuals teach the use of Indo-Arabic numerals and fractions and their calculation algorithms. In the inheritance calculation texts, however, we hardly ever see anything other than spelled out number words. This raises the question of further tensions between various cultures of calculations and various media of inscription. Did inheritance calculators calculate differently than arithmeticians? Were the number words recorded in manuscripts supported by numeral based calculations on dust-boards or other media? The little evidence found in contexts where the two mathematical sub-cultures interacted shows mutual awareness, but, at the same time, no attempt to explicitly harmonize the two approaches.

What we see here, then, is an attempt to generate normative texts that simplify arithmetical practice in a specific context, without thereby resolving some of the major tensions between actual calculation practices, specialized inheritance calculation manuals, and more general arithmetic manuals, giving rise to an interesting kind of a knowledge "trading zone" between various kinds of normative texts and calculation practices.

1.5 Some General Conclusions

The different contributions to this collection highlight various overarching observations. We will formulate five such observations and briefly mention how some of the papers in this volume illustrate them.

First, normative mathematical texts form a complex corpus that may not be as coherent as expected, or one that pretends to be more coherent than it actually is. Lougovaya's discussion of volumetric calculations in Greco-Roman general mathematical treatises and in more occupationally oriented papyri is one example, where "scholarly" solutions do not always cohere with "practical" ones. Friedman's discussion of the various regulations, measurements, and tables in the context of the early modern German yarn processing industry shows the difficulty inherent in homogenizing diverse local occupational practices into a general rationalized arithmetic. Morel and Danna too highlight the experimental and non-systematic nature of emergent new mathematical procedures in early modern commercial and mining mathematics, which lacked strong institutions. And Coulon's discussion of the Būnian and pseudo-Būnian corpus, which attempted to present texts from various sources and adhering to different mathematical norms as a coherent whole, is an example of how past "editorial" practices led later scholars to misinterpret Arabic texts discussing magic squares.

Second, practical considerations shape some aspects of normative texts, but normative texts also enact their own rationality, creating hybrids between overarching mathematical norms and specific practices. One example is Morel and Danna's discussion of the "crystallization" of an early modern mathematical approach from clusters of older problems embedded into the new professional contexts of mining and commerce, implementing an analogy-based rationality. Lugaresi's treatment of eighteenth century Italian hydrology illustrates how the interaction between well-established but insufficient existing mathematical knowledge and new experimental practices generated new intermediary scientific standards and persona. In Kutner's study of accountant education in Roman Egypt, we see the formation of a sort of accountant-mathematical stratum of knowledge that linked, through their education, many accountants, as well as some other related professionals and landowners. Trouillot's discussion of inheritance calculations in nineteenth century North Africa exposes a specialized calculation practice that mediated between a more scholarly arithmetical tradition and the actual, sometimes deviant, practices of inheritance calculations.

Third, the hybrid mathematical norms discussed in the previous paragraph may themselves have only a weak connection to the actual professional practice, so instead of bridging texts and practices, they sometimes yield yet another degree of separation. This is witnessed in Kutner's discussion of the gap between on-the-job accountant training and school texts in the Greco-Roman world. Ashokan also provides various examples of wood measurement evaluation techniques in Kerala that do not fit scholarly mathematical norms and at the same time are unlikely to have had much practical value. And Coulon's analysis shows how texts presented simplified methods for producing and encoding complex magic squares, faulty magic squares, and non-magic squares, whereas in practice only a few standard magic squares were in popular circulation. In Trouillot's paper too, we can see discrepancies between calculations in inheritance manuals and actual divisions of inheritance.

Fourth, normative mathematical texts do not only codify the mathematical aspects of professional activities, but also their social images and the images of mathematics itself, enacting potential ideological interventions. Friedman's discussion of seventeenth century German yarn industry, for example, highlights the relationship between numerate codification of occupational practices and an attempt to endow this industry with a rational aura, especially in light of its contested gender and class background. Lugaresi's analysis of Italian hydrology also emphasizes the way that its educational, theoretical, and practical settings generated new conceptions of division of scientific labor and an accompanying scientific ideology. Finally, Prakash and Sengupta's discussion of South Indian pearl fishery points out how the creation of a supposedly rational practice of pearl evaluation obfuscated the violent extraction of surplus value from the labor of divers who routinely risked their lives.

Fifth, the relations between text and practice are intertwined with the entire ecology of knowledge production that involves human practitioners as well as nonhuman material agencies shaping knowledge and practice. These issues are explicitly theorized in Ashokan's study of how wood measurement and evaluation problems in Kerala intertwined with various material and ecological contexts. In Lougovaya's paper, we can see how the different materiality of various objects and their production led to different Greco-Roman mensuration and calculation practices. Similarly, the materiality of pearls and their extraction studied by Prakash and Sengupta ended up shaping the mathematical practices of their measurement and evaluation.

We hope that these explorations and observations will help motivate further inquiry into the subtle relations between textual normativity and mathematical practices in occupational contexts. The still prevalent image of straightforward relations between mathematical practices and textual norms requires a thorough revision for us to obtain a nuanced understanding of the role of mathematics in occupational contexts.

References

Asper, Markus. 2008. The Two Cultures of Mathematics in Ancient Greece. In *The Oxford Handbook of the History of Mathematics*, ed. Eleanor Robson and Jacqueline Stedall, 107–132. Oxford: Oxford University Press.

Chemla, Karine. 2004. *History of Science, History of Text*. Dodrecht: Springer. https://doi.org/10.1007/1-4020-2321-9.

Cormack, Lesley B., Steven A. Walton, and John A. Schuster, eds. 2017. *Mathematical Practitioners and the Transformation of Natural Knowledge in Early Modern Europe*. Cham: Springer Nature.

Høyrup, Jens. 2007. The Roles of Mesopotamian Bronze Age Mathematics: Tool for State Formation and Administration – Carrier of Teachers' Professional Intellectual Autonomy. *Educational Studies in Mathematics* 66 (2): 257–271. https://doi.org/10.1007/s10649-007-9090-4.

Høyrup, Jens. 2008. What Did the Abbacus Teachers Really Do When They (Sometimes) Ended up Doing Mathematics? An Investigation of the Incentives and Norms of a Distinct Mathematical Practice. In *New Perspectives on Mathematical Practices: Essays in Philosophy and History of Mathematics*, ed. Bart Van Kerkhove, 47–75. Singapore: World Scientific.

Kula, Witold. 1986. *Measures and Men*. Princeton: Princeton University Press.

Lefèvre, Wolfgang. 2021. *Minerva Meets Vulcan: Scientific and Technological Literature—1450–1750*. Cham: Springer Nature.

Long, Pamela O. 2001. *Openness, Secrecy, Authorship: Technical Arts and the Culture of Knowledge from Antiquity to the Renaissance*. Baltimore: Johns Hopkins University Press.

Long, Pamela O. 2011. *Artisan/Practitioners and the Rise of the New Sciences, 1400-1600*. Corvallis: Oregon State University Press. https://muse.jhu.edu/pub/205/monograph/book/11174.

Lugli, Emanuele. 2019. *The Making of Measure and the Promise of Sameness*. Chicago: University of Chicago Press.

Netz, Reviel. 1999. *The Shaping of Deduction in Greek Mathematics: A Study in Cognitive History*. Cambridge: Cambridge University Press.

Omodeo, Pietro Daniel. 2019. *Political Epistemology: The Problem of Ideology in Science Studies*. Cham: Springer Nature.

Roberts, Lissa L., Simon Schaffer, and Peter Dear, eds. 2007. *The Mindful Hand: Inquiry and Invention from the Late Renaissance to Early Industrialisation*. Amsterdam: Koninklijke Nederlandse Akademie van Wetenschappen.

Robson, Eleanor. 2008. *Mathematics in Ancient Iraq: A Social History*. Princeton; Oxford: Princeton University Press.

Scott, James C. 1998. *Seeing like a State: How Certain Schemes to Improve the Human Condition Have Failed*. New Haven, CT: Yale University Press.

Smith, Pamela H. 2022. *From Lived Experience to the Written Word: Reconstructing Practical Knowledge in the Early Modern World*. Chicago, IL: University of Chicago Press. https://press.uchicago.edu/ucp/books/book/chicago/F/bo133038690.html.

Chapter 2
Slavery, School Texts, and Roman Accountants: Gaps and Contexts

Melissa Bailey Kutner

Abstract This paper addresses the gap between the educational and working contexts of accountants in the Western Roman Empire, who were often enslaved, and mathematical educational texts, which come from Roman Egypt where enslavement was less pervasive. Educational mathematical texts in Egypt range from arithmetical tables to complex geometrical problems. While not all of them necessarily reflect the education of every accountant or manager, taken together they suggest a broad and challenging mathematical education, which coincided to some extent with literary training. Evidence from the Western Roman Empire and Italy, though much sparser, suggests a more limited mathematical education with fewer connections to prestigious geometrical problems, much less to literary education. Educational contexts, meanwhile, were more coercive and exploitative, with seemingly less scope for a sense of professional identity and community among accountants. Whether these differences impacted accounting practices remains unclear due to the fragmentary nature of the evidence, but they suggest that accounts from Roman Egypt, which often stand for Roman accounting in general, may not reflect its full variety.

Keywords Accounting · Accountant · Roman · Slavery · Numeracy · Education

2.1 Introduction

The gap between mathematical texts and working contexts is, for many kinds of mathematical practice in the Roman world, and especially in the Latin West, quite large. There are exceptions: the texts of the *agrimensores*, or surveyors, especially when coupled with inscriptions, give a good idea of their methods as well as the social context and self-conception of some of them, at least for the Flavian period

M. B. Kutner (✉)
University of Maryland, Baltimore County, Baltimore, MD, USA
e-mail: mabailey@umbc.edu

Senthil Babu D., R. Wagner (eds.), *Normative Texts and Occupational Practices*, Boston Studies in the Philosophy and History of Science 350,
https://doi.org/10.1007/978-3-032-28419-8_2

onwards.[1] But Vitruvius gives only a very partial idea of the actual practices of architects, and there are no surviving manuals for accounting.[2] School texts, meanwhile, have survived only in Greek. The reasons for these gaps are manifold: the haphazard nature of the survival of texts; the fact that Roman elites had a complex and not always positive view of technical knowledge and professions; and the nature of educational contexts, which were varied and likely (especially at advanced levels) heavily depended on apprenticeship and personal experience.[3]

Contributing to this situation, although obviously not explaining it fully, is the fact that a considerable amount of mathematical work would have been done by enslaved people, especially in the Western Roman Empire and particularly in Italy. This was most true of accounting, but also partially true (more in some times and places than others) of architecture, engineering, and surveying. In textual terms, we seem to have more information the farther the writer was from slavery, and the more the profession itself diverged from slavery: Vitruvius, of course, was freeborn, as seem to have been most or all of the *agrimensores* who wrote the manuals included in the *Corpus Agrimensorum Romanorum*. This means that scholarly attention has disproportionately focused on the non-enslaved as well. We have meanwhile (as mentioned above) no surviving manuals for accounting, the profession that probably remained most tied to slavery.[4] Of course, boundaries were not absolute: enslaved people might be freed, and the working contexts of enslaved, freed, and freeborn overlapped. But the fact remains that professional life was likely in many ways quite different for enslaved workers; this may have extended to their education and training as well.

This paper explores some of these gaps. The primary focus is on the intersection between educational texts and working contexts, and particularly on probing the gap between the context from which the educational texts originate (Roman Egypt), and the training and work of accountants in the Roman West, especially Italy. I focus on accountants partly because their education and methods have received relatively little attention by comparison to those of surveyors and architects, and partly because they were more likely to be enslaved (which of course does not mean that all of them were): focusing on them thus forces us to directly confront the frequent overlap of mathematical practice and slavery in the Roman Empire. Educational texts suggest that managers, agents, and accountants in Roman Egypt might receive a broad mathematical (and, up to a point, literary) training. This education may have had important consequences for their working practices and identities. Comparing these texts with Western Roman contexts, and asking whether managers and accountants there, especially enslaved ones, would receive an

[1] On the *agrimensores* in general the literature is vast; for a start, see Dilke (1971), Chouquer and Favory (2001), Campbell (2000) for a selection of texts in English. Cuomo (2001, Chap. 6) also discusses their self-conception.

[2] Wilson Jones (2000, 33–38) on Vitruvius's text in relation to Roman architecture.

[3] Cuomo (2001, Chap. 6) on Roman elite view of mathematics; Kaster (1983) and Cribiore (2001) on education, which will be discussed much more below.

[4] Kutner (2025) discusses this issue further.

education at all similar, can sharpen our sense of differences across geography and status, even if many questions remain.

2.2 Accounting

Before discussing school texts and training, it will be useful to briefly outline what Roman accounting looked like—what Roman accountants were required to do. First, to state the obvious, Roman accounting varied depending on context. Some enterprises, particularly large estates and very wealthy households, would have used complex accounting; others would have only needed very basic methods, and those keeping the accounts might have struggled more with basic numeracy as well.[5] At its most complex, Roman accounting never involved double-entry bookkeeping or concepts such as depreciation; features such as these, whose absence was considered crucial by earlier scholars such as de Ste. Croix, have been seen as much less significant in recent decades. Current scholars have stressed instead that Roman accounting was complex enough for the needs of Roman organizations: that it was clear and well-organized and allowed for assessments of profitability (even if it is not clear whether Romans always used them for this purpose).[6]

The most complex surviving accounts come from estates in Roman Egypt. The two that have been examined most thoroughly are the third-century CE Appianus estate, in the now-classic analysis by Dominic Rathbone (1991), and the sixth-century Apion estate, in a series of books by Roberta Mazza (2001), Peter Sarris (2006), Todd Hickey (2012), and Ryan McConnell (2017) (the latter two disagree with Sarris in many respects about the overall structure and aims of the estate). Estate managers compiled day-by-day lists as well as monthly and/or yearly accounts organized by ingoing and outgoing payments of money (many of them in credit) and produce (grain, bread, wine, beans and the like). Individual employees of the Appianus estate also had their own accounts. Rathbone argued that the main emphasis of the accounts was on controlling the costs of production, but that they could have been used by the central administration to calculate profitability.[7] The Appianus accounts are not very demanding numerically. Many numbers are round, and there are only a few fractions. The Apion accounts, partly due to the way recording money (in fractions of the gold *solidus*) had changed, partly due to the intersection of rent collection and taxation, are more demanding and involve many calculations with unit fractions. Accounts for both estates, however, are

---

[5] See for example LaGroue (2014, 61–62) on the Theophanes account, an account based on a journey in the fourth century CE.

[6] LaGroue (2014, 10–37) discusses these issues in detail; Rathbone (1991) makes an argument for the complexity of Roman accounting based on the documents of the Appianus estate.

[7] Rathbone (1991, 333–359) on organization of accounts; he makes it clear that the "monthly" accounts were actually compiled all together, probably at the end of the year; on profitability, see Rathbone (1991, 369–377).

organizationally complex, requiring managers to track and categorize large amounts of information and synthesize it into different types of accounts. Rathbone (1991, 349–350) is even able to trace some managers making changes in how they classify information, improving their system over time.

In the Western Roman Empire, there are no surviving examples equivalent to the Appianus or Apion accounts. However, literary evidence, examined in detail by Gérard Minaud (2005), suggests that accounts for large estates or households probably reached similar levels of complexity; at the very least, references to day books that were then synthesized into final accounts are frequent. But specific differences between accounts from different geographic areas, or between Greek and Latin, remain opaque. The largest surviving group of actual accounts are those in the Vindolanda tablets, but it is likely that they are preliminary records.[8] There is thus no single or easy answer to the question of what skills a Roman accountant would have needed. Basic literacy and numeracy would have sufficed for the simplest accounts, but for large estates, taxation, imperial administrators (many of whom would have been enslaved or freed by the imperial family), and similar contexts, skills such as those visible in the Appianus and Apion accounts were likely needed. Some of these skills were probably taught at various levels of schooling, but others—especially specific ways of organizing and synthesizing information—were more likely to be learned on the job, perhaps by beginning at low levels of a hierarchy. The question of what students would have learned early on or in a school setting, vs. the kinds of knowledge absent from these settings and how it, in turn, was acquired, will recur frequently below.

The remainder of this paper juxtaposes school texts with educational and occupational settings for managers/accountants, first in Roman Egypt and then in the Western Empire (particularly Italy, the origin of most of the literary evidence). Even in Egypt, the relationship between educational texts and the work accountants did is not entirely straightforward. Unsurprisingly, the issue only intensifies when attention is shifted to the West. Nevertheless, some contrasts emerge, which shed light on the differing experiences and professional identities of practitioners in each place.

2.3 Accountants and Educational Texts in Roman Egypt

The process of numerical education remains somewhat murky even for Roman Egypt, where at least primary documents on papyri, tablets, and ostraca survive. For Roman education in general, debate exists both about levels (what was taught when, and whether truly distinct levels can be discerned) and whether (and again, when) separate tracks and/or schools existed for people of different statuses and professions.[9] Analyzing primary documents, which contain lists, exercises, tables, and

[8] LaGroue (2014, 62–69) gives a good basic overview of their style.

[9] See Bernard, Proust, and Ross (2014, 44–46) for a recent discussion.

problems, Raffaella Cribiore (2001, 181–183) has pointed out that students seem to have begun to write lists of numbers *after* learning the alphabet, when their handwriting was less awkward. Operations like addition show even better handwriting and were probably taught even later. However, numerical education at this (still elementary) stage may have stressed both memorization and use of the abacus over writing operations down. For most students, basic multiplication and division (use of fractions) might have continued to rely on memorization. After this level, Cribiore relegates most numerical training to specialized contexts in particular schools, largely separate from the teaching of grammarians or other more advanced teachers of literary material. Cribiore assumes this training was necessary for scribes (at least some of whom would keep accounts) and took place in "scribal" schools. Texts and notebooks with multiplication tables, fractions, and metrological conversions generally use high quality scripts, showing practiced writers.

Certain types of exercises on primary documents do seem to show a specifically scribal education. Cribiore (1996, 28–29) calls these "scribal trials," and many have been collected and examined in detail in a recent dissertation by Michael Freeman (2023, Chap. 3). The exercises, which might be literary or administrative (such as model documents, formulas, and so on), seem to emphasize speed of writing, repetition, and memorization of formulas rather than, say, composition or analysis. Where *numerical* education specifically is concerned, however, the notion of scribal schools or separate tracks runs into some difficulties, especially when attempting to discern what type of numerical training would have taken place when—for example, at an earlier (and more general) level of school rather than specifically as part of scribal training.[10] Some of the metrological tables cited by Cribiore as belonging to scribal training, for example, do not seem to have the high-quality scripts she associates with such schools.[11] Freeman (2023, 109–110) notes the common presence of mathematical exercises in "rudimentary hands" and school notebooks, and suggests most numerical education took place in schools *before* students began specialized scribal training.[12] Giuseppina Azzarello (2019a, 245), meanwhile, considers it straightforwardly evident that "that literary and arithmetical curricula were taught in schools at the same time."[13]

Among recent broad overviews of ancient mathematical education, Bernard, Proust, and Ross (2014, 46–47) explicitly associate more complex numerical tables and problems with enslaved people: "Because mastering such skills seemed alien to the spirit and contents of 'secondary' teaching, this aspect of education might be

[10] Even for scribal education more broadly, Cribiore (1996, 28) notes that "the modalities and timing of a specialized scribal education are unclear."

[11] *P.Ryl.* II 64, for example, a short metrological list that according to the editors is written in an "uncultivated hand" with "unusually bad spelling" (Johnson et al. 1915, 4) and *P.Ryl.* III 538, a list of monetary conversions in a "large and unpracticed hand" (Roberts 1938, 180). Both are cited in Cribiore (2001, 183, n. 84). Elsewhere (2001, 40) Cribiore associates metrological tables with elementary schooling instead.

[12] Like Cribiore, however, Freeman spends very little time on numerical material specifically.

[13] Azzarello does not discuss earlier versus advanced levels, however.

plausibly interpreted as the production of advanced slaves trained to be *notarii* (professional scribes) or *calculatores* (accountants) who were completing a professional training."[14] However, many managers and accountants in Egypt were not enslaved, and so while this training *may* have been distinct from elite education it was not limited to the enslaved. Nathan Sidoli (2015), who uses and cites more primary texts than Bernard, Proust, and Ross, gives a somewhat different picture, which stresses overlap between literary and numerical education at least at the primary and secondary levels, but argues that the most advanced "practical" mathematics would have been for professional scribes or architects and surveyors. At this advanced level, there is a vagueness both in terminology (whether titles like *calculator* or *geometer* applied to professionals, teachers or both) and educational settings (a school or an apprenticeship model; what Sidoli (2015, 389–392) terms "studying with a practitioner").[15]

As even the above brief discussion shows, although a general progression can be seen in the mathematical texts, there is considerable ambiguity about the exact context and use of many of them—especially whether, or at what stage, to consider particular texts specialized or relegated to specific groups. The issue is not helped by the general vagueness of the term "scribe," which is often used to include people doing numerical work (like accountants), but of course also applied to a much broader group of people of varied status; or by the scant attention given to mathematical texts specifically in broader overviews of ancient education or scribal training.[16] But the questions of what numerical training happened when, and how much of it overlapped with more general schooling, potentially have large consequences for the self-conception and working contexts of accountants and related professionals. After discussing the various types of mathematical texts more directly, I will return to these broader questions of levels, boundaries, and status.

Several types of texts are typically discussed as belonging to numerical education.[17] At the most basic or elementary level (though, as mentioned above, still seemingly after students had learned to write), these might include lists of numbers alongside lists of words. Then there were tables of addition (rare), multiplication,

---

[14]The "secondary" teaching that Bernard, Proust, and Ross (2014) refer to was undertaken by wealthy or elite students, which they argue would focus (to the extent that it was mathematical at all, rather than literary and rhetorical) on geometry and theoretical mathematics more than "practical" problems.

[15]The texts that Sidoli cites as more specialized are long collections of problems such as *P.Cair. Cat.* 10,758 and *MPER* 1.1, which will be discussed further below.

[16]Assistants, or *boethoi*, is the word preferred by Bagnall (2010, 25) for "estate managers, business agents, tax collectors, especially those who did the actual work for which liturgists were responsible, and others of this kind." This is likely a better term than "scribe" for most of those who would have done accounting and related work—see also Ast (2015), cited further below, who includes town councilors and liturgists—though it does not of course encompass everyone who would have received advanced mathematical training.

[17]An overview and substantial sample of numerical texts are given by Fowler (1999, 222–276). Many examples also appear in *MPER NS* XV. The recently published *P.Math* (Bagnall and Jones 2019) will also be referenced extensively. Additional examples are cited below.

and division (parts, or fractions), which were sometimes memorized, sometimes used as references, and sometimes made by students as exercises.[18] There were also lists of metrological and monetary conversions, and finally, collections of problems. These problems were practical insofar as they involved real-world objects and specific numbers, and they were part of a long tradition that contrasted with that of geometry or theoretical mathematics.[19] I now move to discussing the types of texts in turn and how they may have related to the training of an accountant/steward. I will then return to questions of specialization and status.

2.4 Tables

Arguably the most common type of mathematical text found is the table. Even these are not uniformly present across centuries: Azzarello, in a recent overview of the genre, mentions some 150 published examples, of which fully half date to the fourth to seventh centuries CE: tables are less common in the earlier Roman period and even less in the Ptolemaic. If this pattern is not due to chance, Azzarello (2021) suggests that the reforms of Diocletian may have made conversions more necessary.[20] Changes in the monetary system (especially the use of the gold solidus, which tended to involve extensive use of fractions due to the lack of intermediate denominations between gold and bronze and the instability of the latter) may have had the same effect. Among the tables, those involving addition are rare (perhaps because it was memorized, or performed on an abacus or with fingers), those involving multiplication are more common, and those involving division (or fractions) are most common of all, perhaps because it was most difficult both to perform and to remember.[21]

As mentioned briefly earlier, although these documents are usually referred to as tables, "list" is probably a better term. Though their format has some variations (on which again see Azzarello (2021) for details), they always have two elements, essentially an operation and an answer, lined up in a tabular way. Sometimes, but certainly not always, horizontal separation lines are used between calculations or

[18] See Lougovaya (2022a), for an example of the last. For tables as references or ready reckoners, see Cribiore (2001, 182); for some examples, see Sampson (2022) and Azzarello (2019b), who discusses in detail the features that suggest the table was copied from a model, either as an exercise by a student or a (careless) personal reference by an accountant (69–71).

[19] See Roby (2023, 155–159) for a recent overview (and comparison of the mathematical papyri with the Heronian corpus, a comparison to which I will return); Høyrup (1990) for a more wide-ranging discussion of "sub-scientific mathematics."

[20] Azzarello (2021) also mentions that it is exclusively later texts that appear on wood and wax tablets, and that sometimes pages of these belong to the same notebook but are published under different numbers, inflating the numbers overall. See also Cuomo (2001, 213–214), on mathematical texts in general being more common in this period; she also suggests a possible link with reforms of Diocletian.

[21] See Azzarello (2021) again for these proportions.

sections. The division tables are tables of parts and always use unit fractions. For example:

1 Twelfths
2 The 1/12 (of 6000) = 500
3 Of 1 the 1/12 (= 1/12)
4 Of 2 = 1/6
5 Of 3 = 1/4
6 Of 4 = 1/3
7 Of 5 = 1/3 1/12...[22]

Series of these additions, multiplications, or divisions will then repeat across multiple columns of a papyrus sheet or tablet. One would read or calculate by moving down one document column, then to the top of the next one and down it, and so on. Thus, the actual "table" always consists of only lists of operations, which continue across multiple *document* columns. This is worth stressing since more extensive tables often work quite differently, condensing information in new ways and allowing for forms of cross reference that these lists do not. Andrew Riggsby's (2019) discussion is valuable here, as well as the example for multiplication given by Gierdien (2009, cited by Riggsby), who compares the list format (still used in most primary education to teach multiplication) to a true table that would allow for the examination of more complex relationships and patterns. Gierdien's example is partially reproduced in Table 2.1 (to 5 instead of 12)[23]:

Eleanor Robson's discussions of Mesopotamian tables are also useful in comparison: she contrasts the complex tables used in some administrative contexts (including accounts) with the list-like format used to teach mathematical operations and expresses some surprise at the contrast: "Oddly enough, we find a strong disinclination towards the truly tabular format in later school arithmetic, where one might expect to find tables galore."[24] Part of the reason, she suggests, is that such

Table 2.1 Multiplication Table as in Gierdien (2009)

X	**1**	**2**	**3**	**4**	**5**
1	1	2	3	4	5
2	2	4	6	8	10
3	3	6	9	12	15
4	4	8	12	16	20
5	5	10	15	20	25

[22] *P.Oxy*. XXXIII 2656, as edited and translated by Azzarello (2019a, 247), but modified by me to make it clear that in the Greek "1/12" is not repeated in every line, but only in the first three.

[23] Riggsby (2019, 42–50, and 62 on mathematical lists specifically): "there are no matrices here, much less ones that contribute to meaning." Gierdien (2009, 17) for the full table.

[24] Robson (2003, 30). See also Eddy (2013), who does not really differentiate between table-like lists and true tables in this way, but does discuss the skills needed to make the latter and how the learning of such skills is visible in eighteenth-century Scottish school notebooks.

lists were geared towards repetition and memorization; this emphasis on memorization is also highlighted by Gierdien in contemporary teaching of multiplication.

The Roman/Byzantine tables, similarly list-like and repetitive, were thus not presenting information in ways that allowed for the novel insights or relationships afforded by true tables. Part of the purpose of these list-like tables was probably to foster memorization, and the important role of memorization is suggested by literary references as well, such as Augustine's oft-cited description of children reciting additions in monotonous chants.[25] Even some division lists were probably meant to be memorized. Others may have been kept to serve as references, though the evidence is ambiguous, with scholars often classifying lists as likely being references based only on the fluency of the handwriting or the lack of errors.[26] In still other cases, repetition of the *operations* was part of the point, at least for the more difficult divisions. These show signs of being, in Lougovaya's words, "computational workouts," where students worked through particularly challenging examples as practice or training (Lougovaya 2022a). The "computational workouts" are often more difficult than the kinds of divisions that an actual practicing accountant or scribe would need to work with on a daily basis, and in this respect they have much in common with the problems involving divisions discussed in the next section.

Metrological texts, finally, are also lists, more obviously than the texts discussed above, since they often give relationships as successive sentences without even superficially organizing them into columns (some metrological texts, however, do arrange the same information in a more table-like way, such as *P.Ryl.* II 64). The presentation of the metrological texts in *P.Math.* makes the contrast particularly striking, where on pages 18–19 Bagnall and Jones (2019) condense the relationships expressed in the different metrological texts into actual tables with as many as ten columns and rows. A typical text (e3 in Bagnall and Jones), meanwhile, reads as follows:

1 The surveyor's schoinion [contains] [8] ogdoa, the ogdoon
2 contains 12 cubits, so that the surveyor's schoinion comprises
3 96 cubits, while the linear (schoinion)
4 comprises 100 cubits. The linear cubit is the one…

And so on. As with many of the arithmetical tables, the point here seems to be repetition and memorization. The lists could also have served as references, although they again would not have been as easy to consult as a modern table. Finally, metrological texts heavily resemble one another, and in some cases are identical; such texts seem to have remained standard for centuries (Bagnall and Jones 2019, 16–17). This means that while they included necessary conversions, they obviously did not address actual physical variations that might exist between local measures. The point was instead to internalize, or be able to easily retrieve, the basic conversions

[25] *Conf.* 1.13, cited by Cribiore (2001, 181).

[26] For example, Cribiore (2001, 181); but see Lougovaya (2022a, 663) on the unlikelihood of many tables being ready-reckoners in work contexts (since they often did not reflect divisions commonly used in taxation, etc.).

and equivalences that persisted across time, and to use at least some of these in mathematical problems.

These table-like lists would have shaped future scribes and accountants in several ways. Most obviously and basically, students would have learned to write numbers fluently and memorized or learned how to compute numerical operations, an obvious necessity for compiling accounts, especially the more lengthy and complex monthly and annual summaries that managers for larger estates were clearly required to complete (Rathbone 1991). In some cases, especially with the division tables, students would have even gained numerical skills beyond those strictly required to make accounts. But the lists also more indirectly trained students to think in certain spatial ways and perhaps even to flexibly approach the organization and cross-referencing of information.[27]

Spatially, the "tables" heavily resemble the way that well-organized accounts (especially for large estates) are laid out. These accounts tend to be written as careful two-element lists continuing up and down papyrus columns. An item, person, or task would be listed on the left, and a monetary amount or measurement on the right, in an evenly-spaced way.[28] The resemblance between this layout and that of the mathematical tables is considerably closer than that between accounts and mathematical texts in some Mesopotamian contexts, for example: as referenced above, Eleanor Robson (2003) has shown that complex tables were used in accounting and administrative contexts, but not in mathematical training, in at least some times and places in Mesopotamia. An accountant in Roman Egypt would have been able to easily and seamlessly transfer the spatial organization of one context to another, without the additional training that must have been necessary for these Mesopotamian accountants. While this may have made the organization of information in accounts less spatially sophisticated, it may also have created advantages in terms of flexibility, adaptability, or ease of working in different contexts.

The mathematical lists also contained some features of grouping and cross-referencing that, while not the same as those characterizing the organization of information in accounts, could have helped to facilitate relevant ways of thinking. For example, as Azzarello (2021, paragraph 5) explains, "Tables normally contain all the units, but then only round numbers in tens, hundreds, and thousands," which meant that "to calculate 25 × 4, one would add up the result of 20 × 4 = 80 (to be found in the series of tens) and the result of 5 × 4 = 20 (to be found in the series of units) to obtain 100." Compilers or copiers of such tables would thus internalize that certain information could be condensed or retrieved elsewhere. Operations themselves could also be clustered and organized in different ways, and multiple patterns are visible in the tables.[29]

---

[27] See Eddy (2013) for an analysis of how tables and school texts trained students to do this in the eighteenth century.

[28] See the examples of monthly draft accounts in Rathbone (1991, 426–463), which date to the third century, or *P.Oxy.* LV 3804, an annual account from the Apion estate (sixth century).

[29] Summarized again in Azzarello (2021).

Stewards/accountants working for large estates would have kept preliminary accounts that were probably mostly chronological, but then would need to be reorganized into annual accounts. This required grouping and condensing information in particular ways: merging individual chronological transactions into single entries by type, for example, or deciding how to categorize information (listing loaves as part of wheat, or as its own separate account). Both of these examples come from Rathbone's discussion of the Heroninos archive, where Heroninos can be seen making changes over time in how he organizes information (Rathbone 1991, 347–352), and where final monthly accounts can be compared with various preliminary and intermediate records (Rathbone 1991, 334). The mathematical tables, while they did not directly teach students how to organize accounts, would have laid the groundwork for thinking about how to group, condense, and reference information in tabular lists.

2.5 Arithmetical Problems

In addition to tables and lists, school texts often contained various types of problems. These problems are, again, usually described as practical (in contrast to theoretical), since they are particular cases that concern real-world objects and specific numbers. Many of them seem straightforwardly applicable to a steward or accountant. The collection of problems in *P.Mich.* III 145 (second century CE) is a good example: it contains fractions, monetary conversions (especially between silver and copper drachmas), freight charges, interest payments, and an area problem. Many of the problems require working backwards in a way that probably does not reflect how one would encounter the problem in actual life (being given the interest, a period of time, and a resulting sum, and then being asked to determine the principle, for example, when in practice one would be more likely to start with the principle). They thus require flexibility and the ability to adapt known rules or algorithms. But they do not stray far from basic competencies.

Other arithmetical problems are more difficult or more puzzle-like; though still "practical" in the sense that they deal with relevant skills, they are not straightforwardly utilitarian. For example, problems involving partitions into unit fractions (such as a4, c4, f3, f4, h2, i4 in *P.Math*), like the "computational workout" division tables referenced above (Lougovaya 2022a) and like many problems in *P.Cair. cat.* 10,758 (the Akhmim mathematical papyrus, seventh century CE), are more difficult than the sort of work with unit fractions that stewards would be doing in actual accounts. The problems in *P.Math.* require denominators below 100 and many require a certain number of unit fractions even when the number could be expressed with fewer; they "are puzzles with no immediate practical application" (Bagnall and Jones 2019, 51). Divisors also tend to be more difficult; in accounts, most fractional quantities come from divisions by 2, 3, and their multiples, which is not the case in all the problems in *P.Math.* or similar texts Lougovaya (2022a, 663).[30]

Into the same category—practical situations rendered somewhat more difficult, puzzle-like, or artificial—fall many other arithmetical problems, such as m2 in *P.Math.*, which seems to have involved distributing 300 artabas of wheat among five people so that each person receives eight more artabas than the last (the actual statement of the problem is lost).[31] While workers on an estate might have received different pay grades, it seems unlikely that a manager would start with a fixed amount of pay, rather than simply paying workers whatever total resulted based on the different pay grades over a certain number of days or for a certain task.[32] Or, considerably more removed from the everyday, *P.Vindob.* G 26011e (*MPER NS* 15179), in which the student must determine the shares of the wind, sun, rain, and so on, in the erosion of a statue. Broadly speaking, this playfulness or abstraction from strict practicality is a well-known features of some problem texts. Høyrup (1990, 66–67) discusses it as a feature of much "sub-scientific" mathematics across cultures, and Jones mentions it as a characteristic of mathematical papyri in general.[33] This impracticality also characterized a number of problems in the Demotic Egyptian tradition, problems which Imhausen (2016, 193) calls "suprautilitarian."[34] There are many reasons why a mathematical education might contain such problems, and in fact the persistence of similar problems and setups across centuries and cultures from ancient to medieval might make an explanation specific to Roman or late Roman Egypt an exercise in futility. Sidoli (2015) in fact considers it to be what marks the texts as educational in nature to begin with: mathematical papyri teach general problem-solving skills.[35]

Høyrup (1990, 67), a bit more specifically, sees riddle-like problems in other contexts as a way to form professional identity and demonstrate skill, especially for merchants. Such a purpose may be relevant to the Roman-Egyptian context, but the ambiguities discussed above (especially the degree to which there were specialized or separate schools or tracks for particular professionals) make it difficult to say exactly how. If (as previously suggested and shortly to be argued at greater length) there was in fact at many levels little obvious sign of separation between tracks or between literary and numerical knowledge, then the effect may have been rather

---

[30] Lougovaya's analysis of accounts is not exhaustive but agrees with my own (also not exhaustive) examination of fractions in these texts.

[31] See Bagnall and Jones (2019, 146) for commentary.

[32] As in *SB* 9408.2, lines 60–67, where the task, number of days' work, rate, and resulting total are specified (Rathbone 1991, 429–430).

[33] Specifically, what Høyrup calls "recreational problems" are riddle-like. Jones (2009, 342) writes: "The emphasis on impractical problems in the mathematical papyri raises interesting questions about the goals and rationale of the mathematical education of scribes in antiquity."

[34] This is discussed by Roby (2023, 162–163). Jones (2009, 342) also writes on this type of "riddle" problem.

[35] Many of them are, in his words, "problems that do not correspond to anything that would have been encountered in actual work," (Sidoli 2015, 392).

different than that posited by Høyrup. Instead of instilling identity or pride in a specific group, mathematical education may (along with literary education) have linked people of different backgrounds and future specialties: surveyors with accountants, for example, but also accountants with scribes more generally, enslaved with free accountants or scribes, or even to some degree landowners with those they employed or enslaved. Such an education was by no means universal, but it was not necessarily narrowly professional. The area and volume problems discussed in the next section emphasize this point still further.

2.6 Spatial Problems and *P.Math*

In addition to arithmetical problems, many texts contain spatial problems (area and volume, or plane and solid geometry), sometimes in conjunction with literary texts, and sometimes as part of more lengthy, seemingly specialized collections.[36] Among these, *P.Math.*—as the text that both seems most specific to a business or professional education and contains some of the most challenging mathematical problems—will serve as an example for detailed discussion. *P.Math.* is a fourth century CE papyrus codex containing model contracts, metrological lists, arithmetical problems, and area and volume problems. Problems tend to follow metrological lists in the codex, which Bagnall and Jones (2019, 16) suggest reflects the structure of lessons. The handwriting is varied (as if the writer grew tired over time) and there are many errors in the Greek (as well as in solving the problems), which lead Bagnall and Jones to conclude that the codex was probably written by a student, perhaps as a "fair copy" of tablets used in class. Bagnall and Jones (2019, 55–56) associate the codex with a business education, though not business in the sense of commerce: "skills relevant for the administration of land, rents, and taxes, skills that would have been equally relevant for the agents of liturgists who had to assess and collect taxes and the stewards of the estates of large landowners."[37] The model documents especially seem to link the codex to a land management or administrative context. The difficulty of some of the problems on the one hand, and the lack of literary texts on the other, also might imply a more specialized context than many other educational texts. Even in this more specialized context, however, only the model contracts and the metrological tables are directly and narrowly practical (in the sense described above).

Many of the problems in *P.Math.* are area and volume problems involving a variety of shapes: triangles, rectangles, cylinders, cones, triangular prisms, and some

[36] For example (some of which are discussed further below): *P.Cornell. Inv* 69, *BM Add. MS* 41203, *PSI* III 186, *MPER NS* 1.1, *P.Chester Beatty Library AC 1390, P.Lond.* V 1718, and many examples in *MPER NS* XV.

[37] As the quote suggests, Bagnall and Jones stop short of suggesting that a landowner might have undergone such an education themselves, seeming to associate it, like Bernard, Proust, and Ross as well as Cribiore, with those of somewhat lower social status.

irregular or uncertain shapes. Some of these could have related to a manager's actual work. Certain area problems involve determining amounts to be planted in a field or vineyard, such as f7.[38] Some of the three-dimensional shapes loosely correspond to actual structures needing to be built on an estate: canals or irrigation ditches, cisterns, and perhaps small granaries.[39] Accounts for such construction projects survive,[40] and knowing the capacity of such structures was probably also sometimes necessary. But the problems cannot always be related to a straightforwardly practical context, and in any case, they have parallels in other collections and texts that were doubtless used by people other than stewards. They thus probably had a range of additional purposes, like the exercises discussed above. Socially or professionally, they may have distinguished apt students or professionals through their difficulty. More broadly, they taught skills of visualization and (like the arithmetical problems) finding the correct framework or algorithm to approach a problem.

Interestingly, these spatial problems—especially those involving three-dimensional objects—pose the most difficulties for the student in *P.Math*. The student has no trouble with problems involving basic arithmetic or unit fractions.[41] Either he had learned these problems thoroughly at an earlier stage or he was simply more at ease with them.[42] But the volume (and some of the two-dimensional) problems are different. As Courtney Roby has analyzed in detail, the student frequently fails to visualize problems correctly, chooses improbable dimensions for the objects, or uses the wrong algorithms and procedures. One of the most striking of the diagram failures (o1) is a drawing meant (probably) to be a vaulted granary, which is simply a blob-like shape with a zigzag line on top. Another drawing, o2, should be a quadrilateral based on the problem, but the student has simply drawn a line, with the numbers that should correspond to the four sides above, below, and to either side of the line.[43] An example of procedural failures is n4, where the student, to find the volume of a granary, multiplies the length times the width, then this by a dimension he calls the "vault," and then the two products by one another, resulting in a four-

[38] Bagnall and Jones (2019, 16) try to connect the model lease to area problems, but leases themselves don't show any signs of calculation and might frequently have represented longstanding practice.

[39] Though the type of vaulted granary in n4, for example, does not correspond to the common shapes of various granaries in Egypt as discussed by Rickman (1971, 298–299), nor of course to the large state granaries like those at Karanis (Husselman 1952).

[40] *P.Ant.* I 46, for example; though this does not mean that the maker of the account actually designed the project. *P.Köln* I 53 is an account (for a city council) of wooden beams for the ceiling of a gymnasium that uses cubits and fingers to calculate the volume of materials used; see Lougovaya (2022b, paragraph 27).

[41] Bagnall and Jones (2019, 24–25, 55), as well as commentary on specific problems.

[42] Generally speaking, I use "he," as Roman accountants seem to have been mostly male—though there are two possible female *nummulariae* (money-changers and testers) attested in the imperial *familia* (Swetnam-Burland 2021, 30–36).

[43] In the latter case, the simplified diagram would still have had enough information to solve the problem, but, as Roby points out, the example dimensions given in the problem are impossible, and a more accurate diagram might have made the student realize this (Roby 2023, 168).

dimensional object in a unit Bagnall and Jones (2019, 149) call "hypercubits."[44] Other drawings represent improbably tiny objects, structurally impossible towers, or similar failures (Roby 2023, 178–181).

As Roby explains, the student in *P.Math.* has difficulty finding the correct frame for new kinds of problems. But her analysis makes clear that it is specifically the problems involving *visualizing* and dealing spatially with objects and structures (especially three-dimensional ones) that give the student trouble. Of course, the student may have simply been at an early stage in encountering these sorts of problems. But as Roby also shows in a particularly illuminating comparison, *P.Math.* as a whole is structured in a way that does not make learning easy for the student. While long for a papyrological problem collection, it is very short by the standards of the *Stereometrica* associated with Hero of Alexandria. The *Stereometrica* contains gradual progressions of specific types of problems, moving from simple to complex variations, but *P.Math.* does nothing of the kind: problems are disparate, and a complex problem might appear early in the collection Roby (2023, 175).[45] This arrangement would not make it easy for a student to internalize the correct procedures or use of diagrams, and it might partially explain the difficulties the *P.Math.* solver has. And if, as Bagnall and Jones conjecture, the collection reflects the structure of lessons, it may suggest that the teacher did not understand the best way of teaching this information either.

This is of course quite speculative. We cannot know what components of the lessons were not included in the codex. And as with similar collections of problems, little explanation is given about the solution of the individual problems, so oral interaction/instruction from the teacher is implied in any case; we are certainly missing much of the classroom teaching.[46] Nor do we know whether the student went on to become more proficient and more at ease with visualizing problems and choosing algorithms. But at least at this stage, the student, the teacher, or both had some difficulty in finding the best ways to approach these problems. And it is precisely at the level of the "practical" (connecting the problems to structures one could realistically build or estimate the capacity of) that the student often fails. Nevertheless, the large number of problems suggest a sustained and challenging focus on this type of material, to a degree that may seem surprising for someone probably most concerned with collecting taxes or keeping accounts.

I use the word "surprising" because other material that is often assigned to specialized education of "scribes" stresses memorization, repetition, fast copying, shorthand, and use of formulas—in short, texts that do not suggest much focus on solving problems in new ways, visualization, having to choose frameworks or

[44] See also Roby (2023, 170–171).

[45] Bagnall and Jones (2019, 5) mention that the codex is missing five leaves which were known to be part of it until the early 2000s, and they leave open the possibility of more (saying that it was "at least" 17 leaves in the early 2000s). There are also some uncertainties about the order of the leaves. Even so, it does not seem that the codex could have approached anything like the length or progression of the *Stereometrica*.

[46] See Bagnall and Jones (2019, 21–23) on the style of the problems.

algorithms, or in general flexible thinking.[47] Some of the exercises in *P.Math.*—the model contracts and metrological lists—do seem to emphasize repetition and memorization. But most of the codex is concerned with different skills, and the difficulties of the student are especially interesting to consider in this light.[48] While it is possible to imagine practical applications for some of the problems, the collection as a whole seems aimed at instilling a broader, more flexible, and more creative range of abilities. And it is decidedly *unlike* the most extensive handbook for learning shorthand, the *Commentary*, which groups shorthand symbols into phrases such as "The master utters threats, the slave falls silent" that, as Ella Kirsh (2024, quotes 66 and 72) has argued in a recent extensive analysis, "prepared non-elite students for a life of social disadvantage."[49] The problems in *P.Math.* and other collections do not contain such obvious reminders of status. Who received an education in such problems, and what does this mean for the training of stewards and accountants in particular?

2.7 Flexible Educational Boundaries

I now return to some questions of levels, specialization, and status; and the consequences for the self-conception and working contexts of accountants or managers. First, I stress that primary documents show little separation between literary and numerical education at most levels—especially at what might be called the primary and secondary levels. Second, I discuss texts that *do* seem more specialized (such as *P.Math.*), while nevertheless finding some evidence of blurred boundaries even here. And finally, I discuss what aspects of training (for an accountant in particular) even specialized education contexts seem to lack, and how accountants acquired this training. I argue that these various types of evidence suggest that most accountants in Roman Egypt would have shared much of their educational background with people outside their profession, and that this, in combination with other aspects of their lives and training, would have given them a robust professional identity that was *not* necessarily linked with notions of inferior status.

As discussed earlier, current overviews of mathematical education are often imprecise about whether and when such education became specialized. In fact,

[47] The texts examined and classified by Freeman (2023, Chap. 3) as scribal trials, for example.

[48] The student who wrote the Akhmim Mathematical Papyrus (*P.Cair. cat.* 10,758) had similar difficulties with volume problems, and also made errors in division problems (Smyly 1920; Baillet 1892, 34–35 and commentary on individual problems). Lougovaya (2022a) also details mistakes in the difficult divisions of *O.Petr.Mus.* 64 and 65. See also Lougovaya and Ast (2023) on an arithmetical problem in *P.Bodl.* I 7. Brashear briefly comments on errors in *P.Chester Beatty AC 1390* (1990, 37–39). Students at lower levels would also make errors of course (see Fournet and Gallazi 1996, for example). Still, the frequency of errors in the cited texts may loosely correlate with their advanced level, especially since students so often show mistakes in choosing algorithms or frameworks: in trying to independently apply what they have learned, in other words.

[49] See also Moss (2024, 42).

although many types of mathematical documents are often classed with "scribal education," they do not necessarily overlap with "scribal trials" as such. Examples of most types of mathematical exercises occur in conjunction with literary exercises that are not necessarily scribal in nature. Much mathematical material may thus fit better within a general school setting rather than specialized training. For example, *P.Oxy*. XXXIII 2656, a codex of Menander's *Misoumenos* probably dating to the late third or fourth century CE, also contains tables of fractions from 1/12 to 1/16. The tables seem to have been inserted on pages that were intentionally left blank, and thus the codex may have been some kind of school notebook.[50] *P. Vindob.* G 26740 (*MPER N.S.* XV 178), dating to the second century CE, contains mathematical problems (including area problems), metrological lists, and an excerpt from Homer's *Iliad*. In other cases, as Serafina Cuomo (2020, 102–106) has observed, mathematical and literary texts have been found at the same location: buildings in Tebtunis and Kellis, for example.[51] Cuomo uses this evidence to argue against a strict separation between mathematical traditions (theoretical and practical), but it may also support a general presence of mathematical material in schools.[52] Even Cribiore (2001, 200 and n. 73) cites *P.Oxy*. XXXIII 2656, with its combination of Menander and advanced fractions, in her chapter on the grammarian (whose pupils, the discussion of specialized schools in her previous chapter seemed to imply, would not be concerned with advanced practical mathematics).[53] Such texts imply a broad mathematical and literary foundation shared by some (perhaps many) students.

Certain types of mathematical texts do seem more specialized: notably, large problem collections without literary material. *P.Math.* is one of these (Bagnall and Jones 2019).[54] Other extensive collections of problems without literary material include the seventh-century CE Akhmim mathematical papyrus (*P. Cair. cat.* 10,758); the circa first-century CE *MPER NS* 1.1 (P. Rainer NS 1.1/P.Vindob. [inv.] 19,996); and the fourth-century CE *P.Chester Beatty Library AC 1390*.[55] These collections often also include problems involving volume or three dimensions, which are less common in texts that mix mathematical and literary material (the spatial problems in *P. Vindob.* G 26740/*MPER N.S.* XV 178, for example, all involve area,

[50] See analysis by Azzarello (2019a, especially 244–245). *Bodleian Gr. Inscription [inv.]* 3019, third century CE, also contains language exercises and tables of fractions (Parsons 1970), as does the notebook belonging to Papnouthion (Boyaval 1977).

[51] Cuomo (2020) also discusses *P. Vindob.* G 26740.

[52] The overall picture again resonates with Sidoli (2015), who classifies only a few texts as truly specialized.

[53] This seems to contrast with the discussion on 181–183, although both here and in (1996, 28–29) Cribiore does express doubt about when and after what material a student would begin a specialized scribal education.

[54] Bagnall and Jones (2019, 55) explicitly associate it with a specialized school such as those posited by Cribiore.

[55] *P.Chester Beatty Library AC 1390* in fact also contains an excerpt from the Gospel of John in Coptic; the editors did not think that the two were connected, but rather that the mathematical texts were already written on papyrus that was then purchased for a codex (Brashear et al. 1990, 39).

not volume). The length of these collections and their focus on one type of material make them the best candidates, as suggested by Sidoli (2015) and Bagnall and Jones (2019), for mathematical texts used in professional education—though this education is still somewhat broad, since it could include scribes, architects, surveyors, or the category of people Bagnall (2010, 25) calls *boethoi*.

There is some evidence, however, that even these types of problems might be studied by landowners, town councilors, and other similar "notables." For example, the archive of the poet, landowner, administrator, and notary Dioskoros of Aphrodito (sixth century CE) contains literary educational material at both the grammatical and rhetorical level (Cribiore 2001, 40) as well as *P.Lond.* V 1718. The latter consists of metrological lists and procedures for solving various problems of volume/capacity (of a simple—non-vaulted—granary, ship, reservoir, canal, and the number of bricks needed for a wall).[56] *P.Lond.* V 1718 may have been educational or may have been for Dioskoros' own use; either way, it is in his handwriting and thus shows his direct and close engagement with the material.[57] A more advanced numerical (and literary) education may have been especially useful for someone in Dioskoros' position in late antiquity (Ast 2015).[58] But earlier suggestive examples exist (though not specifically involving geometrical problems): Kronion son of Apion in first-century CE Tebtunis, for example, who ran the *grapheion* (notarial office), made his own accounts in which he can be seen, among other things, converting between silver and copper in ways that would make transactions most advantageous to him (Rathbone 2013, especially 130–134).[59] It might seem obvious that people like Dioskoros or Kronion—who had administrative responsibilities—would benefit from such skills, but given how often numerical education beyond the primary level has been attributed to specialized schools or people of lower status in broader surveys, it is worth emphasizing. It was not just "assistants," but people like Dioskoros and Kronion themselves, who had these numerical interests or skills.

Not everyone, even within professional contexts, would have received such an education to the same degree. And perhaps the educational overlap between different groups of people was especially strong in social categories other than those of the largest landowners, who may have relied more (or exclusively) on assistants. But at least in many parts of Egypt, the implication is that people from different

[56] *P.Lond.* V 1718 is quite similar to *BM Add. MS* 41203A (Skeat 1936), also dating to the sixth century.

[57] Cribiore (2001, 40) suggests that Dioskoros was teaching his own children, local children, or both. Unlike Cribiore, however, I would argue that the mathematical material represents a more advanced level (not primary education as she suggests). Ruffini (2018, 189–190) argues for the practical relevance of the document; the metrological lists contain pervasive errors, however, and would have needed to be corrected first (Rathbone 1981, 267–270).

[58] Especially Ast (2015), paragraph 4.5: "I suggest that many of the petitions, memoranda, accounts, contracts, receipts, etc. that we have from Egypt were generated by councillors and their assistants (the identity and status of these assistants is exceedingly difficult to assess), who had attended local schools."

[59] Rathbone thinks Kronion was likely also a landowner, although the archive does not make this explicit.

groups might share a mathematical education that included numerical skills, skills in spatial layout and organization of information, and familiarity with a range of arithmetical or geometrical problems, varying in difficulty and being at least sometimes puzzle-like or impractical. And relatedly, a number of stewards or managers (along with similar professionals) who received such a mathematical education also seem to have had considerable literary education (Bagnall 2010, 26–27).[60] What the texts as a whole suggest is a mathematical education that loosely linked stewards both with other professionals and with those broadly involved in land, taxation, and administration (including at least some landowners themselves), and which was not necessarily sharply differentiated on the basis of status.[61]

The educational texts also contain many gaps, at least for a steward or accountant. They lack anything directly relating to organizing accounts, to types of accounts (from daybooks to monthly or yearly accounts, for example), or to different types of enterprises (an estate versus a bank, for example). Where did a steward learn these? The best answer based on current evidence seems to be not a school setting but an apprenticeship within a family or simply "on the job." To make a brief comparison, this was the situation in Renaissance Florence, where despite a pervasive accounting culture (and incipient double-entry bookkeeping), the abacus schools that students attended after their elementary education do not actually seem to have taught much accounting practice, as opposed to arithmetic. The treatises associated with abacus schools contained "practical, everyday problems, though, for didactic purposes, of very considerable complexity"; accounting, especially complex or business accounting, was learned as part of a firm instead (Goldthwaite 2015, 619–623, quote 620).[62] In Roman Egypt (much less elsewhere in the Roman world), there is no certain evidence of how the gap between the school texts and fully developed accounts was filled. Apprenticeship contracts survive from Roman Egypt, but only for a limited range of professions (especially weaving); one contract survives for an apprentice learning shorthand, but none for accounting or the more general skills of a steward or manager (Bergamesco 1995, 96–98). Such apprenticeships may still have existed, of course, and evidence is simply lacking.

Dominic Rathbone's brief discussion of the status and background of managers (*phrontistai*) on the third-century CE Appianus estate suggests additional possibilities (Rathbone 1991, 80–81). Some were sons of existing managers and could have learned from their fathers. Others "came up through the ranks," holding a series of

---

[60] The fourth-century Kellis account book and a codex of Isocrates were found in the same room and probably by the same hand, suggesting the *pronoetes* (manager) was also an educator. In the same chapter, Bagnall (2010) gives a few additional examples of accounts juxtaposed with literary texts.

[61] Even if Bagnall and Jones (2019, 55) are correct that *P.Math.* itself was "part of an education that was indeed separate from that of the grammarian and did not presuppose any time in a grammarian's classroom", I would still stress this broader context of overlap between categories as common, and perhaps even the norm.

[62] Goldthwaite describes how a boy who took a job in a firm would begin by tracking the cash box and only gradually learn, and be responsible for, more complex forms of accounting.

jobs before becoming a manager: perhaps the estate, like a Florentine firm, trained its own. In a study of *phrontistai* from second-century CE Tebtunis, Ryosuke Takahashi found that managers were typically villagers who leased land from the wealthy families for whom they worked; these long-term economic relationships might have formed the context for their training and selection. Though Takahashi is less certain than Rathbone, it also seems that generations of the same family might have worked for the same landowner. Finally, at least some managers documented in Tebtunis (though none of the ones in the archive studied by Rathbone) were enslaved or freed. An enslaved manager might have received training more directly from someone in the landowner's household or might have moved through the ranks in a way similar to other managers (Takahashi 2021, 105–118). But the majority were free, and this remained true in later contexts as well (such as the salaried *pronoetai* for the sixth-century Apion estate).[63]

Such managers seemingly could have done *all* their training on the job or within their families, and perhaps many of them did. We have no certain evidence that most *phrontistai* would have used school texts as advanced as *P.Math.* But in places and settings where these school texts and problem collections were used, perhaps they helped differentiate especially apt pupils from others, or the schooling, once completed, served as a badge of mathematical knowledge and problem-solving. It also, as emphasized above, linked people of different social and professional backgrounds. A codex such as *P.Math.* would not have left a student immediately prepared to organize the kinds of accounting practiced on a large estate such as that of Appianus, but would have *overprepared* them for the required arithmetic: as mentioned earlier, the Appianus accounts contain few fractions, and many of the numbers (for example for standard wages) are quite round.[64] The accounts of the sixth-century Apion estate are more challenging, partly because the monetary calculations of the time were done in fractions of the gold solidus, partly because of the extra charges relating to taxes. But the use of unit fractions even in these accounts, while much heavier, is never as challenging as the problems in *P.Math.* and other collections.[65] Overpreparing students in this way—by teaching them especially challenging unit fraction problems, or area and volume problems they might not directly use—could give them a source of pride, as well as a link with those outside their specific work (and perhaps give them additional roles, such as teachers themselves).[66]

Those who shared a similar education would have been aware of sharing it, both across and within groups. In the managers' cases, if they stayed and worked where they had grown up and within the same networks of family and employers, they would have been aware of who had been the most skilled at these sorts of problems

[63] In addition to their salaries, these *pronoetai* seem to have made money through their role as intermediaries in collection of taxes and rent; see McConnell (2017, 56–58).

[64] See examples in Rathbone (1991, Appendix I).

[65] *P. Oxy.* LV 3804 for example.

[66] Bagnall (2010, 25) again (the *pronoetes* at Kellis).

during their education. Rathbone, in his discussion of the *phrontistai* for the Appianus estate, describes the close connections between them: they addressed each other as "friend" and "brother," wrote in familiar tones, sent each other wine, and had parties together. These connections could, of course, have formed for many reasons: Rathbone (1991, 80) mentions in particular the many years they tended to work for the same estate, ensuring they knew each other well. But perhaps the closeness also reflects shared training and educational background.

Both the work that managers did and their educational backgrounds must have varied considerably. I have already mentioned that it is quite possible that not all of them used school texts like *P.Math.* or were trained in school settings at all. Accounts vary widely in level of complexity and accuracy, from the sophisticated Appianus and Apion accounts to the error-riddled travel account of Theophanes. Much of this variation had to do with the type of organization or enterprise—an estate versus a journey or household account, for example—but some of it probably depended on the background and skill of the accountant as well. And just as the educational backgrounds of accountants must have differed, so did their actual work and duties. Takahashi (2021, 109) emphasizes this for the *phrontistai* at Tebtunis: their responsibilities varied widely, and could spread to what looks like household management, as when a manager is instructed to clean clothes and look after children. In contrast, the most upper level administrators (well above the *phrontistai*) of the Appianus estate were wealthy landowners in their own right, who seems to have worked for patronage rather than salaries; below them (and working for them) came salaried assistants, or *boethoi* (Rathbone 1991, 58–71).[67] The training of these *boethoi* perhaps resembled that of the *phrontistai*, was more complex in some way, or was urban- rather than village-based. But evidence is lacking to say for certain how their education may have differed. Finally, it is also unclear whether enslaved managers would have received the same type of education.

We thus cannot use the school texts to speak definitively of a uniform professional identity for all accountants (much less scribes in general, or architects, surveyors, and others who may have learned this type of problem at various stages of their schooling). But the texts suggest that education may have played a role in the formation of professional identity for some. More broadly, the frequent lack of separation between literary and mathematical educational texts suggests links between accountants and those *outside* their specific professional identity—with other specialists as well as with some landowners or local notables. This may have especially been the case in Roman Egypt, where many managers seem to have been freeborn and to have had long, even multigenerational careers in the areas where they were born and grew up.[68] In Roman Italy, to which I now turn, the context was rather different.

[67] Other titles occur as well, but many are basically synonyms for "manager."

[68] And it may have increasingly been the case over time if, as Bagnall suggests, the numbers of "assistants" increased in later centuries due to the increasing importance of cities, or "municipalization" (2010, 30).

2.8 Accountants and Education in the Western Empire

I began this paper with the question of slavery, but have dwelt at length on a context—Roman Egypt—where the overall proportion of enslaved people is thought to have been low, certainly by comparison to Roman Italy, and where many attested managers and accountants seem to have been freeborn.[69] This is, of course, because Roman Egypt is the source of the available school texts, not to mention most accounts (and all complex accounts for large estates). As discussed above, even in Egypt, questions remain about how exactly many of these school texts related to different careers, often lumped together under the term "scribal." Outside Egypt, the situation is even murkier, but here I will attempt to discuss the school texts in relation to accountants and managers in the Western Roman Empire, especially Roman Italy, who were more likely to be enslaved.[70]

Even in parts of the empire where slavery was more pervasive than in Egypt, of course not all managers, scribes, accountants, and related professions were enslaved (or freed). But written sources suggest a strong association between this type of work and slavery, and recent work on the Roman *vilicus*, or overseer, has concluded that most were enslaved.[71] It therefore makes sense to discuss accounting specifically in the context of slavery, while not forgetting that in a number of ways the education and experiences of non-enslaved practitioners might have overlapped. The intersection between accounting/management and slavery has been examined from a number of angles—from the idea that these professionals were usually enslaved because this increased trust in them or power over them (perhaps both), to the ways in which slavery allowed for "depersonalization" in business—but the educational texts from Egypt have not been discussed in relation to slavery and have only briefly been discussed in relation to the education of enslaved people in general.[72]

---

[69] Bagnall and Frier (1994, 70–71) on slavery overall in Egypt: 13.4% in metropoleis and 8.5% in villages. These percentages are for earlier Roman Egypt. Bagnall (1993) argues against seeing a decline in slavery in later centuries, though not particularly an increase either. Where agricultural managers are concerned, he does caution that, with types of documentation such as letters, "If a landowner writes to his *phrontistes*…we do not know the legal status of the steward" (225). As discussed above, detailed studies of these managers (such as Rathbone 1991; Takahashi 2021) tend to find few that were enslaved.

[70] The situation in Egypt may also have differed from that elsewhere in the Eastern Mediterranean, but for now I leave that question aside.

[71] See Kutner (2025) for more on the general association between slavery and numeracy, especially in accounting; Carlsen (2010, 79) on *vilici*, and Carlsen (1995, Chap. 5) on *actores* (more closely associated with finance and accounts), whose duties seem to have become increasingly differentiated from those of *vilici* in the second and third centuries CE. While most of the evidence cited for the Roman West in this chapter dates to the first to third centuries CE, see Harper (2011) for evidence of continued importance of slavery in agriculture (including for managers and overseers of various kinds) at least through the fourth century, especially pages 120–126, 184–191.

[72] Wrenhaven (2015) is a brief recent overview. Hawkins (2016) discusses artisan apprenticeship and treats in detail the situation of apprentices who were enslaved. See Kutner (2025) for references to other aspects of the intersection between accounting and slavery.

What did the education of an enslaved manager or accountant look like, in broad terms? As discussed above, education in general in the Roman world varied, especially depending on urban/rural, center/periphery contrasts. Primary education may have looked similar for many students, but in some places—especially the city of Rome itself—literary evidence implies separate tracks from an early stage, with wealthy students educated either at home or by a higher-status teacher, and others (non-elites or enslaved) already as children going to different schools or being taught by teachers with a more basic, practical, or technical focus.[73] If this is so, then both enslaved and non-elite children would have been on a separate educational path from the start, even if the components of what they learned (literary, basic numbers) did not immediately differ from wealthy children. Literary evidence analyzed by Marrou (1956), Booth (1979), Forbes (1955), and Mohler (1940), which are still fundamental overviews, gives an idea of the contexts within which enslaved people in Rome might have learned. The primary context, unsurprisingly, was the household within which they were enslaved. Large households may have educated enslaved boys together in a school-like setting called a *paedagogium*.[74] Some references also imply that enslaved children might be sent to outside schools for basic education. Finally, some enslaved people underwent formal apprenticeships.

Forbes and the other cited scholars, mostly writing in the 1940s and 1950s, tended to give a rosy view of ancient slavery and slave education in general, insisting that enslaved people were "handled with fairness and even leniency," for example (Forbes 1955, 321). Contemporary scholarship places stronger emphasis on the violence of Roman slavery, which extended to scribes and accountants as well as those performing menial labor (Howley 2025, with copious citations of further literature).[75] But only very recently have such reassessments specifically addressed education of the enslaved.[76] Before discussing educational texts, it is thus worthwhile to focus on these contexts of school or household and attempt to ask what they were actually like. The most salient point is that receiving an education was not some sign that enslaved people were free from exploitation. Education, exploitation, and the enforcing of hierarchies were intertwined.

The atmosphere of exploitation is clear (if not acknowledged by them) in the literary sources used by the above-mentioned scholars. The freedman Trimalchio and his fellow freedmen in Petronius's *Satyricon*, whose extensive conversation

[73] Bernard, Proust, and Ross (2014) emphasize separate tracks for large cities in particular; much of the evidence comes from Rome.

[74] Though see the skepticism of Groen-Vallinga (2022, 189–190), who argues that *paedagogus* simply means "childminder" and therefore a *paedogogium* may not have been an educational setting per se.

[75] Some recent scholars, like Silver (2011), think that large numbers of people sold themselves into slavery specifically to obtain professional training. The widespread Roman associations between slavery, degradation, and violence (see Saller 1994, 134–153) work against this view at least in the early empire, I would argue.

[76] See Moss (2024, Chap. 1) and Kirsh (2024).

about their own backgrounds is often used as evidence for slave education in general, discuss having received a basic mathematical education in addition to (again basic) literacy. In keeping with the two-track view of Roman education and with Petronius's overall tone (which is intended to mock the pretensions of the freedmen), their knowledge of literature is garbled and they admit to never having learned geometry or Homer; their education seems to have been geared especially towards business (Forbes 1955, 327–328; Petr. *Sat.* 58). And indeed, Trimalchio served as *dispensator* for his owner (Petr. *Sat.* 29). At the same time, at various points in the discussion Trimalchio and his dinner guests discuss their exploitation: Trimalchio mentions having been his master's (and mistress's) favorite (the sexual overtones are clear); his own kisses for a boy enslaved by him who is also being given a basic mathematical education suggest the cycle continues (Petr. *Sat.* 74–75). Violence is frequently discussed or implied as well; Trimlachio (often as a joke) threatens to beat people enslaved to him (see Petr. *Sat.* 49 for one of many threatened beatings), and a *dispensator* of his is being prosecuted, although it is unclear why (Petr. *Sat.* 53). Petronius is doubtlessly not invoking sexual exploitation or violence to create any sympathy for the freedmen. The point rather seems to be that these things have permanently tainted them, or even that they were exploiting their *domini* rather than the reverse; Trimalchio mentions having captured his master's mind (Petr. *Sat.* 75).[77] But in any case, the scene suggests that education and exploitation were linked.

Other literary sources, though briefer, support the same picture. Take the lines in Martial, for example, which Booth discusses as implying the existence of schools outside households for enslaved people: the word for the students in the poem is *capillati*, later followed by *delicatae...chorus mensae*, both of which Booth (1979, 11–13) plausibly argues refer to young, long-haired (*capillati*) enslaved pages who waited at banquets (see also Mart. *Epig.* 10.62.1–5). The teachers mentioned are a *calculator* and a *notarius*, making it clear that the enslaved boys are learning calculation and shorthand or related skills. But *capillati*, in its allusion to long hair and to the carefully cultivated physical appearance of enslaved boys who waited at tables or performed similar kinds of household service, alludes, if obliquely, to the same kind of sexual exploitation that Petronius discusses more openly.[78]

One context outside these literary sources provides some additional insight: the physical space of a *paedagogium*, or school for enslaved boys. Peter Keegan's (2013) rich analysis of the graffiti in the Palatine Paedagogium in Rome, which was used to educate children enslaved in the imperial household in the second and third centuries CE, offers an opportunity to escape the purely literary record, although the evidence is of course fragmentary. Some graffiti refer to hard, monotonous labor, such as a representation of a donkey turning a mill with the words, "Toil, little donkey, in the manner which I have toiled, and it will benefit you" (Keegan 2013, 77).

[77] Mohler (1940, 268–269) blandly asserts that "even here we must remember that the victims were favorites."

[78] See Keegan (2013, 74) for a discussion of appearance in general (sumptuous clothes, cosmetics), long hair ("these coiffures are very much like those of women"), and words such as *capillati, comati,* and *criniti* that referred to this last characteristic.

None of the boys in the *paedagogium* would have actually performed hard physical labor such as turning a mill, but as Candida Moss (2024, 29–30) has shown, those who served as scribes might have suffered from arthritis, tendonitis, and back and neck pain, and would have been as subject to physical or sexual violence as any other enslaved person. Other graffiti give more clues about the general circumstances in which they lived, notably a number of sexual terms. The words include *pedico* (related to *pedicare*, to penetrate anally), its Greek equivalent πυγίζω, *pedicans,* and πάθος (a man who takes a passive role during sex). As Keegan points out, by themselves (or attached to names), these terms "provide no clear indication of an emotive register" (Keegan 2013, 88), but the practices invoked are those associated by Roman writers as a matter of course with the sexual exploitation of young male slaves. The graffiti may have been intended to provide a humorous outlet, to insult particular slaves, or to claim a sort of power (*pedico* having an active connotation) but they reflected a potential reality for many of the students in the *paedagogium* (Keegan 2013, 85–89). Finally, not all graffiti allude to labor or exploitation: a number refer to gladiators and chariot racing, suggesting a love of those quintessential Roman entertainments without a hint of the stigmatization shown by elite writers (Keegan 2013, 89–91).

What kind of education were boys in the *paedagogium* receiving, there or elsewhere?[79] To write, certainly; Keegan proposes that the wide variation in epigraphic technique reflects the different education levels of the writers, with most of the graffiti reflecting the style of beginning writers—the capitals characteristic of formal inscriptions—and only about 10% in the miniscule cursive used in documents (Keegan 2013, 75–76). Otherwise, the graffiti provide only a few possible indications of professions: *ianitor* and *custos*, doorkeepers or more broadly people who controlled access and provided protection; *opifer*, someone who provided medical aid; *episcopus*, a guardian or supervisor; *perfusor*, perhaps someone who poured water over bathers; and *pueri* and *iuvenes*, boys or young men, terms associated with close personal attendance; and *Capillatus*, discussed above (seemingly a personal name here) (Keegan 2013, 79–83).[80]

Mohler, Forbes, and the other scholars cited earlier tended to assume that the enslaved boys working as "pages" were also being trained for other things, including business, accounting, and administration, and Martial seems to corroborate this. Mohler (1940, 269) goes so far as to compare an enslaved Roman serving as a waiter to a contemporary college student waiting tables in the summer. Obviously this is characteristic of his overall rosy view of the situation, but he is also compensating for the overall lack of specific or specialized training mentioned in the sources. It is interesting that the imperial *paedagogium* reflects a similar gap. We know—from literary sources, from epigraphy, from abundant evidence in general—that enslaved people and freedmen of the imperial family had many roles as

[79] Again, Groen-Vallinga (2022, 189–190) doubts that education itself took place in the *paedagogium.*

[80] A number of words for fine clothing might also suggest a *vestiarius,* someone responsible for a wardrobe, and several personal names suggest service at banquets.

administrators, accountants, and managers, often in an organized hierarchy (Weaver 1972). If specialized education should be evident anywhere, the *paedagogium* seems like the correct place. But the graffiti give little sense of this; the titles and names cluster around roles of close personal service, and none of the graffiti allude to numbers or calculation. One graffito does refer to an imperial freedman as a *procurator*—a manager or administrator, though we do not know of what (Keegan 2013, 80). Here we are closer to the kind of mathematical or accounting training I have been discussing in this paper. But since the person named is a freedman, he would not have been a current student in the *paedagogium*, and his connection to the current students is unknown.

Students in the *paedagogium* may have received a mathematical education and simply never made graffiti about it, of course. Or perhaps only some of them received it, at least in an extensive way. The low percentage of graffiti using miniscule cursive (10%) may simply show that older students wrote less graffiti. Or it might mean that only a subset of students learned to write in this way—those destined for secretarial or administrative careers. Perhaps extensive numerical training was similarly undertaken only by a few. A final possibility is that such training primarily took place beyond the *paedagogium*. Particular students chosen for more administrative roles may have done most of their learning on the job (as indeed seems to have been the case with their counterparts in Egypt), as part of something like an apprenticeship, or simply by beginning in very low-level positions.

As school-like settings, *paedagogia* (possibly) and the sorts of schools mentioned by Martial (perhaps lower-level, used by those who did not enslave enough people to run schools within their houses) seem like the most likely contexts in which tables and problems like those found on papyri in Egypt would have been used. References to specific problems or types of exercises in Roman educational settings are rare, however, which makes it even more difficult than in Egypt to discern stages of learning and who was taught what. The most specific and oft-cited passage comes from a poem of Horace (*Ars Poetica* 325–330), who wrote that:

> *Romani pueri longis rationibus assem*
> *discunt in partis centum diducere. "Dicat*
> *filius Albini: si de quincunce remota est*
> *uncia, quid superat?... Poteras dixisse. — Triens. — Eu!*
> *Rem poteris seruare tuam. Redit uncia, quid fit?*
> *«Semis».*
> Roman boys learn through long calculations how to divide the *as* into a hundred parts.
> [Teacher]: Now let the son of Albinus say—if from five-twelfths one-twelfth [uncia] is taken, what remains? You could have already said.
> [Pupil]: One third.
> [Teacher]: Ah!—you'll be able to guard your wealth. One-twelfth is added—what is it?
> [Pupil]: One half.[81]

[81] The translation is my own.

The problems are simple, but the division into parts and the use of unit fractions suggest exercises similar to the fraction problems in the Greek texts from Egypt. It is likely that students memorized and worked with addition, multiplication, and division tables, and with unit fractions, especially though not exclusively according to the duodecimal system relevant for much of Roman calculation (including, as Horace alludes to here with the mention of guarding wealth, finance and inheritance). The fractions were written differently in Latin than in Greek, of course, but were still mostly unitary.[82] Horace's mention of *pueri*, and the dialogue with a teacher, obviously suggest a school setting. Despite the associations some Roman authors make between numerical education and low status, Horace's passage is not specific about status, and its overall context (contrasting Romans' sordid love of wealth with the loftier pursuits of the Greeks) seems applicable to any Roman. Problems like these were probably a feature of any school, whether for enslaved or free students.

Murkier in the Roman context are the more difficult problems from the papyri—those concerned with measuring areas and especially volumes, or those which are more puzzle-like. Authors such as Horace and Petronius do not mention these sorts of problems, either in relation to enslaved/formerly enslaved people or to freeborn people. Petronius makes clear that the mathematical education of a freedman such as Hermeros included fractions and stopped short of theoretical geometry (and most literature), but gives no hint of problems that centered on measuring real objects. Actual Latin collections of problems for students, meanwhile, are late. The earliest is *Propositiones ad acuendos juvenes*, attributed to Alcuin of York and thus to the eighth to ninth centuries CE. Some types of problems in it resemble those in the Greek papyri; many are new (Hadley and Singmaster 1992). Most are arithmetical problems or riddles, but there are also geometrical ones: these deal with areas (round, triangular, rectangular, quadrilaterial), not volumes. Problems such as those in *P.Math.* that deal with cylinders, granaries, and similar objects are absent. Alcuin's collection might mean that arithmetical and some geometrical problems circulated in Latin and were taught in schools much earlier, especially since some of the problem types and methods are familiar from Egyptian and Greek examples (Butzer 1995, 22).[83] Some of the treatises of the *agrimensores* contain problems, though problems are not the primary focus of most of them. But since these are specialist texts, they do not answer the question of whether such problems would have been part of a general mathematical education for someone who was not training specifically to be a surveyor.

The Roman agricultural writer Columella engages with several area problems in his treatise *De Re Rustica*, and at least one of these problems matches (except for the choice of numbers) a problem in the surveyors' text *De iugeribus metiundis*. Columella takes a tone similar to that found in educational texts, showing the reader

[82] See Bailey (2013, 177–178) for more discussion of this passage in relation to fractions and the Roman monetary system.

[83] Problems 9, 10, 21–25, and 27–31 use a formula for the area of a quadrilateral that was used by Egyptians and Roman *agrimensores*, for example.

the procedures to follow in measuring various shapes (not all of them logical shapes for fields). As Daniel Bertoni (2017) has shown in a perceptive analysis, Columella positions himself somewhat in between authors such as Hero of Alexandria and the surveyors with their more practical methods (though rather closer to the latter). While Hero in the *Metrica* includes geometrical proofs as well as alternative methods for calculating various problems, Columella adopts a tone and method of presentation closer to the surveyors, using the first person when solving the problem, giving only one way to solve it, and not including proofs. When showing how to measure the area of a circle segment, Columella also makes the algorithm less clear than Hero, introducing ambiguity that may have given rise to errors in later traditions (Bertoni 2017, 542–545).[84] Columella's style ultimately resembles school texts more than it does Hero, and he may have been drawing on manuals or collections of problems.

But Columella, even while showing his competence at such problems, distances himself from the technical knowledge overall, insisting that it is properly the domain of *geometrarum*, not Roman landowners (Bertoni 2017, 545–547). Columella's use of such problems thus again does not allow us to place them within a general mathematical education with any certainty, either for the freeborn or enslaved. And problems involving *volume* are more uncertain still. Columella and other Roman writers situate these problems in the context of technical expertise and specific professions. Presumably, the problems were seen as useful or necessary for certain professionals: surveyors, architects, engineers, builders. Some of these professionals were enslaved, though the number would have varied depending on the context and time period.[85] But how much their education overlapped with that of accountants, or how much of it took place in settings like *paedagogia*, is an open question. Some of them probably at least began in similar contexts in large Roman households or the imperial *familia* and then perhaps moved on to specialized training under individuals. Others would have been enslaved to begin with *by* professionals—by builders or contractors, for example, who might themselves have been freed slaves.[86] They might have spent little or no time in school-like settings, instead undergoing more

[84] Hero always makes clear which numbers correspond to base and height; Columella uses 4 for the height without further specification. The author of *De iug. met.* then multiplies by 4 in his algorithm, even though the height in his example is 5. In other words, he has mistaken the number 4 for being part of the algorithm, rather than being a specific height in Columella's computation.

[85] See Dilke (1971, 35–46) for surveyors; the situation in the Republic is somewhat unclear, due to a reference of Cicero's that implies large numbers of equestrian-ranked surveyors, but see Classen (1994), who thinks it unlikely these equestrians were performing the actual labor themselves. Later it is clear that surveyors were increasingly professionalized and that many were independent and had been trained in the military. Anderson (1997, 3–67) argues that the most prestigious architects were freeborn, but a number of others were enslaved/freed, mainly attested by epigraphic evidence (47–48). More builders were enslaved or freed, though again not all.

[86] DeLaine (2000, 120–122) discusses some of the evidence for the building industry in general: some builders and a *mensor* who seem to have been trained in aristocratic households, others who were trained by builders or carpenters. The tools depicted on tombs—compasses, levels, measuring rules, and so on—could belong to either architects or builders.

focused training from the start. But for those who came from wealthy Roman households—as most accountants or managers would have—did their early, shared education include problems such as those in *P.Math.*, especially the ones concerned with area and volume?

On balance, the evidence seems to suggest no. Direct mentions of problems in educational settings for boys in general, and especially for enslaved ones, focus on numerical skills, as discussed above. Meanwhile, writers such as Columella locate problems of measurement and area within a professional context, not a general educational one. *Propositiones ad acuendos juvenes*, our only actual problem collection, is too late to be of much help, but its inclusion of area problems perhaps complicates the impression left by Columella and other Roman writers. However, even this collection contains no problems involving three-dimensional objects. Finally, while the graffiti in the imperial *paedagogium* are merely suggestive—since the argument is an argument from absence and this is, after all, only one context—they do not reference mathematical education in any way, or even include any numbers. Some ambiguity lingers, of course. It is still possible to imagine these problems being taught in a school or *paedagogium*, if only to help identify talented students and assign them to particular kinds of training or professions, just as at least some students in the imperial *paedagogium* learned professional writing. But on balance, the evidence suggests that measurement and visualization problems played a minimal role in shared education; when they were taught at all, it may have been at later stages in focused, professional settings.

This scenario, if accurate, leaves little room for accountants, specifically, to have learned these sorts of problems—in other words, makes it unlikely that a text such as *P.Math.* would have been compiled or used by one in Rome. While there are, of course, many gaps once again, the textual references to the training of accountants or managers make no mention of such problems. The freedmen at Trimalchio's dinner, some of them explicitly accountants, give no hint that their education included anything similar (and the reference to having not studied geometry, while it might apply only to Euclidian geometry, may rule out geometrical problems as well). Columella discusses the training of managers in agricultural contexts, and expects them to do accounts,[87] but is more concerned that they have extensive experience of farming: he envisions them being trained in specific skills by different workers so that they themselves can evaluate work, and cautions against choosing an urban enslaved person to be a *vilicus* (which may suggest the latter was common practice, although this cannot be demonstrated for certain) (Carlsen 2010, especially 83–86).[88] Elsewhere, when Columella discusses his measurement problems, he gives no hint that this is a skill a manager should have (even when he discusses practical applications such as determining how much seed to plant in a field): the methods are for

[87] Despite elsewhere suggesting a *vilicus* might be illiterate; overall, however, he seems to expect otherwise.

[88] Other job titles, especially *actor* and *dispensator*, are less connected to direct farm management and more to accounting/finance, but their education still probably did not include the problems under discussion, for the reasons mentioned above.

geometers, or (with caveats) for landowners such as himself (Bertoni 2017). Sometimes, Roman writers suggest that trustworthiness is more important than accuracy for an accountant or manager in any case, downplaying mathematical skills overall (Gianella 2014, 140–141).[89]

I suggested earlier in this chapter that even if problems were not always directly relevant to accountants and managers in Egypt, this type of mathematical education may have had important consequences. It may have conferred some general prestige—since these types of problems were associated with surveyors, architects, and other skilled professionals, and were more related to geometry than the purely arithmetical problems—as well as individual prestige on the most skilled students. It seems to have overlapped with the education of at least some landowners and liturgists themselves. And finally, it perhaps created a certain amount of solidarity, as well as shared professional identity and pride. The specific managers studied by Rathbone and Takahashi had many other reasons for forming close ties, including their long careers for the same estates and their long residences in the same places. But education may have been another one.

Much of this was perhaps unavailable to enslaved Roman accountants in the West. They seem to have had few ways of forming a shared identity with professions that were more prestigious, such as those of the surveyors and architects (partly because fewer of their practitioners were enslaved). It seems likely that the education of the enslaved accountants diverged from that of other such professionals, perhaps early on, and was also perhaps more sharply differentiated than in Egypt from that of landowners or freeborn wealthier Romans. Literary sources situate the accountants firmly within the household and within the context of slavery in ways that suggest this status marks and defines them, perhaps more than it does the other professions (even when, again, members of these other professions were enslaved). This was perhaps because of the emphasis placed on trust (even sometimes above competence) in managerial positions, and the worry on the part of enslavers that this trust would (from their perspective) be abused. Petronius illustrates this when he has Trimalchio discuss how he became his master's favorite and then, as a result, effectively controlled his master as *dispensator*: the sexual exploitation of the enslaved by the master is turned on its head, perhaps a persistent fear for some Romans. Columella, in his agricultural treatise, is much more circumspect, but perhaps has something similar in mind when he cautions landowners not to appoint urban enslaved people (whom they may have known best and therefore trusted) as *vilici*.

In short, accountants and managers are depicted as *close*—raised within the household, exploited, trusted, feared, and, when freed (according to Petronius) carrying the taint of that close exploitation. The graffiti in the Palatine *paedagogium* suggest that these associations persisted among those enslaved by the imperial family, at least some of whom (based on the cursive handwriting and the mention of at least one *procurator*) became administrators. An inscription from a *columbarium*

[89] Some legal texts stress that good intentions were more important than accuracy.

(collective tomb of an elite household) commemorates a freedman *a cubiculo et procurator*, a bedchamber attendant and manager, again stressing the overlap in personal and managerial roles (Groen-Vallinga 2022, 205). Finally, both the sexual graffiti and the names given to the enslaved in the imperial *paedagogium* suggest ongoing reminders of their status, a continual sort of "putting in their place" or process of enslavement.[90]

Enslaved accountants may have, and probably did, form various kinds of solidarity with one another or other enslaved people, and they may have had professional pride as well. Those with complex posts clearly trained under others and often moved within a hierarchy that was professional, not only that of enslaver/enslaved. Professional pride may have developed more in advanced or on-the-job training than in school-like settings. The worries displayed by wealthy Romans that the accountants they enslaved would cheat, manipulate, or deceive them suggest an expertise that was at times inaccessible to the wealthy Romans (and thus, again, a source of pride for the accountants themselves). But on the whole, both the educational and working contexts for such accountants seem more fragmented and segregated than those suggested by the educational papyri in Egypt.

In Egypt, the fact that many managers and accountants were not enslaved had practical consequences for their well-being and physical autonomy. But this paper has also stressed educational differences with the Western Empire, and in particular the existence of a complex and challenging mathematical education that could generate professional identity and pride, and that overlapped more with early literary education than is sometimes acknowledged. In Egypt, it was not just that fewer accountants were enslaved, but that aspects of their education were shared with those of varying status—including town councilors and local notables. Whether this blurring of educational boundaries extended to Egyptian accountants who *were* enslaved is less clear. But at the very least, the educational mathematical texts, unlike the shorthand manual the *Commentary*, do not dwell on hierarchies, but instead on puzzles, visualization, and algorithms that would have challenged any student, enslaved or free.

References

Anderson, James. 1997. *Roman Architecture and Society*. Baltimore: Johns Hopkins University Press.

Ast, Rodney. 2015. Writing and the City in Later Roman Egypt. Towards a Social History of the Ancient 'Scribe' CHS Research Bulletin 4 (1). http://nrs.harvard.edu/urn-3:hlnc.essay:AstR.Writing_in_the_City_in_Later_Roman_Egypt.2016.

Azzarello, Giuseppina. 2019a. Μισούμενα on the Misoumenos: Neglected Tables of Fractions in P.Oxy. XXXIII 2656. *Trends in Classics* 11 (2): 241–255.

[90] See Keegan (2013, 84) on the names, many of which evoke objects or work, emphasizing the enslaved people as "speaking tools."

Azzarello, Giuseppina. 2019b. SB XX 15071 + P.Mich. III 146: The American Journey of a Parted "Table of Parts". *Archiv für Papyrusforschung und verwandte Gebiete* 65 (1): 52–98.
Azzarello, Giuseppina. 2021. Arithmetical Tables. In *Oxford Classical Dictionary*, 4th ed. https://doi.org/10.1093/acrefore/9780199381135.013.8523.
Bagnall, Roger. 1993. Slavery and Society in Late Roman Egypt. In *Law, Politics, and Society in the Ancient Mediterranean World*, ed. Baruch Halpern and Deborah W. Hobson, 220–240. Sheffield: Sheffield Academic Press.
Bagnall, Roger. 2010. The Readers of Christian Books: Further Speculations. In *I papiri letterari cristiani: atti del convegno internazionale di studi in memoria di Mario Naldini*, ed. Guido Bastianini, 23–30. Firenze: Istituto Papirologico G. Vitelli.
Bagnall, Roger, and Bruce Frier. 1994. *The Demography of Roman Egypt*. Cambridge: Cambridge University Press.
Baillet, J. 1892. Le papyrus mathematique d'Akhmîm. Paris: Libraire de la Société Asiatique.
Bagnall, Roger, and Alexander Jones. 2019. *Mathematics, Metrology, and Model Contracts: A Codex from Late Antique Business Education (P.Math.)*. New York: New York University Press.
Bailey, Melissa. 2013. Roman Money and Numerical Practice. *Revue belge de philologie et d'histoire* 91 (1): 153–186.
Bergamesco, Marco. 1995. Le διδασκαλικαί nella ricerca attuale. *Aegyptus* 75 (1/2): 95–167.
Bernard, Alain, Christine Proust, and Micah Ross. 2014. Mathematics Education in Antiquity. In *Handbook on the History of Mathematics Education*, ed. Alexander Karp and Gert Schubring, 27–53. New York: Springer.
Bertoni, Daniel. 2017. Geometry and Genre in Columella. *The American Journal of Philology* 138 (3): 527–554.
Booth, Alan D. 1979. The Schooling of Slaves in First-Century Rome. *Transactions of the American Philological Association (1974-2014)* 109:11–19.
Boyaval, B. 1977. Le Cahier de Papnouthion et les autres cahiers scolaires Grecs. *Revue Archéologique, Nouvelle Série* 2:215–230.
Brashear, William, Wolf-Peter Funk, James M. Robinson, and Richard Smith. 1990. *The Chester Beatty Codex Ac. 1390. Mathematical School Exercises in Greek and John 10:7-13:38 in Subachmimic*. Leuven: Peeters.
Butzer, P. L. 1995. Mathematics in Egypt and Its Connection with the Court School of Charlemagne. *Contemporary Mathematics* 190:1–30.
Campbell, Brian. 2000. *The Writings of the Roman Land Surveyors: Introduction, Text, Translation, and Commentary*. London: Society for the Promotion of Roman Studies.
Carlsen, Jesper. 1995. *Vilici and Roman Estate Managers until AD. 284*. Rome: "L'Erma" di Bretschneider.
Carlsen, Jesper. 2010. Recruitment and Training of Roman Estate Managers in a Comparative Perspective. In *Bulletin of the Institute of Classical Studies. Supplement No. 109, By the Sweat of Your Brow: Roman Slavery in Its Socio-Economic Setting*, ed. Ulrike Roth, 75–90. Oxford: Oxford University Press.
Chouquer, G., and F. Favory. 2001. *L'Arpentage Romain: histoire des textes, droit, techniques*. Paris: Errance.
Classen, C. Joachim. 1994. On the Training of the Agrimensores in Republican Rome and Related Problems: Some Preliminary Observations. *Illinois Classical Studies* 19:161–170.
Cribiore, Raffaella. 1996. *Writing, Teachers, and Students in Graeco-Roman Egypt*. Atlanta: Scholars Press.
Cribiore, Raffaella. 2001. *Gymnastics of the Mind: Greek Education in Hellenistic and Roman Egypt*. Princeton: Princeton University Press.
Cuomo, Serafina. 2001. *Ancient Mathematics*. New York: Routledge.
Cuomo, Serafina. 2020. Thinking in Cases in Ancient Greek Mathematics. In *Thinking in Cases: Ancient Greek and Imperial Chinese Case Narratives*, ed. Markus Asper, 97–107. Berlin: De Gruyter.

DeLaine, Janet. 2000. Building the Eternal City: The Construction Industry of Ancient Rome. In *Ancient Rome: The Archaeology of the Eternal City*, ed. John Coulston and Hazel Dodge, 119–141. Oxford: Oxford University School of Archaeology.

Dilke, O. A. W. 1971. *The Roman Land Surveyors: An Introduction to the Agrimensores*. Newton Abbot: David & Charles.

Eddy, Matthew Daniel. 2013. The Shape of Knowledge: Children and the Visual Culture of Literacy and Numeracy. *Science in Context* 26 (2): 215–245.

Forbes, Clarence A. 1955. The Education and Training of Slaves in Antiquity. *Transactions and Proceedings of the American Philological Association* 86:321–360.

Fournet, Jean-Luc, and Claudio Gallazi. 1996. Une tablette scolaire mathématique de Tebtynis. *Bulletin deml'Institut Français d'Archéologie Orientale* 96:171–176.

Fowler, David. 1999. *The Mathematics of Plato's Academy: A New Reconstruction*. Oxford: Clarendon Press.

Freeman, Michael A. 2023. The Hands That Write: Life and Training of Greco-Roman Scribes. PhD diss., Duke University.

Gianella, Nicole Julia. 2014. The Mind of the Slave: The Limits of Knowledge and Power in Roman Law and Society. PhD diss., University of Southern California.

Gierdien, Faaiz. 2009. Musings on Multiplication Tables and Associated Mathematics and Teaching Practices. *Pythagoras* 70:16–31.

Goldthwaite, Richard. 2015. The Practice and Culture of Accounting in Renaissance Florence. *Enterprise & Society* 16 (3): 611–647.

Groen-Vallinga, Miriam. 2022. *Work and Labour in the Cities of Roman Italy*. Liverpool: Liverpool University Press.

Hadley, John, and David Singmaster. 1992. Problems to Sharpen the Young. *The Mathematical Gazette* 76 (475): 102–126.

Harper, Kyle. 2011. *Slavery in the Late Roman World, AD 275-425*. Cambridge: Cambridge University Press.

Hawkins, Cameron. 2016. Contracts, Coercion, and the Boundaries of the Roman Artisanal Firm. In *Work, Labour, and Professions in the Roman World*, ed. Koenraad Verboven and Christian Laes, 36–61. Leiden: Brill.

Hickey, Todd. 2012. *Wine, Wealth, and the State in Late Antique Egypt: The House of Apion at Oxyrhynchus*. Ann Arbor: University of Michigan Press.

Howley, Joseph. 2025. Despotics. In *Writing, Enslavement, and Power in the Roman Mediterranean, 100 BCE–300 CE*, ed. Jeremiah Coogan, Joseph Howley, and Candida Moss, 24–43. Oxford: Oxford University Press.

Høyrup, Jens. 1990. Sub-Scientific Mathematics: Observations on a Premodern Phenomenon. *History of Science* 28 (1): 63–87.

Husselman, Elinor M. 1952. The Granaries of Karanis. *Transactions and Proceedings of the American Philological Association* 83:56–73.

Imhausen, A. 2016. *Mathematics in Ancient Egypt: A Contextual History*. Princeton: Princeton University Press.

Johnson, J. D. M., Victor Martin, and Arthur S. Hunt. 1915. *Catalogue of the Greek Papyri in the John Rylands Library, Manchester*. Vol. II. Manchester: Manchester University Press.

Jones, Alexander. 2009. Mathematics, Science, and Medicine in the Papyri. In *The Oxford Handbook of Papyrology*, ed. Roger S. Bagnall, 338–357. Oxford: Oxford University Press.

Kaster, Robert A. 1983. Notes on 'Primary' and 'Secondary' Schools in Late Antiquity. *Transactions of the American Philological Association* 113:323–346.

Keegan, Peter. 2013. Reading the 'Pages' of the *Domus Caesaris: Pueri Delicati,* Slave Education, and the Graffiti of the Palatine Paedagogium. In *Roman Slavery and Roman Material Culture*, ed. Michele George, 69–98. Toronto: University of Toronto Press.

Kirsh, Ella. 2024. 'Etched Into the Soul': The Education of Shorthand Writers in Late Antiquity. *The Journal of Roman Studies* 114:61–84.

Kutner, Melissa Bailey. 2025. Numeracy. In *Writing, Enslavement, and Power in the Roman Mediterranean, 100 BCE–300 CE*, ed. Jeremiah Coogan, Joseph Howley, and Candida Moss, 106–121. Oxford: Oxford University Press.

LaGroue, Lance Elliott. 2014. Accounting and Auditing in Roman Society. PhD diss., University of North Carolina at Chapel Hill.

Lougovaya, Julia. 2022a. Computational Workout: Division Tables as Training Exercises. In *Proceedings of the 29th International Congress of Papyrology*, ed. Mario Capasso, Paola Davoli, and Natascia Pellé, 661–672. Lecce: Centro di Studi Papirologici dell' Università del Salento.

Lougovaya, Julia. 2022b. On Cubic and Other Volumetric Cubits and Fingers. *Pylon* 1.

Lougovaya, Julia, and Rodney Ast. 2023. Between Ahmes and Alcuin: P.Bodl. 1 7 Revisited. *Pylon* 3.

Marrou, H. I. 1956. *A History of Education in Antiquity*. New York: Sheed and Ward.

Mazza, Roberta. 2001. *L'archivio degli Apioni: terra, lavoro, e proprieta' senatoria nell'Egitto tardoantico*. Bari: Edipuglia.

McConnell, Ryan. 2017. *Getting Rich in Late Antique Egypt*. Ann Arbor: University of Michigan Press.

Minaud, Gérard. 2005. *La comptabilité à Rome: essai d'histoire économique sur la pensée comptable commerciale et privée dans le monde antique romain*. Lausanne: Presses polytechniques et universitaires romandes.

Mohler, S. L. 1940. Slave Education in the Roman Empire. *Transactions and Proceedings of the American Philological Association* 71:262–280.

Moss, Candida. 2024. *God's Ghostwriters: Enslaved Christians and the Making of the Bible*. New York: Little, Brown and Company.

Parsons, P. J. 1970. A School-Book from the Sayce Collection. *Zeitschrift für Papyrologie und Epigraphik* 6:133–149.

Rathbone, D. 1981. The Weight and Measurement of Egyptian Grains. *Zeitschrift für Papyrologie und Epigraphik 53* (1983): 265-275.

Rathbone, Dominic. 1991. *Economic Rationalism and Rural Society in Third-Century AD Egypt: The Heroninos Archive and the Appianus Estate*. Cambridge: Cambridge University Press.

Rathbone, D. 2013. Village Markets in Roman Egypt: the case of first-century AD Tebtunis. In Kauf, Konsum und Markte : Wirtschaftswelten im Fokus: van der romischen Antike bis zur Gegenwart, ed. Monika Frass, 123-144. Wiesbaden: Otto Harrassowitz.

Rickman, Geoffrey. 1971. *Roman Granaries and Store Buildings*. Cambridge: Cambridge University Press.

Riggsby, Andrew. 2019. *Mosaics of Knowledge: Representing Information in the Roman World*. Oxford: Oxford University Press.

Roberts, C. H. 1938. *Catalogue of the Greek Papyri in the John Rylands Library, Manchester*. Vol. III. Manchester: Manchester University Press.

Robson, Eleanor. 2003. Tables and Tabular Formatting in Sumer, Babylonia, and Assyria, 2500 BC-50 CE. In *The History of Mathematical Tables: From Sumer to Spreadsheets*, ed. Martin Campbell-Kelly, Mary Croarken, Raymond Flood, and Eleanor Robson, 19–47. Oxford: Oxford University Press.

Roby, Courtney. 2023. Learning from Mistakes: Constructing Knowledge in Late Antique Mathematical Texts. In *Knowledge Construction in Late Antiquity*, ed. Monika Amsler, 155–187. Berlin: De Gruyter.

Ruffini, Giovanni R. 2018. *Life in an Egyptian Village in Late Antiquity: Aphrodito Before and After the Islamic Conquest*. Cambridge: Cambridge University Press.

Saller, Richard. 1994. *Patriarchy, Property and Death in the Roman Family*. Cambridge: Cambridge University Press.

Sampson, C. Michael. 2022. New Light on P.Mich 16 186. *Pylon* 2.

Sarris, Peter. 2006. *Economy and Society in the Age of Justinian*. Cambridge: Cambridge University Press.

Sidoli, Nathan. 2015. Mathematical Education. In *A Companion to Ancient Education*, ed. W. Martin Bloomer, 387–400. Malden: Wiley Blackwell.

Silver, Morris. 2011. Contractual Slavery in the Roman Economy. *Ancient History Bulletin* 25:73–132.

Skeat, T. C. 1936. A Greek Mathematical Tablet. *Mizraim* 3:1–8.

Smyly, J.G. 1920. Some Examples of Greek Arithmetic. Hermathena 19: 105–114.

Swetnam-Burland, Molly. 2021. Women's Work? Investors, Money-Handlers, and Dealers. In *Women's Lives, Women's Voices: Roman Material Culture and Female Agency in the Bay of Naples*, ed. Brenda Longfellow and Molly Swetnam-Burland, 29–49. New York: University of Texas Press.

Takahashi, Ryosuke. 2021. *The Ties That Bind: The Economic Relationships of Twelve Tebtunis Families in Roman Egypt*. London: Institute of Classical Studies, University of London.

Weaver, P. R. C. 1972. *Familia Caesaris: A Social Study of the Emperor's Freedmen and Slaves*. Cambridge: Cambridge University Press.

Wilson Jones, Mark. 2000. *Principles of Roman Architecture*. New Haven: Yale University Press.

Wrenhaven, Kelly L. 2015. Slaves. In *A Companion to Ancient Education*, ed. W. Martin Bloomer, 464–473. Malden: Wiley Blackwell.

Chapter 3
Metrology of Volume in Greco-Roman Egypt: Evidence from the Papyri

Julia Lougovaya

Abstract Methods for calculating the volume of solids expounded in Greco-Roman mathematical texts may lead one to assume that the volume of 'real life' objects would be calculated according to those methods, too, with the shape of the object determining the method. Texts stemming from occupational contexts, however, reveal that in practice not only the shape of the object, but also the material which it consisted of influenced the way its volume was computed. Largely relying on papyrological texts that either recorded actual measurements or else were meant to train practitioners how to perform mensuration, this paper offers for the first time a survey of practical metrology of volume in Greco-Roman Egypt. It demonstrates consistency in the choice of computational procedures for quantifying three types of materials—stone, earth, and wood—and examines possible reasons for apparent deviations from the 'normative' geometry of solids.

Keywords Ancient metrology and mensuration · Volume computation · Practitioners' mathematics · Greco-Roman Egypt · Problems in papyri

3.1 Introduction

In the Greco-Roman world, just as in other ancient cultures, conventions used for quantifying three-dimensional entities depended on the nature of materials being measured. Metals, for example, were quantified by weight, while cereals, pulses, oil, wine and some other commodities that do not by themselves form a fixed three-dimensional object were measured with the help of measuring containers in units of

J. Lougovaya (✉)
Universität Heidelberg, Heidelberg, Germany
e-mail: lougovaya@uni-heidelberg.de

Senthil Babu D., R. Wagner (eds.), *Normative Texts and Occupational Practices*, Boston Studies in the Philosophy and History of Science 350,
https://doi.org/10.1007/978-3-032-28419-8_3

dry or liquid capacity.[1] Three types of materials, however, were measured in terms of the amount of space occupied by the object they formed. These were stone, earth, and wood that had been worked into simple geometric solids, such as blocks of stone, heaped or excavated earth, and pieces of timber or trunks of felled trees. The spatial extent of these materials was determined not by means of standardized containers or weights, but by computations based on linear dimensions of the objects made of them, and expressed in units of volume.[2]

Greek writings on metrology and mensuration expounded methods for calculating the volume of many solids, which may lead one to assume that the volume of 'real life' objects would be calculated according to these same methods, with the shape of the object determining the method. Papyrological evidence, however, indicates that practical conditions of working with actual materials, that is the occupational context in which practitioners engaged directly with objects made of stone, earth, or wood, determined the methods used for quantifying them. This paper offers for the first time a survey of practical metrology of volume in Greco-Roman Egypt, which is based largely on papyrological sources. It demonstrates consistency in the choice of computational procedures for quantifying stone, earth, and wood and examines possible reasons for apparent deviations from the 'normative' geometry of solids.

3.2 Sources on Practical Metrology of Volume

The evidence for the volume computation of objects defined by their shapes, magnitudes and materials—as opposed to abstract geometric solids—is furnished largely by two kinds of textual sources. First, there is the corpus of metrological and mensurational texts transmitted in manuscripts. Many of these texts, which have been assembled by Heiberg under the headings *Definitiones*, *Geometrica*, *Stereometrica* I, *Stereometrica* II and *De mensuris* and published in volumes IV (1912) and V (1914) of his edition of Heron of Alexandria, are conventionally referred to as the pseudo-Heronian corpus.[3] They contain definitions of geometrical shapes and of units of measurement, metrological conversions, and mensurational

[1] The two-volume edition of Greek and Latin metrological texts in Hultsch 1864–1866 and the chapter on metrology in Greco-Roman Egypt in Wilcken 1899, 1:738–780, remain indispensable. For a brief summary of the subject, see Bagnall 2009, 185–189. Berg 2020 offers a good survey of the visual representation of measuring containers with a focus on modii, and measuring grain is illustrated in the mosaics of the so-called Hall of the Grain Measurers in Ostia, for which see, for example, https://www.ostia-antica.org/regio1/19/19-1.htm. There has been, however, no treatment of volume metrology of materials in the form of solids, the common assumption being that their volumes were computed as if for abstract geometrical solids.

[2] Volumes of granaries were determined similarly to the volume of excavations, but then converted to the units of capacity to express the amount of grain they contained, see below.

[3] On the corpus of Greek metrological writings, its constituent parts, and its relation to Heron of Alexandria and to 'the practitioners' tradition', see Høyrup 1996, esp. 7–13; Vitrac 2010; Acerbi and Vitrac 2014, 430–516.

problems with instructions on how to solve them. Problems concerned with solids tend to be grouped according to the shapes of the objects. Most do not make provisions for material-specific mensuration but expound the normative methods for determining the volume of solids according to their shape. Those rare exceptions that are concerned with measuring objects made of specific materials tend to be corrupt in the manuscripts. This is exemplified best by a small treatise ascribed to Didymus of Alexandria and transmitted in two different versions under the titles *On Measurement of Marble and All Sorts of Timber* or *On Measurements of All Sorts of Timber*, both of which are ridden with mistakes and omissions.[4] The reason for the poor transmission of texts related to material specific metrology is likely owed to the fact that surviving manuscripts are based on collections of earlier texts. Some of these texts were normative in the sense that they focused on properties of abstract geometrical solids, while others must have come from occupational contexts and were aimed at instructing actual practitioners in how to measure concrete physical materials with the choice of material-specific units and computational procedures being included. For a medieval copyist, material-oriented mensurational problems may have looked like geometrical problems, the setting for which was little more than a rhetorical device, but which seemed flawed in their use of units or computational procedures. Consequently, material-dependent specifications could be omitted in the process of copying and computational procedures or values of variables 'corrected', with the result that texts concerned with practical metrology in the manuscript tradition "survived in a state of deep confusion and disorder" or disappeared completely.[5]

The second type of sources on practical volume metrology comes from papyrological evidence, that is, unique texts preserved by the dry climate of Egypt, not in medieval manuscripts. These texts, which were produced as a result of actual measuring of physical objects or else as teaching material to instruct practitioners in how to measure such objects, include accounts, receipts or contracts which mention quantified materials (though not the calculations that produced the stated quantities), metrological conversions, and mensurational problems. Combined, they shed light on such particularities of material-specific volume mensuration as the types and shapes of objects, in the form of which the material was measured, units of measurement, and computational procedures. The prevalence of papyrological sources means that the metrology of volume gleaned from them pertains primarily

[4] The first title (μέτρα μαρμάρων καὶ παντοίων ξύλων) is attached to an apparently more corrupt version of the work preserved in Cod. Parisin. Gr. suppl. 387, fol. 105r–107v, dated to the fourteenth c. (codex *C*), and the second title (περὶ παντοίων ξύλων τῆς μετρήσεως) heads a shorter and somewhat more coherent version in Cod. Const. Pal. vet. 1, fol. 64r–66r. The latter codex (*S*, also called *Seragliensis* G I 1), which used to be dated to the eleventh c., has been identified as the work of Ephrem, a copyist known to be active in the middle of the tenth century, and consequently a date in the 960s has been argued for its production; cf. Prato 1993, 279–281, Acerbi and Vitrac 2014, 91–92. The transcription, photographs and translation of codex *S* can be found in Bruins 1964; for a possible reconstruction of the text on the basis of primarily codices *S* and *C*, cf. the critical edition in Heiberg 1927.

[5] Bernard et al. 2014, 51.

to Greco-Roman Egypt, even if evidence for some similar conventions can be found outside Egypt, too. It is thus on the practical metrology of volume in Greco-Roman Egypt that I focus here while also drawing on metrological texts transmitted in manuscripts.

3.3 Occupational Context

One of the crucial points in understanding the practical metrology of volume and its apparent deviation from 'normative' geometry lies in its intrinsic connection to occupational contexts. Papyrological documents record data that was derived from actual measurements and computations related to such projects as the construction of dikes, the quarrying of stone, and the building of houses, to name just a few, even if they usually contain little or no details about the computational methods used. Mensurational problems, however, are often presented as concrete examples of such projects and consequently the methods of computation that they demonstrate in order to determine the volume of a given object are not abstract but are chosen to fit the circumstances in which the object was quantified. These problems served as training exercises and likely circulated in specialized schools or else were copied by individuals aspiring to occupations that required basic computational skills and competence in trade-specific metrology.[6] While we are not well informed about these schools, and the course and content of elementary and practical mathematical education remain elusive,[7] problems in volume metrology are indicative of the conscious adjustment of elementary mathematical skills to occupational contexts, which are traceable at two levels. First, the physical, practical and economic conditions related to working with stone, earth, and wood likely influenced the way these materials were quantified, whether or not practitioners such as builders, carpenters, or quarrymen recorded the measurements or performed the calculations themselves. Secondly, those in charge of (ac)counting and recording—clerks, overseers or other kinds of administrative agents—would have to be well versed both in elementary

[6] On such specialized schools see Cribiore 2001, 182–184, who refers to them as 'scribal.' Bagnall and Jones (2022, 55–56), however, question the term 'scribal' and envisage the environment as a school devoted to the training of business agents or similar professionals, which may have been the context in which *P.Math*., a papyrus codex dated to the fourth century CE and containing a collection of problems in practical mathematics along with metrological texts and model contracts, was produced.

[7] Greek literary sources were pointedly not interested in the acquisition of elementary or practical skills, whether in literacy or in numeracy, with the consequence that "sources for the *idealized* curricula are much richer and substantial for Greco-Roman antiquity" than for training in elementary and practical mathematics, cf. Bernard et al. 2014, esp. 43–47 (quoted p. 47). For mathematical education in Greco-Roman antiquity, see Jones 2012, Bernard et al. 2014, Sidoli 2015. Thanks largely to the study of Greek school texts on papyri by Rafaella Cribiore (1996), the *ordo docendi* for teaching how to read and write in Greco-Roman Egypt is now well mapped (cf. too Johnson 2015), but that for numeracy remains a desideratum, see Netz 2022, 265.

calculations and in different trade-specific computations, and it would have been part of their training to learn which method to select for determining the volume of an object according to the material it was made from. Since accounting and overseeing could be performed in Greco-Roman Egypt at various levels and by different people both in their private or official capacity, all depending on the period, place as well as the nature and scale of projects and transactions, it is difficult to associate practical mensuration with a certain social stratum. The difficulty is exacerbated by the fact that the evidence for training in practical metrology is not chronologically or geographically homogeneous. Thus, the number of problems and metrological texts in the papyri increases in the second part of the fourth century CE, which might correspond to changes of administrative and taxation practices that led to an increased demand for people capable of carrying out practical calculations.[8] It is conceivable that in the earlier period a larger part of practical computation was performed by agents or officials trained for the purpose in a way that, for some reason, left less of a paper trail.[9]

3.4 Basic Principles

Two basic principles of practical volume mensuration can be immediately observed in papyrological evidence. First, most volume computations were based on 'averaging approximation'. This means that any solid with a tapering shape would be considered equal in volume to a prism or a cylinder having as its base the cross-section half-way between the lower and upper surfaces of the original solid. Correspondingly, its volume was computed as the product of the area of the mean cross-section and the height (or depth). The averaging-approximation procedure would be followed even when a method producing a more accurate result could be as easily employed, yet it seems to have been avoided for measuring one particular material, stone.

Secondly, the system of volumetric units and subunits used to express volume in practical mensuration differed from the one described in the metrological corpus. In the latter, units and subunits of volume (or area) were commonly defined in terms of units of length applied to three (or two) dimensions, as exemplified by the following definition of the foot:

[8] See, for example, Cuomo 2001, 214–215. The increase in evidence is particularly pronounced in Upper Egypt, the reasons for which are yet to be understood.

[9] There is no reason to assume that, at least in Greco-Roman Egypt, calculations were performed by slaves trained for the purpose. While it is of course perfectly possible that there were plenty of slaves trained in practical mathematics and metrology, the papyri furnish no evidence that associates specifically slaves with performing professionally any kind of calculating or accounting in general. Cf. Kutner in this volume (p. 24), who rightly notes that "many managers and accountants in Egypt were not enslaved, and so while this [sc. more advanced practical] training *may* have been distinct from elite education it was not limited to the enslaved."

ὁ πούς ὁ εὐθυμετρικὸς ἔχει παλαιστὰς δ, δακτύλους ις, ὁ δὲ ἐπίπεδος ἔχει παλαιστὰς ις, δακτύλους σνς, ὁ δὲ στερεὸς πούς ἔχει παλαιστὰς ξδ, δακτύλους Δϙς.
The linear foot has 4 palms, 16 fingers; the area foot has 16 palms, 256 fingers; the solid foot has 64 palms, 4096 fingers. (Pseudo-Heron, *Definitiones* 132, ed. Heiberg 1912)[10]

In practical metrology, however, the relations between the base volumetric units and their subunits were the same as between the corresponding units and subunits of length. Thus, the volumetric foot, which was conceptualized as a cube with the edge equaling 16 fingers, just as in the normative texts, comprised 16, not 4096, volumetric fingers, while the volumetric cubit was divisible into 24 volumetric fingers, because one linear cubit contained 24 fingers. Consequently, these 'fingers', which can be envisaged as one-finger-thick 'slices' of the cubes of the corresponding volumetric units, were relational and not absolute units: a finger of the volumetric foot differed in size from a finger of the volumetric cubit. The 'absolute' volumetric subunits, i.e. those defined by the subunit of length applied to three dimensions, could be used to express the size of volumetric units and subunits, but never to express the volume of an object in practical mensuration. These basic principles of practical metrology, however, could play out differently depending on what material was measured.[11]

3.5 Survey of Material-Specific Metrology of Volume

Since the focus of my survey is on practical metrology, it is organized not by the shapes of geometrical solids and computational procedures used to determine their volumes, but by the materials, which were measured in terms of volume. For ease of comparison, three aspects of mensuration for each material are considered: (a) the circumstances, or occupational contexts, that called for its quantification and the shapes that objects made of it tended to have; (b) the linear and volumetric units and subunits; (c) and the computational procedures used for determining the volume.

3.5.1 Stone

Circumstances and Shapes

Stone was used primarily for architectural projects and sculpture and must have been measured routinely. Often, however, only linear dimensions of a quarried or

[10] Unless specified otherwise, all translations are mine.

[11] Both basic principles of Greco-Roman practical metrology find parallels in mensurational practices of other ancient cultures. For the averaging approach to computing the volume of tapering solids in Egyptian, Mesopotamian, Indian, or Chinese volume computations, cf. a survey in Friberg 1996. For the concept of volumetric units as surfaces raised to a height rather than as cubes in Mesopotamia and China, cf. Høyrup 2019, 210–211; Chemla and Ma 2020, 245–259; Proust 2022, esp. 205–206.

worked piece were needed and written down, but occasionally in documentary evidence and in some mensurational problems the volume of stone blocks was computed, too. Since stone was quarried in the shape of rough cuboids or columns, a rectangular prism and a cylinder were the most common shapes, while mensurational problems also feature pyramids, cones, and their frusta.

Linear and Volumetric Units

The basic linear unit for measuring stone in Greco-Roman Egypt was the foot. A length smaller than a foot could be expressed either as a fraction of the foot or in fingers and/or palms with the usual ratio:

1 foot = 4 palms = 16 fingers

A first-century CE papyrus roll, *MPER* I 1, which contains problems concerned with finding the volume of various stone solids, specifies in the introductory part that the unit of length for measuring stones is the foot of four palms, and not a schoinion or a cubit.[12] Although occasionally cubits did appear in accounts or problems concerned with measuring stones, these instances remained exceptional to the end of Greco-Roman antiquity.[13] Mixing cubits and feet in measuring the same object was avoided.[14]

The volume of a stone object was usually expressed in feet, which were sometimes specified as 'solid' feet (στερεοὶ πόδες), with a solid foot corresponding to a cube with the edge of one foot. Quantities less than one solid foot could be rendered as its fractions or with the help of a subunit called a 'finger' or sometimes a 'solid finger' (στερεὸς δάκτυλος), which equaled 1/16th of the (cubic) foot.[15] This solid finger can be visualized as a 'slice' of a (cubic) foot measuring 1 foot × 1 foot × 1 finger. A solid palm, which correspondingly equaled one-fourth of the solid cubit, could also be used to express volume.

Whereas volumetric 'solid' units and subunits were used in actual mensuration of objects, metrological texts provide their equivalencies expressed in absolute unit called a 'simple' or 'ordinary' finger (ἀγελαῖος δάκτυλος) which corresponds to a cube with the edge of one finger, i.e. what we would call a cubic finger. These units were used only notionally to express the size of volumetric units: the solid foot

[12] Cf. col. 1, l. 2–col. 2, l. 4: ὁ ποὺς ὧι μετρεῖται τὰ λ̣[ι]θ̣ι̣κὰ ἔργα τετραπάλαιστος καλεῖτε (lege καλεῖται)… ταῦτα (sc. λιθικὰ ἔργα) γὰρ οὔτε σχοινίωι οὔτε πήχει μετρεῖται, "the foot in which stone works are measured is called the four-palm foot … for stone works are measured neither in schoinia nor in cubits."

[13] Cf. *P.Chester Beatty AC 1390*, page 2, lines 1–6.

[14] When both units occur in the measurements of the same object, as, for example, in Problem 4 of P.Berol. inv. 11529 verso (second c. CE), one of them (cubit in this case) is almost certainly a copying mistake. The papyrus is edited in Schubart 1916, but the edition needs revising.

[15] The designation 'cubic' refers to the size of the unit, while the parentheses indicate that the qualifier is a modern addition.

equaled in size to 4096 ordinary fingers and its subunit solid finger equaled to 256 ordinary fingers.[16]

The concept of expressing the volume of stone blocks in volumetric subunits that stand in the same relation to the volumetric units as their linear counterparts is documented both earlier in Egypt and elsewhere in the Greek world. It can be illustrated by Papyrus Reisner I, a hieratic papyrus dating to the nineteenth century BCE, which contains records of linear dimensions along with the volume of stone blocks used for a construction project.[17] In this papyrus, as was common in Egyptian metrological practice, the main unit of length is a cubit comprised of seven (not six as in the Greco-Roman period!) palms.[18] Linear dimensions smaller than a cubit are expressed either as a fraction of a cubit or by the subunits of a palm and a finger. The volume is correspondingly computed and recorded either by the number of cubits, including the fractional part, or in cubits, palms and fingers. The ratio of volumetric units to subunits remains the same as that of the corresponding units of length: the (volume) cubit is subdivided into seven (volume) palms or 28 (volume) fingers. Their visualization is presented in Fig. 3.1.

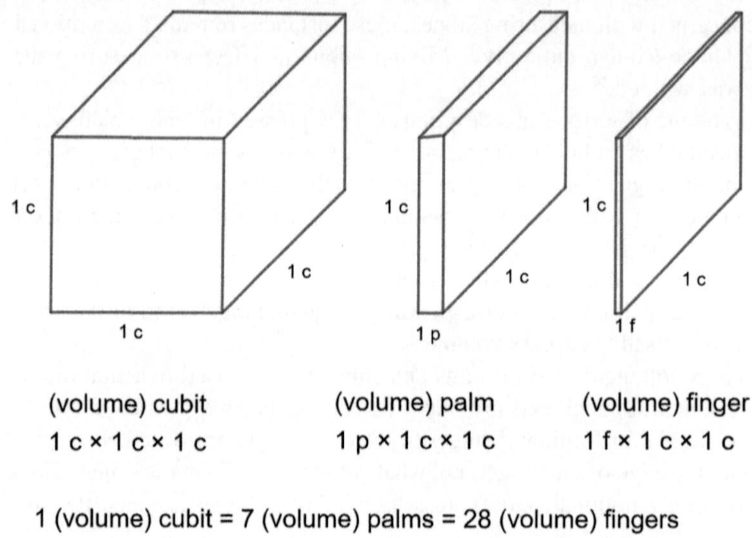

Fig. 3.1 Visualization of the Egyptian volumetric cubit and its subunits after Papyrus Reisner I

[16] Cf. the discussion of metrological units in Gerstinger and Vogel in *MPER* I 1, 52–53, and for the nomenclature see below n. 39.

[17] For an analysis of computations in this papyrus in the context of architectural practices, see foremost Imhausen and Rossi 2009; for a detailed discussion of possible computational techniques used to compute the volume in this papyrus, cf. also Gillings 1972, 218–234.

[18] A survey of metrological systems in Pharaonic Egypt can be found in, for example, Imhausen 2016, 41–51.

Outside Egypt an analogous way of measuring stone can be seen in building accounts pertaining to construction work in the sanctuary of Apollo at Didyma in Asia Minor. Inscriptions spanning over a century display annual records of work, and several of them list the number and the corresponding volume of limestone and marble blocks used in a certain course of the building or transported to the sanctuary. Since no linear dimensions of the blocks are given, it is impossible to ascertain how the volume was computed, but the consistently small number of volumetric subunits in recorded quantities indicates that they were relational, not absolute, as can be illustrated by the following excerpt from an account dated to ca. 170 BCE (*SEG* 26.1235, lines 23–8):

> ὑπὸ δὲ τῶν ὑπὸ Ἀπολλᾶν λίθοι λευκοὶ ἑβδlομήκο[ν]τα καὶ ὀκτώ, ὧν αὐχένιαι ἓξ κεφαλαί, ἐννέαl διάτọ[ι]χοι, δύο ψευδοδιατοίχο<υ>ς, κυματιαίους δεκlαπέ[ν]τε, ἀγελαίους τριάκοντα ἐννέα, ὧν πόδες στlερε[οὶ] δισχίλιοι ὀκτακόσιοι ἑξήκοντα ἓξ δῶρον δάlκ̣τυλοι δύο …
>
> Work performed under (sc. the foreman) Apollas: seventy-eight marble blocks, including six pilaster capitals, nine bonding blocks, two facing (?) bonding blocks, fifteen blocks with cyma molding, thirty-nine plain blocks, which amount to *two thousand eight hundred sixty-six solid feet, a palm, two fingers*.[19]

Computational Procedure

Most of the evidence for finding the volume of stone comes from just one papyrus, *MPER* I 1, but the 30+ problems it contains furnish a spectrum of geometrical solids which include cylinders, prisms, triangular and rectangular pyramids and their frusta, cones and conical frusta. Computational procedures employed in the papyrus produce accurate results,[20] but some of them seem to be more complicated than necessary. Thus, the procedure for computing the volume of a truncated square pyramid has the following form, where a_1 is the length of the edge of the bottom and a_2 of the top faces:

$$V = \left[\left(\frac{a_1 + a_2}{2} \right)^2 + \frac{1}{3} \left(\frac{a_1 - a_2}{2} \right)^2 \right] h$$

The cumbersome form of this computational procedure is puzzling and its origin has been disputed, especially because a simpler method was already known in

[19] In his translation, Wolfgang Günther 1969/70, 246, the editor of the inscription, chooses to render the last figure as 2866 3/8 (cubic) feet, with the fractional part derived from the correspondence of one palm to one-fourth and two fingers to one-eighth of a foot.

[20] That is, apart from the usual approximations of π to 3 and $\frac{\sqrt{3}}{4}$ to 1/3 1/10.

Pharaonic Egypt.[21] It is used in Problem 14 of the so-called Moscow mathematical papyrus (MMP),[22] in which the volume of a square truncated pyramid is computed by a procedure that one still learns in school:

$$V = \frac{1}{3}\left(a_1^2 + a_2^2 + a_1 a_2\right)h$$

It is difficult to believe that this method was forgotten or, even if forgotten, that it could not have been easily devised anew.[23] It is more plausible that the use of the seemingly more awkward computation may have been a choice influenced by mensurational practices. Indeed, the procedure used for determining the volume of a square truncated pyramid in *MPER* I 1 can be viewed as a correction, or adjustment, by the addition of $\frac{1}{3}\left(\frac{a_1 - a_2}{2}\right)^2 h$ to the result which would have been produced by means of the averaging approximation, $\left(\frac{a_1 + a_2}{2}\right)^2 h$. It is thus possible that what looks to us like a more complicated computation may have been easier to remember for an ancient practitioner than a 'simpler' one and was thus preferred in the context of practical mensuration. Another advantage of this procedure may have been that it also gave the practitioner a sense of when an approximation or a more precise computation was in order.

Similarly structured computational procedures are used in *MPER* I 1 also for finding the volume of other tapering solids,[24] whereas their predecessors can be found in Mesopotamian mathematics, in which both approximate and precise methods for the volume of tapering solids were used. Particularly interesting is Problem 28 of tablet BM 85194, in which the volume of a moat that has the shape of a square truncated pyramid is calculated with the help of a correction to the volume yielded

[21] Vogel 1930 offers a survey of the procedures used to compute the volume of a truncated pyramid in Egyptian mathematics, including the evidence from the Greco-Roman period. For an overview of volume computations of whole and truncated square pyramids in ancient mathematics, see Friberg 1996.

[22] For the principal edition of the MMP, which is dated to the first half of the second millennium BCE, see Struve 1930. For discussion of Problem 14, one of the two most famous problems of the papyrus, see, e.g., Gillings 1972, 187–193; Imhausen 2016, 73–75.

[23] Although the 'clunkier' procedure for computing the volume of a truncated pyramid and cone is used and expounded in several places in Heron's *Metrica* and pseudo-Heron's *Stereometrica*, the fact that the 'simpler' formula for the volume of a truncated cone is recorded in *Stereometrica* I 16 (ed. Heiberg 1914) makes it highly unlikely that the analogous computational procedure for a truncated pyramid would not have been known. These examples are discussed in Vogel 1930, 245–246. For a survey of algorithms used to compute the volume of truncated solids in the pseudo-Heronian corpus, cf. now the notes to *Metrica* II. 6–10 in Acerbi and Vitrac 2014, 262–279.

[24] For a survey of procedures used for volume calculations in this papyrus, see Gerstinger and Vogel in *MPER* I 1, 39–40.

by the application of the averaging procedure. The correction term has the form of $\frac{1}{12}(a_1 - a_2)^2 h$, which is essentially identical to the one used in the Greek papyrus.[25] The similarity of the procedures is perhaps owed to the same basic approach of considering tapering solids to be *approximately* equal in volume to solids built on a mean cross-section, with the possibility to achieve a more precise quantification by the use of a correction term, if needed. It is remarkable, however, that in Greek mensuration the use of procedures producing accurate results is found either in metrological works that are indifferent to the material measured or in practical mensuration of stone, but not in practical quantification of earth and wood.

3.5.2 Earth (and Granaries)

Circumstances and Shapes

Just as in Mesopotamia, agriculture in Egypt depended on irrigational infrastructure, the maintenance of which was implemented since Pharaonic times by a system of compulsory work. The fulfillment of corvée as well as of paid labor could be accounted for by the number of days or by the amount of excavated earth and the extent of the cleaned water reservoirs and canals.[26] Among geometrical shapes associated with irrigation work were the quadrilateral prism, cylinder, cone, pyramid, and conical and pyramidal frusta.

Computing the volume of granaries followed the same procedures used for measuring earth, the similarity possibly owed to granaries being large receptacles, not unlike cisterns or ditches, with the same basic shapes. The purpose of measuring a granary was, however, to determine the amount of grain it contained, and since grain was quantified in units of capacity, the volume of a granary would be converted to its capacity.

---

[25] For the text of the problem with computational procedures explained, see Friberg 2005, 243–244; Robson 2007, 136. For the averaging approach in computing volume in Mesopotamian mathematics, cf. Robson 2007, 123, and for a survey of what the author calls "texts concerned with both correct and incorrect computations of volumes of whole or truncated square pyramids," see Friberg 1996 §1.5 (quotation from Friberg 2005, 243).

[26] For documentary evidence for work on dikes and canals, see Bagnall 1993, 15–23; Michel 2005; for dike corvée cf. also Westerman 1925. In the imperial Roman period (first through early third centuries), the obligation of spending 5 days a year on irrigational work is attested by the so-called penthemeros-receipts, hundreds of which survive, cf. Ferretti et al. 2023.

Linear and Volumetric Units

The main units for linear dimensions of excavations and embankments were the cubit and xylon, but fingers, palms, hammata, and schoinia could occasionally be used, especially in late antique documents and exercises.[27] The linear units have the following ratios:

1 cubit = 6 palms = 24 fingers
1 xylon = 3 cubits
1 schoinion = 8 hammata = 96 cubits

The volumetric unit for measuring earth was almost exclusively the naubion, which in the Roman period was equivalent to the amount of space occupied by a cube with a side of one xylon, or three cubits, thus: 1 naubion = 27 (cubic) cubits = 1 (cubic) xylon. If the volume amounted to less than one naubion, it was usually left in (cubic) cubits.

Linear dimensions of granaries were measured in cubits, and their volume was computed in (cubic) cubits. The number of (cubic) cubits was then multiplied by 3 1/4 1/8 to convert it to the number of artabas, which were the units of dry capacity. The conversion factor was based on the correspondence of one artaba to one (cubic) foot and the equivalency of one (cubic) cubit to 3 1/4 1/8 (cubic) feet.[28]

Computational Procedure

Determining volume of excavations or of objects made of earth seems to have been invariably based on averaging approximation. Although documentary texts never describe how a recorded volume was computed, the procedure can be deduced if both the volume and the linear dimensions were written down. A papyrus documenting measurements of an excavation for a cistern serves as a good example (*P. Oxy*. LXVI 4537, sixth–seventh c., lines 7–10):

ἄνω πλάτος πήχ(εις) κδ
κάτω πλάτος πήχ(εις) κβ
βάθος πήχ(εις) ς
εἰς ναύει(α) πη ς´

Upper width 24 cubits,
lower width 22 cubits,
depth 6 cubits;
converted to naubia, 88 1/6.

[27] For example, some problems inscribed on the wax tablets *T.Varie* 71–75 (sixth c. CE) ask to find the volume of an excavation the area of which is given in (square) hammata and the depth in fingers.

[28] Although there were different standards of artabas, this is the only volume-to-capacity equivalency attested in the papyrological evidence; cf. Wilcken 1899, 1:753; Shelton 1981, esp. 100–102, and now the conversion tables in *P.Math*. E verso and M recto.

The cistern had the shape of a conical frustum with the upper diameter of 24, the lower diameter of 22 and the depth of 6 cubits (diameters are called 'width' here). The recorded number of naubia, 88 1/6, shows that the volume of the cistern was computed as that of a cylinder the diameter of which equaled the mean between the upper and the lower diameters.[29] What is impossible to know in this case, however, is whether the volume was first calculated in (cubic) cubits and then converted to naubia through a division by 27 (i.e. $\frac{3}{4} \times \left(\frac{24+22}{2}\right)^2 \times 6 = 2380\ 1/2$, then 2380 1/2 ÷ 27 = 88 1/6), or whether the factor 1/27 was worked directly into the computation of the volume (i.e. $\frac{1}{36} \times \left(\frac{24+22}{2}\right)^2 \times 6 = 88\ 1/6$), as both procedures are found in mensurational problems.

That the volume of an excavation in the shape of a conical frustum was determined with the help of averaging approximation comes as no surprise. It is somewhat astonishing, however, to see the same procedure used to find the volume of an excavation in the form of a whole cone. Yet this is precisely how one is instructed to proceed in Problem g2 in *P.Math.*:

> A circular tapering excavation, the circumference of which is 20 cubits, depth 1 1/2 1/4 cubits. Find out how much it contains. We do it this way: I take half of the circumference, the result is 10. 10 multiplied by itself is 100. Of this, 1/12, the result is 8 1/3. Multiplied by the depth, 1 1/2 1/4 cubits, the result is 14 1/2 1/12 (cubic cubits).[30]

The description 'a circular tapering excavation' for which the length of only one circumference is given means that it has the shape of a cone. Its volume is computed as that of a cylinder with a circumference half the circumference of the base: since the cone has no top-face, the mean between the base and the top is half the base. The application of this procedure leads to the volume of a cone approximated to one-fourth of the volume of the cylinder with the same height and base, even though it must have been known that the volume of a cone equaled one-third of the volume of such a cylinder! The method used in the papyrus, however, is entirely consistent with the general approach to measuring the volume of solids of earth by means of averaging approximation, absurd as it might look if viewed as an exercise in solid geometry.

Granaries were evidently measured the same way as excavations except that the volume calculated in (cubic) cubits was converted to capacity expressed in artabas. The process can be illustrated by another problem (n2) in *P.Math.*:

[29] Since in practical mensuration the value of π was assumed to be 3, the area of a circle with the diameter d was computed as $\frac{3}{4}d^2$ or, if the circumference c was given, as $\frac{1}{12}c^2$.

[30] I do not retain in my translation the square brackets, which indicate missing and restored parts of the Greek text; for the Greek, consult the edition in Bagnall and Jones 2022, 88–90, with Lougovaya 2023 §9.

> A triangular (sc. pyramidal) granary, the length of which is 20 cubits, width 30 cubits, depth 6 cubits. To find out the artabas. We do it this way: the width multiplied by length, 20 times 30, it is 600. Take a quarter of it, it is 150. Multiply by the depth of 6 cubits, it is 900. Multiply 900 by 3 1/4 1/8, it is 3037 1/2 (artabas). This way for similar cases.[31]

The granary has the shape of a rectangular pyramid, but its volume is computed not as one-third of the prism with the same base and height, which would be the accurate procedure well attested in geometrical works and used for mensuration of stone,[32] but as one-fourth of it. The method ought to have been owed to the same use of averaging approximation that we saw above in the computation of the volume of the conical excavation.

A notable feature in the mensuration of granaries is the persistence with which the cubit is used as the unit of measurement. Had granaries been measured in feet, as stone was, computation of the volume as the product of linear dimensions would have produced (cubic) feet, the number of which would have corresponded directly to the number of artabas. This apparent simplification, however, does not seem to have been ever adopted, as all surviving mensurational problems record measurements of granaries in cubits and instruct one to multiply the computed volume by 3 1/4 1/8 to convert it to artabas.

3.5.3 Wood and Timber

Circumstances and Shapes

Quantifying wood in terms of volume seems to have been needed mostly in the context of construction projects and trade. Timber and wood measured in these circumstances tended to reflect the shape of a tree in that its length was significantly greater than the linear dimensions of its cross-section, no matter the precise shape. Most objects were beams, circular or rectangular in cross-section, that either were a cylinder or prism or else could be approximated to these solids.[33] Despite the simplicity of shapes, computing and expressing the volume of wood was perceived as requiring special attention. This is evidenced by the treatise on measuring wood

[31] For the Greek text see Bagnall and Jones 2022, 98, with comments on interpretation in Lougovaya 2023 §10–12.

[32] Cf., for example, Heron's *Metrica* II 5 (ed. Acerbi and Vitrac 2014, 260): … πᾶσα πυραμὶς τρίτον μέρος ἐδείχθη τοῦ στερεοῦ τοῦ τὴν αὐτὴν βάσιν ἔχοντος αὐτῇ καὶ ὕψος ἴσον, "… it has been shown that any pyramid equals in volume to one third of the solid having the same base and height." For the accurate procedures used to compute volume of stone in the shape of rectangular or triangular pyramids, cf. *MPER* I, 39–40.

[33] A notable exception is a wooden rim of a wheel in Problem a3 in *P.Math*. For the purpose of determining its volume, it is approximated to a prism with the length equaling the mean of the outer and inner circumference of the rim and with the cross-section determined by its width and thickness; cf. the reedition of this problem in Bagnall and Jones 2022.

ascribed to Didymus of Alexandria,[34] as well as by a relatively high number of mensurational problems concerned with the volume of a wooden object.[35] A peculiar approach to measuring wood apparently persisted even after the Arab conquest of Egypt, since Abu Kamil criticized the 'people of Egypt' for their way of measuring wood in his *Book on Mensuration.*[36]

Linear and Volumetric Units

For linear dimensions of a wooden object or a piece of timber cubits were used for expressing the length and fingers for the cross-section, with one linear cubit equaling 24 fingers.

Volume was expressed also in cubits, with one volumetric cubit comprising 24 fingers, just as the volumetric foot used for quantifying stone comprised 16 (not 4096!) fingers. Unlike in stone mensuration, however, the volumetric cubit in which wood was measured came in at least two varieties. They differed in size from each other and equaled either a half or a third of a cubic cubit, respectively. Mensurational problems never specify them by name,[37] but a metrological papyrus, *P.Oxy.* XLIX 3455 (third–fourth c.), identifies three volumetric cubits as 'solid', 'Ptolemaic', and 'Nicomedian', and describes each as a cuboid with the length and width of 24 fingers, but with the thickness of 24, 12, and 8 fingers, respectively. The cubits called Ptolemaic and Nicomedian evidently correspond in size to the volumetric cubits in which the volume of wood could be expressed. Although it is impossible to know whether this nomenclature was used in association with measuring wood or for other materials quantified in a similar way, for the sake of convenience I use it to refer to the peculiar cubits in which the volume of wood was expressed.[38] *P.Oxy.* XLIX 3455 also gives the sizes of the three volumetric cubits; they are expressed in

[34] One version of the work considers measuring both wood and marble, while the other focuses on wood only, see n. 4 above.

[35] In addition to problems transmitted in Didymus and the pseudo-Heronian corpus, there are five problems in published papyri: *P.Math.* (a3, b5, c1, and g4) and *P.Harris* 50 (third c. CE), for which cf. Lougovaya 2022a; a few more are preserved among unpublished Oxyrhynchus papyri.

[36] Measuring of wood is the subject of chapter 34, where he writes, "People of Egypt, for measuring their wood, proceed according to something [for which there is no one preceding them] which is measurement neither of volume nor of surface. I have explained to you in the preceding part of my book how they measure, how it is ridiculous and laughable, and I have explained what it involves for them in terms of deficiency, weakness, lack of common sense," translation from Sesiano 2014, 398. The 'preceding part' of the book, to which Abu Kamil refers, does not seem to be transmitted in the surviving text of the treatise.

[37] In an unpublished papyrus from Oxyrhynchus the cubit used for expressing the volume of wood is called ξυλικὸς πῆχυς, 'wood cubit'. It equals half a cubic cubit and its fingers are called 'wood fingers'.

[38] Paul Tannery, who identified in the treatise assigned to Didymus the smallest of the three volumetric cubits as a peculiar unit used for measuring wood, described it as "un parallélépipède rectangle ayant pour base une coudée carrée et une hauteur de deux paumes" (Tannery 1881, 159); on this analogy, the Ptolemaic cubit could be called a three-palm cubit; cf. Lougovaya 2022a.

Table 3.1 Three kinds of volumetric cubit with subunits and equivalencies

Designation in *P.Oxy.* XLIX 3455	Terminology after Tannery	Ratio to 1 (cubic) cubit	Solid (στερεοί) fingers	Mixed (χυδαῖοι) fingers	Ordinary (ἀγελαῖοι) fingers
στερεός Solid		1	24	576	13824
Πτολεμαικός Ptolemaic	Three-palm cubit	1/2	24	288	6912
Νικομηδικός Nicomedian	Two-palm cubit	1/3	24	192	4608

units equaling one (cubic) finger and called ordinary fingers (ἀγελαῖοι δάκτυλοι) and in units equaling 1 finger × 1 finger × 1 cubit and called 'mixed fingers' (χυδαῖοι δάκτυλοι), cf. Table 3.1.[39]

Since the volumetric cubit, no matter its size, always comprised 24 fingers, these fingers differed in size depending on the cubit. In expressions of volume, they were sometimes called 'solid fingers' (στερεοὶ δάκτυλοι), but often simply 'fingers', with no further specification, and for smaller quantities fractions of a finger would be used (cf. the example cited below).

Computational Procedure

All wooden objects that figure in mensurational problems either are a cylinder or a prism or are approximated to a cylinder or a prism by the same averaging method used for computing volume of earth.[40] The reason that computing volume of wood required special attention was thus likely due to the potentially confusing range of units one had to operate with.

It was probably the prevalent shape of wood and timber that made it convenient to measure its length in cubits, but the cross-section in fingers, and to follow a computational procedure that would not require converting these dimensions to the same unit. The method was to calculate first the volume on the basis of the numerical values of linear magnitudes, disregarding the difference in units; this produced a result in units 1 finger × 1 finger × 1 cubit, which were either not named at all or were called mixed fingers (χυδαῖοι δάκτυλοι). This quantity, which served only as

[39] Since elsewhere both adjectives ἀγελαῖος and χυδαῖος tend to have a similar meaning of 'simple', 'ordinary' or 'common', it is not immediately clear how to translate them when they are used in the same context with evidently different meanings. I have opted for 'ordinary' for ἀγελαῖος to designate the basic nature of the unit; χυδαῖος, which derives from the verb χέω, could mean 'heaped-up' or 'massed together'; cf. LSJ, χέω II.2, which has suggested to me 'mixed', possibly in reference to the different dimensions of the unit.

[40] Note that the volume of a wooden pyramid in a problem preserved in Didymus 41 (ed. Heiberg 1927) is computed as one-fourth of the product of the base area by the height.

an interim result, was then converted to one of the two volume cubits used for quantification of wood and corresponding solid fingers. The conversion factors differed depending on which cubits the final result was to be expressed in, because they had a different number of mixed fingers. Thus, for Nicomedian cubits, the number of mixed fingers was divided by 192, with the remainder divided by 8 to produce the corresponding fingers, and for Ptolemaic cubits the factors were 288 and 12 respectively.[41] Problem c1 in *P.Math.* illustrates the computational procedure for finding volume of a piece of timber:

> A piece of new wood, 28 cubits (sc. in length), width from the roots 16 fingers, from the foliage 12 fingers, thickness from the roots 8 fingers, from the foliage 6 fingers. Find out how big the timber is. I do it this way: I add the width, 16 and 12, the result is 28, half of which is 14. I add the thickness, 6 and 8, the result is 14, half of which is 7; (multiply) by 14, the result is 98; (multiply) by the length of 28 cubits, the result is 2744. This I divide by 288 and (to convert) the remainder into fingers, (I divide by) 12, the result is 9 and 12 2/3 fingers. So it holds for similar cases.[42]

As the first step, the linear dimensions and then the area of the mean cross-section are determined (with explanations and implied units added in parentheses): the average width is $\frac{16+12}{2}=14$ (fingers), the average thickness is $\frac{6+8}{2}=7$ (fingers), and the area of the mean cross-section is $7 \times 14 = 98$ (area fingers). The value of the area is then multiplied by the length, $98 \times 28 = 2744$ (mixed fingers) and the product is divided by 288 (to convert to Ptolemaic volumetric cubits), with the remainder of the division divided by 12 (to convert to volumetric fingers). The fact that the text of the problem does not explain the use of the conversion factors 288 and 12 was possibly owed to a presumption that it would have been obvious in the occupational context in which the piece of wood was measured.

While scarcity of explanation is characteristic for problems concerned with practical mensuration in general, it is particularly detrimental for our understanding of texts dealing with quantification of wood. Surviving problems never state explicitly which cubit is meant, let alone why it should be. Nor do they contain any nomenclature for the cubits and fingers in which the result is expressed except occasionally calling both or either of them στερεοί, 'solid', no matter the type of cubit.[43] Consequently, it is only the conversion factor applied to the interim result that reveals which of the cubits is at work in a given problem. The choice may have correlated to the type of wood and/or its purpose, for example, with the wood of young trees or timber for shipbuilding being measured differently from other kinds of wood. Or perhaps it was determined by the locality. There is some indication that the cubit equaling in size a half-cubic cubit was associated with Egypt: not only *P.Oxy.* XLIX 3455 calls it Ptolemaic, but it also appears in all but one of the

[41] No mensurational problem on papyrus can be shown to express the volume of wood in 'solid' cubits equaling a cubic cubit in size. If it were used, the conversion factors would be 576 and 24.

[42] The translation is adopted from the reedition of the problem in Lougovaya 2022b.

[43] Only in the unpublished Oxyrhynchus papyrus mentioned in n. 37 is the cubit explicitly called ξυλικός, 'wood'.

volume-of-wood problems preserved in papyrological evidence, all of which come from Egypt. Yet the fact that one problem found on a papyrus (Problem b5 in *P.Math.*) computes the volume using the smaller cubit, which is called Nicomedian in *P.Oxy.* XLIX 3455, suggests that it must have been known in Egypt, too, even if it were less widespread.[44] Didymus uses only that cubit, but since nothing is known about the authorship of the treatise attributed to him, it is impossible to locate it.

3.6 Conclusion

Documents and especially mensurational problems in the papyri demonstrate clearly that practitioners chose and adjusted measuring procedures to the material they were dealing with (summarized in Table 3.2). The choice was likely influenced by the practical and economic circumstances of working with the actual physical objects made of stone, earth or wood, about which we can make some suppositions, even if ancient sources furnish us with no explanations. Thus, the reason for choosing a computational procedure that produces accurate results to determine the volume of stone could be related to the high cost of extraction, working, and transportation of this material.[45] Furthermore, the reference in the work ascribed to Didymus to measuring not any stone but marble in particular might indicate that the

Table 3.2 Summary of metrological conventions for measuring stone, earth and wood

	Stone	Earth	Wood
Linear measurements	foot = 16 fingers = 4 palms	cubit = 1/3 xylon = 1/12 hamma = 1/96 schoinion	cubit = 24 fingers = 6 palms
Base volumetric unit	(cubic) foot	naubion	-Ptolemaic cubit (= 1/2 cubic cubit) -Nicomedian cubit (= 1/3 cubic cubit)
Composition of the base volumetric unit	1 foot = 16 (volume) fingers	1 naubion = 1 (cubic) xylon = 27 (cubic) cubits	1 cubit = 24 (volume) fingers, no matter the size of the volume cubit.
Common format of expressing the volume	Number of feet + parts of a foot; or number of feet + number of fingers	Number of naubia + parts of a naubion; or number of naubia + number of cubits	Number of volume cubits + number of volume fingers + parts of a volume finger.
Common use of approximation for computing the volume	no	yes	yes

[44] The cubit equaling a half-cubic cubit, aka Ptolemaic, features in Problems a3, c1, and g4 of the same papyrus codex.

[45] For the high price of stone, cf. Bagnall 1993, 42 and 49; Russel 2013, 8–36.

accuracy of quantification may have been preferred specifically for more expensive stones such as marble.

Computational procedures used to determine the volume of objects made of earth and wood resulted in underestimation unless the objects were a prism or cylinder. In the case of earth, which was a bulky material, perhaps even the linear dimensions could not be determined with too much precision and consequently the volume would be necessarily not precise, too, all together suggesting to practitioners that a slight underestimation was a safe option. On a somewhat more cynical view, the fact that computed volume of earth was often smaller than the actual volume might have been in the interest of those who commissioned the work, just as an overestimation of the area of a not-rectangular quadrilateral field computed by the surveyor's formula is sometimes viewed as being in the interest of tax collectors.[46]

The reason for using approximation when computing the volume of wood, which was usually quantified when shaped into long beams or logs and thus, if tapering, in the shape of a truncated cone or pyramid, may have been the perceived insignificance of the difference. Indeed, in most problems asking to find the volume of a wooden beam or log the differences in the linear dimensions of the base and the top are relatively small. Furthermore, should the differences be more substantial, including when the beam or log had the shape of a pyramid or cone, the narrow end of the piece would probably be of little use and consequently a computation leading to an underestimation of its volume may have made sense when viewed in terms of its utilitarian value.

Determining volume of objects in occupational contexts was not a task in solid geometry, even if it did make use of solid geometry. Consequently, mensurational problems were not mathematical exercises, strictly speaking, but instructions for practitioners on how to quantify certain materials. When viewed in these terms, the problems become a window into occupational practices and conventions that people followed for reasons that may have had little to do with geometry.

Acknowledgments I am grateful to the anonymous reviewers for their suggestions and to Rodney Ast for many a fruitful conversation on the topic of Greek metrology and for his comments on the draft of this paper. It is also with gratitude that I acknowledge the kind support of the Ralph E. and Doris M. Hansmann Member Fund at the Institute for Advanced Study, where I spent the fall semester 2023 working on a project devoted to mensurational and mathematical papyri, of which the current article forms a part.

Abbreviations for Editions of Greek Papyri

MPER I = Hans Gerstinger and Kurt Vogel, *Mitteilungen aus der Papyrussammlung der österreichischen Nationalbibliothek in Wien, Neue Serie* I, *Griechische literarische Papyri* I. Vienna: Österreichische Staatsdruckerei, 1932.

P.Chester Beatty Library AC 1390 = William M. Brashear, Wolf-Peter Funk, James M. Robinson, and Richard Smith, *The Chester Beatty Codex Ac. 1390. Mathematical School Exercises in Greek and John 10:7–13:38 in Subachmimic*. Leuven: Peeters, 1990.

P.Harris = J. Enoch Powell, *The Rendel Harris Papyri of Woodbrooke College, Birmingham. Nos. 1–165*, Cambridge: At the University Press, 1936.

[46] For this view, see, for example, Grenfell et al. 1902, 385–386; Asper 2012, 57.

P.Math. = Roger S. Bagnall and Alexander Jones, *Mathematics, Metrology, and Model Contracts*, New York: New York University Press, 2019.

P.Oxy. XLIX = Adam Bülow-Jacobsen and John E.G. Whitehorne, *The Oxyrhynchus Papyri*. XLIX, *Nos. 3431–3521*. London: The Egypt Exploration Society, 1982.

P.Oxy. LXVI = Nikolaos Gonis, Juan Chapa, Walter E.H. Cockle, Dirk Obbink, Peter J. Parsons and J. David Thomas with contributions by 8 others, *The Oxyrhynchus Papyri*. LXVI, *Nos 4494–4544*. London: The Egypt Exploration Society, 1999.

T.Varie = Rosario Pintaudi and Pieter J. Sijpesteijn, *Tavolette lignee e cerate da varie collezioni* (Papyrologica Florentina XVIII). Florence: Edizioni Gonnelli, 1989.

Bibliography

Acerbi, Fabio, and Bernard Vitrac, ed. 2014. *Héron d'Alexandrie, Metrica*. Pisa: Fabrizio Serra.

Asper, Markus. 2012. 'True' and 'False' Errors in Ancient (Greek) Computation. In *Productive Errors: Scientific Concepts in Antiquity*, ed. Markham Geller and Klaus Geus, (TOPOI—Dahlem Seminar for the History of Ancient Sciences, Preprint Max-Planck-Institut für Wissenschaftsgeschichte 430), 57–66. Berlin: Max-Planck-Institut für Wissenschaftsgeschichte. https://www.mpiwg-berlin.mpg.de/Preprints/P430.PDF.

Bagnall, Roger S. 1993. *Egypt in Late Antiquity*. Princeton: Princeton University Press.

Bagnall, Roger S. 2009. Practical Help: Chronology, Geography, Measures, Currency, Names, Prosopography, and Technical Vocabulary. In *The Oxford Handbook of Papyrology*, ed. Roger S. Bagnall, 179–196. Oxford: Oxford University Press. https://doi.org/10.1093/oxfordhb/9780199843695.013.0008.

Bagnall, Roger S., and Alexander Jones. 2022. P.Math. Leaf A Verso, Mathematical Problem a3 Revisited: A New Algorithm in Greek Mensurational Mathematics, *Pylon 2*, December 2022. https://doi.org/10.48631/pylon.2022.2.92969.

Berg, Ria. 2020. Iconography of the Modius Measure in Ostia: The Visualizing of Economic and Cultural Exchange. In *Trade and Commerce in the Harbour Town of Ostia*, ed. Alice Landskron and Claudia Tempesta, 79–106. Graz: Unipress Graz.

Bernard, Alain, Christine Proust, and Micah Ross. 2014. Mathematics Education in Antiquity. In *Handbook on the History of Mathematics Education*, ed. Alexander Karp and Gert Schubring, 27–53. New York: Springer.

Bruins, Evert M. 1964. *Codex Constantinopolitanus Palatii Veteris No. 1*, Parts I–III. Leiden: Brill.

Chemla, Karine, and Biao Ma. 2020. The Use of Volume in the Measurement of Grain in Early Imperial China. In *Mathematics, Administrative and Economic Activities in Ancient Worlds*, ed. Cécile Michel and Karine Chemla, 239–279. Cham: Springer.

Cribiore, Raffaella. 1996. *Writing, Teachers, and Students in Graeco-Roman Egypt* (American Studies in Papyrology 36). Atlanta: Scholars Press.

Cribiore, Raffaella. 2001. *Gymnastics of the Mind: Greek Education in Hellenistic and Roman Egypt*. Princeton: Princeton University Press.

Cuomo, Serafina. 2001. *Ancient Mathematics*. London: Routledge.

Ferretti, Lavinia, Susan Fogarty, Elisa Nury, and Paul Schubert. 2023. *Description of Greek Documentary Papyri: Penthemeros Receipt*. Geneva: Grammateus Project. https://doi.org/10.26037/yareta:cpqm3et4bbefhgjus3uoprmmzu.

Friberg, Jöran. 1996. *Pyramids and Cones in Cuneiform and Other Mathematical Texts: New Hints of a Common Tradition in Ancient Mathematics: A Paper Presented at the Workshop on the Cultural History of Mathematics and Informatics, Freie Universität, Berlin, July 10–12, 1995* (Preprint Max-Planck-Institut für Wissenschaftsgeschichte 29). Berlin: Max-Planck-Institut für Wissenschaftsgeschichte.

Friberg, Jöran. 2005. *Unexpected Links Between Egyptian and Babylonian Mathematics*. Hackensack, NJ: World Scientific.

Gillings, Richard J. 1972. *Mathematics in the Time of the Pharaohs*. Cambridge, Mass.: MIT Press.

Grenfell, Bernard P., Arthur S. Hunt, and J. Gilbart Smyly. 1902. *The Tebtunis Papyri I*. London: The Egypt Exploration Society.

Günther, Wolfgang. 1969/70. Eine neue didymeische Bauinschrift. *Istanbuler Mitteilungen* 19/20: 237–247.

Heiberg, Johan, ed. 1912. *Heronis Alexandrini opera quae supersunt omnia*, Volumen IV. *Heronis definitiones cum variis collectionibus. Heronis quae feruntur geometrica*. Leipzig: Teubner.

Heiberg, Johan, ed. 1914. *Heronis Alexandrini opera quae supersunt omnia*, Volumen V. *Heronis quae feruntur stereometrica et de mensuris*. Leipzig: Teubner.

Heiberg, Johan, ed. 1927. *Mathematici graeci minores*. Copenhagen: Hovendkommissionœr.

Høyrup, Jens. 1996. *Hero, Ps-Hero, and Near Eastern Practical Geometry: An Investigation of Metrica, Geometrica, and Other Treatises* (Filosofi og videnskabsteori på Roskilde Universitetscenter, 3. række: Preprints og reprints). Roskilde. https://rucforsk.ruc.dk/ws/portalfiles/portal/4340101/1996_Hero.pdf.

Høyrup, Jens. 2019. *Selected Essays on Pre- and Early Modern Mathematical Practice*. Cham: Springer. https://doi.org/10.1007/978-3-030-19258-7.

Hultsch, Fridericus. 1864–1866. *Metrologicorum scriptorum reliquiae*. 2 vols. Leipzig: Teubner.

Imhausen, Annette. 2016. *Mathematics in Ancient Egypt. A Contextual History*. Princeton: Princeton University Press.

Imhausen, Annette, and Corinna Rossi. 2009. Architecture and Mathematics in the Time of Senusret I: Sections G, H and I of Papyrus Reisner I. In *Beyond the Horizon: Studies in Egyptian Art, Archaeology and History in Honour of Barry J. Kemp*, ed. Salima Ikram and Aidan Dodson, 440–455. Cairo: American University in Cairo Press.

Johnson, William A. 2015. Learning to Read and Write. In *A Companion to Ancient Education*, ed. W. Martin Bloomer, 137–148. Malden: Wiley Blackwell.

Jones, Alexander. 2012. Mathematics, Science, and Medicine in the Papyri. In *The Oxford Handbook of Papyrology*, ed. Roger S. Bagnall, 338–357. Oxford: Oxford University Press. https://doi.org/10.1093/oxfordhb/9780199843695.013.0015.

Lougovaya, Julia. 2022a. On Cubic and Other Volumetric Cubits and Fingers, *Pylon* 1, July 2022. https://doi.org/10.48631/pylon.2022.1.89333.

Lougovaya, Julia. 2022b. A Pyramidal Frustum Volume Algorithm in P.Math. Leaf C Recto, *Pylon* 2, December 2022. https://doi.org/10.48631/pylon.2022.2.92970.

Lougovaya, Julia. 2023. Notes on Mathematical and Mensurational Papyri, *Pylon* 4, December 2023. https://doi.org/10.48631/pylon.2023.4.101704.

Michel, Nicolas. 2005. Travaux aux digues dans la vallée du Nil aux époques papyrologique et ottomane: une comparaison. In *L'agriculture institutionnelle en Égypte ancienne: État de la question et approches interdisciplinaires* (Cahiers de Recherche de l'Institut de Papyrologie et d'Egyptologie de Lille 25), 253–276. Lille: Université Charles-de-Gaulle Lille.

Netz, Reviel. 2022. *A New History of Greek Mathematics*. Cambridge: Cambridge University Press.

Prato, Giancarlo. 1993. Due postille paleografico-codicologiche. In *Symbolae Berolinenses für Dieter Harlfinger*, ed. Friederike Berger, Christian Brockmann, Giuseppe De Gregorio, Irene Ghisu, Sofia Kotzabassi, and Bent Noack, 279–291. Amsterdam: Hakkert.

Proust, Christine. 2022. Volume, Brickage and Capacity in Old Babylonian Mathematical Texts from Southern Mesopotamia. In *Cultures of Computation and Quantification in the Ancient World*, ed. Karine Chemla, Agathe Keller, and Christine Proust, 197–264. Cham: Springer.

Robson, Eleanor. 2007. Mesopotamian Mathematics. In *The Mathematics of Egypt, Mesopotamia, China, India, and Islam. A Sourcebook*, ed. Victor Katz, 57–186. Princeton: Princeton University Press.

Russel, Ben. 2013. *The Economics of the Roman Stone Trade*. Oxford: Oxford University Press.

Schubart, Wilhelm. 1916. Mathematische Aufgaben auf Papyrus. *Amtliche Berichte aus den Königlichen Kunstsammlungen* 37 (8): 5–9.

SEG = Supplementum Epigraphicum Graecum. 1923. Leiden: Brill.

Sesiano, Jacques. 2014. Abū Kāmil's Book on Mensuration. In *From Alexandria, Through Baghdad: Surveys and Studies in the Ancient Greek and Medieval Islamic Mathematical Sciences in Honor of J.L. Berggren*, ed. Nathan Sidoli and Glen van Brummelen, 359–408. Heidelberg: Springer.

Shelton, John. 1981. Two Notes on the Artab. *Zeitschrift für Papyrologie und Epigraphik* 42: 99–106.

Sidoli, Nathan. 2015. Mathematical Education. In *A Companion to Ancient Education*, ed. W. Martin Bloomer, 387–400. Malden: Wiley Blackwell.

Struve, Wasili W. 1930. *Mathematischer Papyrus des staatlichen Museums der schönen Künste in Moskau* (Quellen und Studien zur Geschichte der Mathematik, Astronomie und Physik, Abteilung A: Quellen 1). Berlin: Springer.

Tannery, Paul. 1881. Les mesures des marbres et des divers bois de Didyme d'Alexandrie. *Revue Archéologique N.S.* 41: 152–164.

Vitrac, Bernard. 2010. Héron d'Alexandrie et le corpus métrologique: état des lieux. In *Géométrie(s), pratiques d'arpentage et enseignement: quels liens et dans quel contexte?*. Paris: March 2010. hal-00473981.

Vogel, Kurt. 1930. The Truncated Pyramid in Egyptian Mathematics. *The Journal of Egyptian Archaeology* 16: 242–249.

Westerman, William Linn. 1925. Dike Corvée in Roman Egypt. *Aegyptus* 6: 121–129.

Wilcken, Ulrich. 1899. *Griechische Ostraka*, Vol. 2. Leipzig: Giesecke & Devrient.

Chapter 4
Arithmetic Manuals as Crystallization of Mathematical Practices. Evidence from Italian and German Texts (14th–16th c.)

Raffaele Danna and Thomas Morel

Abstract This chapter explores early modern European practical arithmetic, focusing on mathematical manuals written in fifteenth-century Pisa and sixteenth-century Saxony. It delves into the complexities of how we can interpret mathematical problems within the context of contemporary socio-economic dynamics and educational practices. It argues that we can better understand the relationship of mathematical texts with contemporary professional practices by studying their contents together with the social conditions of their authors and the professional contexts in which these texts were used.

By analyzing and comparing Italian and German sources, the chapter highlights similarities and differences in arithmetic methods and their adaptation to local settings. Case studies from late medieval Italian city states and the mining rush in the Ore Mountains illustrate the intertwining of arithmetic with pedagogical and professional activities. The chapter also discusses the role of practical arithmetic manuals in disseminating numeracy, transmitting lived mathematical skills, and teaching how to adapt and apply them to the multifaceted contingencies of the world of practice.

Building on a specific case study about alloying problems, the chapter further explores how economic changes facilitated the recombination or "crystallization" of mathematical knowledge to address emerging needs. We argue that practical arithmetic texts can be understood as fairly close to contemporary practices, and thus inform us about them, but that these texts were never meant to be merely written records. They should be viewed as complex devices reflecting a combination of transcription, prescription, and contextual adaptation, rather than as static repositories of knowledge.

R. Danna
Scuola Superiore Sant'Anna, Pisa, Italy
e-mail: rd533@cantab.ac.uk

T. Morel (✉)
Bergische Universität Wuppertal, Wuppertal, Germany
e-mail: morel@uni-wuppertal.de

Senthil Babu D., R. Wagner (eds.), *Normative Texts and Occupational Practices*, Boston Studies in the Philosophy and History of Science 350,
https://doi.org/10.1007/978-3-032-28419-8_4

Keywords Practical arithmetic · Early modern Europe · Pisa · Annaberg · Educational and economic practices · Italian city states · Saxony

4.1 Introduction

This chapter deals with some developments in early modern practical arithmetic, its teaching and professional use. Its two complementary case studies are set in fifteenth-century Pisa and in the Ore Mountains of Saxony at the turn of the sixteenth century. Although textbooks, manuscript and printed, play an important role in this study, our main focus is on the relationship between these texts, their social conditions, and the various practices of arithmetic. It is important to underline that we are not dealing with the mathematics practiced in late medieval and early modern universities, but rather with a mathematics practiced in the world of the vernacular crafts, a professional mathematics that was not concerned with the research of higher, abstract truths.

In his study of the *Rechenmeister* Johannes Faulhaber, Ivo Schneider has sketched some of the characteristics of this "professional mathematics" (*Berufsmathematik*):

> the values and traditions of a professional mathematics, which was bound to the conditions of the market, of supply and demand, and was partly organized like a craft [...] the mathematics of the *Rechenmeister* consisted of a largely unstructured mass of problems, whose importance, however, was gauged from the interplay of supply and demand on the market (Schneider 1993, xii).

How can we better understand the relationships between the texts written by practical mathematicians, such as the *maestri d'abaco* and *Rechenmeister*, and the "conditions of the market" mentioned by Schneider? How did the activities of these masters—teaching and mercantile—articulate? How did this knowledge relate to the dynamics of socio-economic conditions? In other words, how should we read and interpret the mathematical problems preserved in texts of practical arithmetic? Do they reflect faithfully the professional activities of the time, or do they provide an abstract distillate of practical experiences?

These are complicated issues that can hardly be unequivocally answered. This chapter tries to tackle these questions in several ways: first, we study both Italian and German sources, something that has rarely been done in spite of the deep influence that Italian traditions had on Germany.[1] This comparative approach allows us to highlight similarities at a general level and to identify common patterns. Second, this study provides a focus on the social world of reckoning masters, figures that are still little known in the historiography, at least in the history of science. We will show that practical arithmetic masters contributed to the broad diffusion of basic

[1] The topic is briefly touched upon in (Høyrup 2008).

numeracy with the new (Indo-Arabic) numerals[2] and that their activities had a wide circulation at the time. Reckoning masters developed problem-solving methods that set demonstrations, theoretical issues, and, in many cases, generalizations aside. These methods played a significant role in the development of a wider set of professional practices (Denzel 2002, 23–30). In both our examples, we discuss how texts (printed or manuscript) interacted with the practices that surrounded them—pedagogical and professional.

In the first part, a general outline of the commercial revolution will be sketched, emphasizing how this period of deep economic change coincided with the adoption of practical arithmetic, with Indo-Arabic numerals among economic agents, and with its institutionalization. The second part focuses more closely on fifteenth-century Pisa, studying an exceptional manuscript written in the 1440s by Cristofano di Gherardo di Dino that contains both a practical arithmetic and the related curriculum of a *scuola d'abaco*. The third part turns to the Ore Mountains, and describes how the mining rush of the 1470s brought mining officials and scholars in contact. The fourth section focuses on the *Rechenmeister* Adam Ries, exploring how his teaching activities articulated with his duties as mining official.

In order to produce fruitful comparisons, both case studies proceed from general to particular and ultimately focus on a specific cluster of problems, namely the computations related to assaying, minting, and alloying. The comparative approach of the paper allows to observe how mathematical procedures were adapted to local contexts, revealing broad similarities in arithmetical methods based on proportionality between the commercial Italian city and the German mining region. Even as the methods were mathematically similar, the environments were rather different: not only are various metrological units used, but the aims are also largely divergent. In the city-states of Italy, merchants used arithmetic mainly to change money and test their metallic fineness. In the mining states of Germany, mining officials followed their rulers' orders, mostly minting and selling coins at fixed rates.

Our examples show that the relationship between books and professional practices was not a straightforward one. An important reason is that both merchants and mining officials were part of highly developed written societies, in which practical arithmetics were but one specific kind of a large set of practical texts, and therefore provided only one perspective on a broader body of practical knowledge.[3] Moreover, these texts did not develop in a vacuum. They drew from the traditions of previous mathematical texts, as well as from the problems that emerged in practical experience, and arithmetical problems related to mining can indeed be found in Babylonian, Greek, and Arabic traditions.[4] Practical arithmetic texts are precious sources to explore these exchanges, as they had a degree of proximity to emergent practices

[2] Following the recommendation by Daston (2022, 289), we use the term 'Indo-Arabic' to avoid the implicit religious connotations of the expression 'Hindu-Arabic'.

[3] Italian merchants used *Pratiche di mercatura*, but also (from the sixteenth century) ready reckoners or lists of weights and measures. Mining officials had *Rezessbücher*, *Gegenbücher* and leaflets advertising the revenue of mining shares. For a general reference, see (Hoock and Jeannin 2001).

[4] A short overview is provided in (Tropfke 1980, 565–572).

while belonging to a clear tradition of mathematical texts. From the late Middle Ages onwards, a cluster of problems related to alloying exhibiting new features, such as presenting the degree of finesse for gold and silver using 24 carats and 16 lots, and testifying of concrete assaying procedures, can be found in Italian and later German practical arithmetic manuals (Tropfke 1980, 566).

Their authors, in fact, were often directly engaged with practices, both educational and mercantile. Our argument is that we should read these texts neither as the direct description of contemporary practices nor as the mere repetition of traditional methods, but rather as what we have come to describe as the "crystallisation" of clusters of problems that drew from both past experience and present needs. This reading allows to account for the fact that these texts include a combination of problems belonging to long mathematical traditions, fictitious problems, and problems that appear to be related to the business practices that surrounded them. The latter are particularly interesting, as they provide evidence of how mathematical procedures responded to emerging needs.

Economic change set the conditions for new experimentations and recombinations of mathematical knowledge. New knowledge could be produced when various bodies of knowledge interacted through an interplay between tried and tested methods, local rules and practices, and emergent problems. This new knowledge, in turn, contributed to socio-economic change through its applications. If we understand practical texts as the "crystallisation" of this liquid process, we can read them, so to speak, as snapshots of a wider collective process of experimentation.

These moments of interplay between established knowledge and emergent problems can be thought of in terms of windows of opportunity that opened up when different bodies of knowledge, different practices, different problems, and different bodies of practitioners/professionals met, interacted, and recombined their expertise to address emergent needs. From this point of view, alloying is a meaningful case. It allows to observe two moments in which these windows opened up, bringing forth new possible solutions, in fifteenth-century Italy and sixteenth-century Germany. During these windows of opportunity, previous knowledge unsettled and 'liquified' until finding new forms that better suited the changed needs of its society. As the case of German cities of the Ore Mountains shows, these windows could only last a couple of generations before winding down into new institutionalised forms.

From this perspective follows that we read these texts as complex devices. They were not merely the receptacle of external practices, nor were they exclusively prescriptive. For example, texts authored by persons who engaged directly in teaching activities were likely a complicated mix of transcription and prescription. Moreover, similar texts could be used in different ways, and respond to different needs, once contextual conditions changed. Texts are indeed tricky things, as they can be used in many ways! The only efficient method to shed light on these complex dynamics is to read these sources together with a study of the practices in which they emerged. This shows that manuals of practical arithmetic were not accurate or exhaustive representations of their surrounding practices; at the same time, they were not completely speculative or performative in nature.

Texts of practical arithmetic describe, record, guide through, and prescribe a codified set of practices, mostly in vernacular, and always from the point of view of a local actor. In order to understand when these texts describe, or when they record, guide, or prescribe, we need to step—as much as this is possible from available contextual information—into the point of view of the writer, and understand for each text or passage what the writer was intending to do. These texts were just one of the several tools available to these people, and we need to read them as tools, not as depositaries of codified and fixed knowledge.

As instructional texts aimed at the vernacular arts, practical arithmetic manuals did not aim to develop general systems of knowledge. Rather, they were marketed as providing would-be practitioners with a set of tools that could be applied to the ever-changing circumstances of practice (Smith 2022, 58, 130–137). Rather than demonstrating a system of theorems, these texts aimed at teaching a mathematical attitude, an ability to resort to a set of mathematical tools to address problems stemming from practice. This mathematical attitude was achieved not through deductive reasoning, but by identifying analogical similarities between large samples of problems. This attitude was trained by making readers go through long lists of problems in which the same mathematical methods were applied to an array of circumstances (Daston 2022, 58). If we accept this reading, we can solve the apparent contradiction due to the fact that these texts include both problems that had a degree of proximity to contemporary practices and problems that were not related to their surrounding context: both sets of problems could be used to train a mathematical attitude towards practice. By identifying analogies between these large samples of examples, would-be practitioners acquired a practice of mathematics, or—to use an expression by Fibonacci—turned a *scientia* into a *habitus*.[5]

4.2 Practical Arithmetic in Italy

4.2.1 The Commercial Revolution of the Thirteenth Century and the Emergence of Italian Practical Arithmetic

The so-called "commercial revolution of the thirteenth century" was a period of profound transformation in European commercial practices (De Roover 1953, 80–85). During this century, a number of fundamental business techniques were developed among southern European merchants, and especially among Italians. The first international commercial enterprises were founded in this period, often relying on the new corporate structure of the *compagnia*: a contract that was not linked to a single commercial venture, and allowed to create business organisations that often had multiple branches located in different trading hubs (Goldthwaite 2009, 64–66). These were the first international commercial-banking enterprises in Europe, and

[5] See below, note 9.

their appearance coincided with the emergence of the first sedentary merchants, whose job diverged substantially from the traditional figure of the wandering merchant. Rather than moving between markets along with their wares, these merchants acted as coordinators of a large variety of businesses, relying on a network of agents scattered in the main commercial hubs of Europe and the Mediterranean. These merchants have attracted the attention of generations of economic historians, but they generally do not feature in histories of science, and even less frequently in traditional histories of mathematics. Leonardo Pisano (Fibonacci) is the exception that confirms the rule, as, even though we do not have direct evidence of his commercial enterprises, it is likely that he was also active as a merchant.[6] However, as these were the first European practitioners to make extensive use of Indo-Arabic numerals and of proto-symbolic algebra, they may also interest the historian of mathematics. To understand why merchants adopted these advanced mathematical tools, we need to take a closer look at their business practices.

International merchants were literate and highly numerate. The written word had an essential role in their profession, as the management of their companies could only be achieved through a continuously updated flow of information. Because of this pressing need, international merchants were among the first European economic agents to take to the written word, mainly relying on their everyday language. As a consequence, some of the oldest evidence of Italian vernaculars comes from their documents (Castellani 1958, 19–95). The complexity of their business led these merchants to develop important innovations in commercial practices. For example, their account books show the incremental experimentation in accounting practices that led to the development of double-entry bookkeeping—a fundamental principle of accounting that is still practiced today.[7] These accounts also show the complex mathematical operations these merchants had to face.

Particularly interesting are the calculations related to exchange. Because of the high political fragmentation of late medieval Italy, units and measures varied considerably between marketplaces. Also monetary units varied greatly, making it necessary to calculate the exchange rate between them. As a result of inter-city trade, several currencies circulated on the same marketplace. In the middle of the thirteenth century, moreover, the Italian monetary system became trimetallic, with bullion, silver, and gold coins in circulation (Goldthwaite 2009, 48–52; Lopez 1956, 219–40). The relative price of these currencies varied depending on the trends in the value of silver and gold, i.e., on the supply and demand of silver and gold on the European scale: dynamics that no city state could control (Goldthwaite 2009, 609–11). As a result, the relative price of coins in circulation—even of those minted by the same city state—varied continuously. This implies that operations of exchange were also necessary to navigate local monetary systems. This means that

[6] There is also a growing interest among historians of mathematics in reconstructing the mathematical practices of the world of practitioners. See, for example (Rowe and Dauben 2024).

[7] General references include (De Roover 1956; Goldthwaite 2015; Sangster 2016).

operations of monetary exchange concerned not only the small élite of international merchants, but also economic agents active on the local marketplace.

This emergent need to calculate operations of exchange faced by a considerable share of Italian practitioners helps to explain their early adoption of Indo-Arabic numerals. While these numbers had been circulating in European centres of learning since the twelfth century (Nothaft 2014; Folkerts 2001; Folkerts and Kunitzsch 1997; Karpinski 1921), Italian city states were probably the first European vernacular contexts to adopt the arithmetic developed in the Arabic world (Van Egmond 1976; Heeffer 2011, 109–30). As Italian merchants often had to calculate operations of exchange between incommensurable units (of measurement and of value), they faced an increasing need to handle rational numbers. These operations were facilitated by the adoption of Indo-Arabic numerals. This is because Indo-Arabic numerals introduced into Europe a positional numeral system that made written calculations possible and, more importantly, because Indo-Arabic numerals were introduced into Europe together with the symbolism for fractions we still use today—i.e., with the numerator placed above the denominator, separated by a horizontal line. In comparison with the Roman duodecimal subdivision of the unit (Maher and Makowski 2001, Yeldham 1927), this symbolism allowed to represent and to calculate with rational numbers in a more synthetic and more flexible way (Danna 2026).

While direct evidence of the early circulation of Indo-Arabic numerals among Italian merchants is scant, they had surely adopted this numerical notation by the late thirteenth century (Ulivi 2008, 404). Starting from this period, an increasing number of Italian city states founded schools specifically dedicated to teaching Arabic arithmetic to prospective merchants and other practitioners. These are the so-called "abacus schools". In late medieval Italian vernaculars, the word "abacus" was used to indicate any kind of reckoning tool, and not just the counter abacus. As a result, the mathematics developed with the commercial revolution, which was based on the adoption of Arabic arithmetic, was called "abacus mathematics". Abacus schools secured the intergenerational transmission of a body of practical knowledge that had developed incrementally during the thirteenth century and, by the end of the century, was considered an essential part of commercial training.

The Commune of Pistoia, for example, decided to establish a similar school stating explicitly that, without a sound training in abacus mathematics, merchants and artisans would not have been able to run their businesses: "sine scientia abaci mercatores et artifices utiliter et bene se exercere non possunt" (Grendler 1989, 22). Abacus schools were supported by public funds in most Italian centres, except for Florence and Venice, where the demand for this kind of training was sufficiently strong to make public financing of abacus schools unnecessary (Ulivi 2002, 133). Abacus schools were run by lay masters ("abacus masters"), who gave an essential contribution to the first codification of this mathematics. These reckoning masters are in fact the authors of the practical arithmetic manuals that, starting from the late thirteenth century, were written in increasing numbers across Italian city states (Danna 2021). Unlike Fibonacci's *Liber abaci*, abacus manuals were written in the vernacular, and were clearly addressed to a readership of practitioners (Bocchi 2017, Høyrup 2024).

As the mathematics attested in these texts was developed in a close entwinement with commercial practices, abacus manuals are particularly significant sources to study the relationship between mathematical texts and the practices that surrounded them. In terms of contents and of textual structure, abacus manuals are rather standardised. They often open with a synthetic section that introduces Indo-Arabic numerals, and shows how to read them. This is followed by sections illustrating the algorithms to carry out the fundamental arithmetical operations: addition, subtraction, multiplication, and division. The bulk of these manuals then follows, with long lists of worked examples in which arithmetic is applied to an extensive range of practical problems. These sections are arranged by the kind of application, such as problems of exchange, of calculation of interest rates, of barter, of division of profits and losses, and of alloying. Some advanced manuals also include a treatment of algebra, and a section on practical geometry (Van Egmond 1983, Høyrup 2024). The fundamental mathematical process used in these texts is the so-called "rule of three", i.e., a proportion where the unknown term is calculated from the values of three known terms.

Contrary to the tradition of the arithmetic primers used in universities—the so-called *algorismi*, which were mostly written in Latin—the tradition of abacus manuals is not dominated by a few influential authors. While Latin *algorismi* are often copies of Alexander of Villedieu's *Carmen de algorismo* or Sacrobosco's *Algorismus*, abacus manuals were written by a large number of abacus masters and anonymous writers, without any name dominating the tradition. This high level of standardisation in the absence of a clear textual model reflects both the comparatively lower level of hierarchy that characterised vernacular schools and their close ties with the world of late medieval arts and crafts. The standardisation of abacus mathematics occurred from the bottom-up, in a similar way to how business practices were standardised through conventions that were often institutionalised in late medieval guilds.[8] Significantly, abacus schools were also identified with the term "bottega" (e.g., *bottega d'abaco*): the same word used to indicate the shops of craftsmen.

While these elements allow us to examine how commercial practices influenced abacus manuals, it is also possible to explore how abacus manuals were, in turn, linked to the practices that surrounded them. Given the context of the abacus school, the first relationship we need to discuss is between abacus manuals and educational practices. As surviving abacus manuals were primarily written by abacus masters who held schools, these texts probably circulated mainly in educational settings. We know of more than one case in which abacus manuals were passed down across generations of abacus masters, as part of the tools necessary to run an abacus school. For example, in 1367 Paolo dell'Abaco—a prominent Florentine master—stated in his will that the use of his school, its furniture, and its books, should pass down to his pupil Michele di Gianni (Van Egmond 1977; Murano 2015). Later in the century, Giovanni di Bartolo inherited the school of Antonio Mazzinghi, and Antonio's abacus books appear to have been part and parcel of the inheritance (Black 2007, 239; Ulivi 2002, 70–71; Arrighi 1965).

[8] In some cases, abacus masters themselves founded their own guilds (Van Egmond 1977, 13).

As such, the contents of abacus manuals were likely related to the classes taught by abacus masters. The style and structure of these manuals provide some clues about the educational practices used in abacus schools. The theoretical sections of these manuals are kept to the bare minimum, and rules are often expressed in a synthetic way. For example, the rule of three is expressed with a concise formula that is attested already in the oldest abacus manual: "If we were told to solve a problem in which there were three things, we need to multiply the thing we want to know by the thing which is of a different kind, and to divide into the other".[9] Similar formulations of the rule of three can be found in most abacus manuals, providing a standard that favours simplicity and memorability of a mathematical rule over its rigorous demonstration.

These characteristics suggest that in abacus schools mathematical rules, rather than being demonstrated, were simply given by the master. The pupil would then assimilate these rules by applying them in a long series of examples. Rote learning and long series of repetitions were a characteristic of late medieval instruction, and abacus mathematics was probably no exception (De Munck et al. 2007). Abacus manuals are mostly made of long lists of solved examples in which arithmetic is applied to a variety of practical problems. As these lists are mostly arranged by domain of application (e.g., exchange, interest rates, profits and losses, etc.), they provide repositories of problems that abacus masters could resort to while teaching classes dedicated to that kind of application. In his *Liber abaci*, Fibonacci presents a passage in which the combination of rote learning and constant practice is praised as the way to turn a *scientia* into a *habitus*: "Thus, who may wish to master the practice of this science should exercise it with enduring and constant practice, because through practice science is turned into a habit".[10]

It is therefore likely that abacus masters made students memorize synthetic formulations of mathematical rules, and then guided them through long series of examples in which these rules were applied to a variety of problems. In this way, mathematical rules were not "demonstrated" with the rigour of classical mathematics, but were rather "corroborated" by their effective applications, in a sort of quasi-experimental proof of their validity (Field 2016, 197–98). Also from this point of view, abacus mathematics reveals its proximity to the epistemology of the crafts, whereby each mathematical rule is considered as a sort of tool, and is used for a specific purpose. As suggested by Fibonacci's passage, these texts did not aim to develop a systematic body of codified knowledge, but rather to teach a set of mathematical skills, a mathematical attitude—a *habitus*—that could be applied to a variety of problems.

[9] Florence, Biblioteca Riccardiana, ms. 2404, f. 1r, 1–8, edited in: "Se ce fosse dicta alchuna ragione e lla q(ua)le se proponesse tre chose, sì devemo m(ultip)licare quilla chosa che noie volemo sap(er)e (contra) q(ui)lla che non è de quilla medessme, a pa(r)ti(r)e nell'altra"

[10] "Unde qui per eum huius scientie praticam bene scire voluerint, oportet eos continuo usu et exercitio diuturno in eius praticis perstudere; quod scientia per praticam versa in habitum". (Giusti and D'Alessandro 2020, 3).

4.2.2 A Fifteenth-Century Manuscript Manual from Pisa: Written Word and Wider Practices

A unique source to document the relations between abacus manuals and educational practices is a manuscript written in Pisa in 1442 by the abacus master Cristofano di Gherardo di Dino. The manuscript is kept in the Biblioteca Riccardiana of Florence (ms. Ricc. 2186), and it is a well-preserved paper volume comprising 131 folios. The manuscript includes a fairly standard abacus manual, but it has the exceptional characteristic of opening with a few folios in which Cristofano outlined in detail the syllabus adopted in his school, reproduced in Fig. 4.1. Both texts were written by the same hand, which is most likely the autograph hand of Cristofano. Starting from folio 9r, the school curriculum is followed by the abacus manual itself (fols. 9r–80r), by a brief section on astronomy/astrology (fols. 81v–82v), a few blank pages (fols. 83r–91v), and a section on practical geometry that is introduced as a vulgarisation of Fibonacci's *Practica Geometriae* (fols. 92r–131r). The final page (fol. 131v) reports a few personal notes (*ricordi*). While the bulk of the manuscript is still unpublished, the syllabus of Cristofano's school has been edited by Arrighi (Arrighi 1966; Franci 2003).

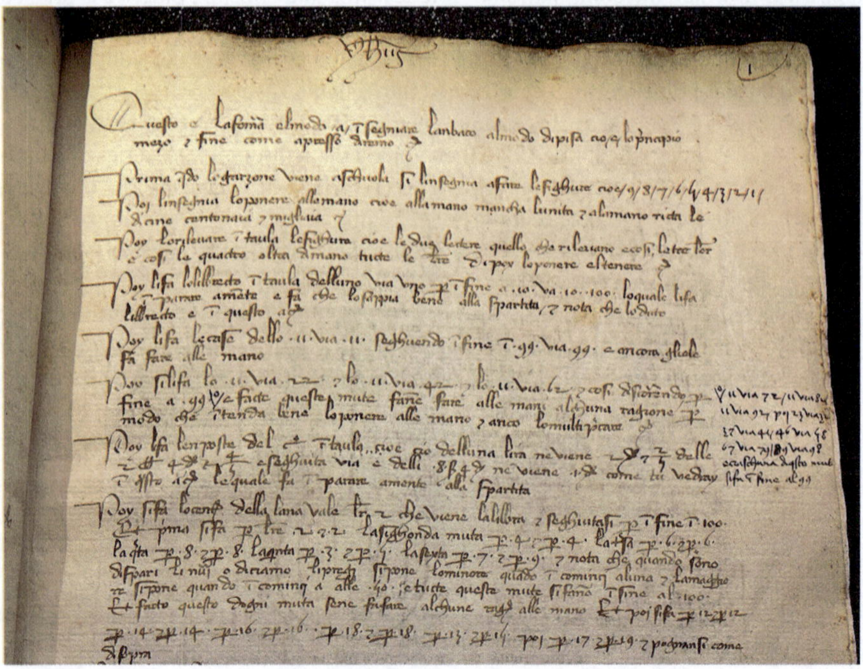

Fig. 4.1 Biblioteca Riccardiana, Firenze, ms. Ricc. 2186 c.1r, su concessione del Ministero della Cultura

As can be seen in Fig. 4.1, Cristofano opened his curriculum with a short introduction: "This is the form and the way to teach the abacus in Pisa, i.e., the beginning, the middle, and the end, as will be shown below".[11] The curriculum is divided into units, called "mute" (literally "relays/teams"). The first unit is dedicated to learning to read the ten Indo-Arabic figures and to the so-called technique of "finger reckoning", and to what appears to be the first steps in writing numbers with Indo-Arabic figures. On the third line, we can read:

> First when the *garzone* comes to school he is taught to make the figures, e.g. 9 8 7 6 5 4 3 2 1. Then he is taught to place on the left hand the units and on the right hand the tens, hundreds, and thousands. Then [he is taught] to place figures on a table, i.e., [to understand] what two figures mean, and what three, and what four, and what all other letters, and then to place and hold.[12]

Finger reckoning was a technique to represent numbers on one's hands, and pupils had to learn it in order to report partial remainders and other quantities on their hands while engaged in written calculation. It is interesting to notice that the pupil is identified with the same word used for apprentices, i.e., "garzone". Pupils were most likely only boys. This first preliminary module was followed by a dense list of multiplication tables that needed to be learned by heart: "Then one does the multiplication table of 1 by 1 until 10 by 10 100, which needs to be learned by heart, and we make sure that [the pupil] knows it well".[13] Multiplication tables are followed by an extensive list of examples of multiplications, with an increasing level of difficulty. This is followed by modules on division, in which different algorithms are used. The curriculum then covers fractions ("rotti", literally "broken numbers") and operations with fractions. This theoretical foundation is then applied to a variety of practical problems: interest rates (simple and compound), actualisation, alloying, and exchange. This is followed by a list of miscellaneous problems of exchange between different measures and currencies and of division of profits and losses (fols. 2r–8v).

A passage on the first folio suggests that the duration of the course depended on the progress made by each pupil, as they were promoted from one "relay/team" to the following one when the master considered them ready to do so: "each of the aforementioned [problems] are done until the master judges [the pupil] to be proficient".[14] The curriculum also reports that it was customary for the master to give some homework to

[11] Biblioteca Riccardiana, ms. Ricc. 2186, fol. 1r.: "Questo è la forma e 'l modo a insegnare l'anbaco al modo di Pisa cioè lo pricipio mezo e fine come apresso diremo".

[12] Biblioteca Riccardiana, ms. Ricc. 2186, fol. 1r.: "Prima quando lo garzone viene a schuola si l'insegnia a fare le fighure cioè 9 8 7 6 5 4 3 2 1. Poi l'insegnia lo ponere alle mano cioè alla mano mancha l'unità e a la mano ricta le dicine centoniaia e migliaia. Poy lo rilevare in taula le fighure cioè le due letere quello che rilevano e così le tre letere e così le quactro oltra di mano tucte le letere di poy lo ponere e 'l tenere."

[13] Biblioteca Riccardiana, ms. Ricc. 2186, fol. 1r.: "Poy li fa lo libbrecto in taula dell'uno via uno per infine a 10 va 10 100 lo quale li fa imparare a mente e fa che lo sappia bene a lla spartita".

[14] Biblioteca Riccardiana, ms. Ricc. 2186, fol. 1v.: "di ciascuna delle sopradicte imposte tanto quanto pare al maestro ch'egli vi sia bene scorto e presto".

his pupils, and suggests that the school may have operated full time: "and note that this is a general rule: each evening [the master] gives pupils, according to their team/relay, a few problems to solve that they have to bring back solved the following morning".[15] At the end of most of the short paragraphs outlining the curriculum, Cristofano left some empty space where he planned to insert the references to the specific folios of the rest of the manuscript where each item of the syllabus was dealt with. Even though these spaces were then not filled, they suggest a close relationship between the subjects covered in the curriculum and those included in the abacus manual.

The text of the manual opens with an introduction in which Cristofano reported his name, the date, and the place of composition of his manuscript: "I, Cristofano di Gherardo di Dino, Pisan citizen of the chapel of Saint Sebastian in the quarter of Kinzica, have started to write this abacus book today, first of May 1442, in the name of god and of the salvation".[16] The manual then introduces directly the standard formulation of the rule of three, followed by a series of problems illustrating how to operate with fractions (fols. 9v–12r). The rest of the manual reports sections covering various applications of mathematics: division of profits and losses, exchange, and miscellaneous applications of the rule of three (fols. 20v–26r), interest rates, simple and compound (fols. 26v–27v), algebra (fols. 28r–32r), a short section on geometry (fols. 32v–37v), actualisation (fols. 38r–40v), a miscellaneous section (fols. 41r–67v), a section on alloying (fols. 68r–73r), a further miscellany of problems (fols. 73v–78r), and multiplication tables (fols. 78v–81r). The final folios include a brief section on astronomy/astrology (fols.81v–82v), some blank pages (fols. 83r–91v), a vernacular readaptation of Fibonacci's *Practica geometriae* (fols. 92r–131r), and a few personal notes (*ricordi*) (fol. 131v).

As can be seen from this outline, there are some differences between the curriculum and the contents of the actual manual. While the curriculum presents a detailed outline of the initial classes in which pupils were taught how to make calculations with Indo-Arabic numerals, the manuscript provides little on these theoretical foundations of the course. A possible explanation for this difference may be that the master did not need the help of an aide memoir to teach these foundational classes, for which he may have relied just on the text of the curriculum and on examples designed on the spot. When it comes to the applications of mathematics, however, the curriculum and the manual broadly cover the same matters. This suggests that Cristofano may have used the manual as a repository of problems, to be consulted selectively to find problems relevant to the specific class he was teaching. The manuscript also includes some advanced sections, such as the one about algebra, which were not mentioned in the curriculum, and which may have been reserved for pupils who progressed to the most advanced classes.

---

[15] Biblioteca Riccardiana, ms. Ricc. 2186, fol. 2r. "Et nota che questa è reghula generale: ogni sera dare loro le ragioni a ciaschuno sighondo le mute loro che le denno recare facte la mactina rivegniente".

[16] Biblioteca Riccardiana, ms. Ricc. 2186, fol. 9r.: "Io Cristofano di Gherardo di Dino citadino pisano della cappella di sancto Bastiano in Chinsica quartiere di Pisa, Oggi questo dì primo di maggio 1442 col nome di dio e di sarvamento cominciai a scrivere lo presente libbro d'anbaco".

Therefore, Cristofano's abacus manual appears to have been in a close dialogue with the educational practices that surrounded it. While it did not record the foundational contents of the abacus curriculum, it provided a rich repository of problems on the practical applications of abacus mathematics, which were probably used as examples during Cristofano's classes.

More complex is assessing the relations between the problems discussed in abacus manuals and practices beyond the abacus school. Surviving manuals suggest that, with time, abacus manuals increasingly circulated not just among masters, but also among a wider readership that mainly comprised practitioners. In a similar way to abacus masters, these practitioners may have used abacus manuals as repositories of tools and techniques to address quantitative problems. As manuals were often passed down across generations of readers—both in the case of masters and in the case of practitioners—it appears that readers were more interested in the mathematical techniques included in these texts rather than in the accuracy of the contextual information they recorded. For example, information about the value of coins, commercial customs, and wares were usually recorded in so-called "pratiche di mercatura" ("mercantile manuals"), and not in abacus manuals. There is a number of documents that blur the two genres, especially from the earlier period of the tradition, but with time the two genres tended to be increasingly distinct. An example of these early texts that combine the characteristics of the abacus manual and the *pratica di mercatura* is the *Zibaldone da Canal*, a Venetian text written around 1370 (Stussi 1967; Danna and Locatelli 2026).

The standard applications included in abacus manuals—such as exchange of moneys and measures, division of profits and losses, barter, interest rates, and alloying—have a clear relevance for practitioners active in market exchange. However, one should not jump to the conclusion that the text of these manuals reflects faithfully what happened in the market. As argued above, these were texts aimed at providing the reader (or the listener) with the mathematical tools necessary to navigate the late medieval economic system. If we accept this reading, the purpose of these texts was to instruct, and not to represent faithfully. At the same time, instructional material needed to have a degree of proximity to the economic reality with which it engaged. Cristofano's manuscript provides ample material to show this. For example, Cristofano reported both problems that include what may be information taken directly from the marketplace and problems that are clearly imaginary.

An example of the former kind of problems is the following: "the *torsello de panni* [a unit of length], which is equivalent to 26 canes or 104 arms, is sold for 67 *lire*, 13 *soldi* and 9 *denari*. How much will 157 *torselli* and 64 $\frac{1}{4}$ soldi be?"[17] As it references any possible merchandise, the following can be considered as an example of an imaginary problem: "the kantar [a unit of weight] of whatever merchandise

[17] Biblioteca Riccardiana, ms. Ricc. 2186, fol. 5r.: "Lo torsello de panni lo quale è canne 26 cioè braccia 104 è venduto lb 67 s 13 d 9 che varranno li torselli 157 s 64 $\frac{1}{4}$?"

is sold for 27 *tari* and 13 *gramme*, what will be the value of 93 kantars and $23\frac{1}{2}$ rolls?"[18]

From the point of view of an instructional text, the faithful adherence of its examples to contemporary practices is less important than the transmission of mathematical procedures that disciples could apply to a variety of problems once they left the school. The two examples given above are clearly two analogous problems, and they are both solved with the same mathematical technique, i.e., the rule of three. These problems could be included in the same sections of the text even if they had very different degrees of proximity to commercial practices. This can again be explained if we consider that the primary goal of these manuals was to teach the reader (or the listener) how to apply a set of mathematical tools. The second example shows that, even though they remained firmly grounded in their commercial settings, the examples of abacus manuals also gestured towards some form of generalisation.

As stated above, a key context in which these mathematical tools were applied was the monetary market. Problems of exchange provided the tools necessary to translate values between the several currencies in use at the time. Sections on alloying addressed a further need related to the monetary system of late medieval Italy (Travaini 2000). This need emerged, again, from the fractured structure of late medieval Italy. Several cities formed independent city-states and minted their own currencies. As a result, a variety of coins were in circulation, which made calculations of exchange necessary. At the same time, further problems arose from the material aspect of these currencies. Coins were recurrently subject to debasement—i.e., to a slow reduction in their content of precious metal. The value of coins also tended to decrease with time because of clipping and of wear due to everyday use. Moreover, a variety of counterfeit coins struck by unofficial mints circulated at the time (Spufford 1988). As a result, the intrinsic value of coins in circulation could vary, which affected significantly their price and purchasing power. This meant that not just mint masters, but also regular merchants and anyone else that regularly dealt with monetary values, had to be acquainted with assaying procedures to gauge the fineness of the coins they had at hand, and eventually with minting practices. The sections dedicated to alloying that can be commonly found in abacus manuals addressed this widely shared social need to deal with these operations.

Cristofano's manuscript is no exception, as it includes a number of problems related to solving these issues. A first series of problems of alloying can be found in the first applications of the rule of three. In these problems, the reader is instructed on how to calculate the value of a given alloy, i.e., the operations necessary to essay a coin. For example: "an ounce of pure silver is worth 29 *soldi* and 4 *denari*. I ask how much an ounce of silver alloyed at $9\frac{1}{4}$ ounces will be worth" or "a pound of

[18] Biblioteca Riccardiana, ms. Ricc. 2186, fol. 2v.: "Lo cantaro di qualunque mercantia si sia è venduto teri 27 gramme 13, che varranno le 93 cantari e ruotuli $23\frac{1}{2}$?"

pure gold is worth $101\frac{1}{4}$ ducats. What will be the value of gold alloyed at $21\frac{1}{3}$ carats?"[19]

The manual also presents a further section entirely dedicated to alloying, which is found on folios 68r–73r of the manuscript. This section is mainly concerned with the operations necessary to obtain an alloy of a desired fineness. It includes a series of problems of the following kind: "One has 7 *lire* of pure silver, out of which he wants to make an alloy that includes $3\frac{1}{4}$ of silver per pound. I ask how much will be the entire amount of metal, and how much copper will be needed".[20] This problem is solved with the rule of three. As one *lira* was equivalent to twelve ounces, seven *libre* equalled 84 ounces, the problem could be solved with the following proportion: $\frac{x}{84}=\frac{1}{3+\frac{1}{4}}$.

Cristofano reports several variations of this kind of problems, for example by giving the reader an initial amount not of pure metal, but of an alloy. For example: "one has 12 pounds of an alloy that includes $2\frac{1}{2}$ ounces of silver per pound. He wants to make an alloy with $3\frac{1}{3}$ ounces of silver per pound".[21] Other problems ask the reader to combine two existing alloys in the correct proportion in order to make a new alloy of a specific fineness: "one has an alloy that has 7 ounces of silver per pound, and has (another) alloy that has 11 ounces of silver per pound, and he wants to make 20 pounds of an alloy that has 8 ounces of silver per lira. And he wants to add neither copper nor silver. I ask how much of each alloy he will have to use".[22]

This mathematical knowledge, which emerged in Italy with the commercial revolution, spread to the rest of the European continent in the following centuries. Germany offers a particularly interesting comparative study, because it allows us to see both how a body of mathematical knowledge was transmitted from Italy to other European areas, and to see how it reacted to local conditions. The booming mining

[19] Biblioteca Riccardiana, ms. Ricc. 2186, fol. 23r.: "L'uncia dell'ariento fine vale s 29 d 4. Adomando che varrà l'uncia dell'ariento a legha di uncie $9\frac{1}{4}$" (note that the expression "$9\frac{1}{4}$" is equivalent to our modern expression "$9+\frac{1}{4}$"). "La libbra dell'oro fine vale ducati $101\frac{1}{4}$. Che varrà la libra dell'oro che sia a carati $21\frac{1}{3}$?"

[20] Biblioteca Riccardiana, ms. Ricc. 2186, fol. 68r.: "Uno (h)a libbre 7 d'ariento fine del quale vuole fare moneta che tegnia uncie $3\frac{1}{4}$ per libbra d'ariento. Adimando quanto si è la somma di tucto il consolare, et quanto sarà lo rame che vi bisognia".

[21] Biblioteca Riccardiana, ms. Ricc. 2186, fol. 69v.: "Uno ae libbre 12 di moneta che tiene d'ariento uncie $2\frac{1}{2}$ per libbra, del quale vuole fare ariento a unce $3\frac{1}{3}$ cioè vuole fare moneta. Adimando quanto ne consolerò e quanto ariento v'agiungerò".

[22] Biblioteca Riccardiana, ms. Ricc. 2186, fol. 72r.: "Uno ae moneta che tiene uncie 7 d'ariento per libbra et (h)a moneta a uncie 11 per libbra et egli vuole fare libbre 20 di moneta che tengha uncie 8 d'ariento per libbra. Et non ci vuole avere ad agiungere né rame né ariento. Adimando quanto vi mecterà di ciascuna di quelle 2 monete".

industry in the Ore Mountains offers a particularly illuminating case to observe how the mathematics codified in practical arithmetic manuals was indeed related to the practices that surrounded these texts, and how it responded to changes in their socio-economic context.

4.3 The Arithmetic of Mining and the *Rechenmeister* Tradition

Mercantile arithmetic in the Holy Roman Empire was to a large extent an Italian import. It is possible to track its development from early contact in South Germany, in the rich Swabian and Franconian free towns of Augsburg, Ulm, and of course Nuremberg (Van Egmond 2005). From the late fourteenth century onward, rich merchant families would send their sons to Florence, and especially to Venice (to the *Fondaco dei Tedeschi*, literally the "warehouse of the Germans"). Young people would there be trained in the new mercantile arithmetic—and labelled it the *welsche Praktik*—the Italian practice (Braunstein 2016, 407–450). From there, this knowledge quickly diffused through many parts of the Empire, eventually reaching the northern Hanseatic towns where it mixed, not without difficulties, with the existing accounting and computing traditions (Jenks and North 1993).

These exchanges suddenly took a new dimension in the late fifteenth century, when the discovery of rich metallic veins in the Ore Mountains of Saxony set up a second silver rush, the so-called *Berggeschrey*.[23] One person played a decisive role in the process: Martin Römer († 1483). Romer was a powerful merchant, whose family came either from Nuremberg or from the Saxon region. From the 1450 onward, he was involved in the spice business and became extremely wealthy, eventually being in the position to lend money to Albrecht, Duke of Saxony.[24] He then diversified his investments in textile and mining operations, from copper to silver, all of which "required flexibility, international networks and a high level of investment capital to start with", amassing hundreds of thousands of florins (Kahleyß 2013, 155).

From the early 1460s, Römer and a handful of associates set up exploring missions in the mountains close to Zwickau, encouraged by tax exemptions from the duke. From 1470 onward, a massive amount of silver was found and attracted thousands of miners, leading to the foundation of new cities such as Schneeberg and Annaberg (Dietrich 1991, chap. 1). Eager to lure investors and skilled workers, the Saxon Duke and his counterparts in other mining regions enacted and printed mining laws (*Bergordnungen*). These public documents stipulated anything from the size of concession to the way they were measured, and, most importantly, how profits were allocated and coins minted (Ermisch 1887). Although we know little about

[23] General references include (Laube 1974; Dietrich 1991)

[24] Biographical information about Martin Römer can be found in (Herzog 1865; Kahleyß 2013)

the training of this first generation of investors, we will see that they used the new Indo-Arabic numerals fluently and possessed an operative arithmetical knowledge which came to play a decisive role in the success of mining operations.

Martin Römer was indeed much more than a shrewd investor, and soon became *Zehntrechner*, a leading mining official in Schneeberg, responsible for computing and collecting the tenth tax on mining profits. These fiscal documents—*Zehntrechnungen*—have been studied in detail by Karl Hahn. Perusing through these repetitive administrative documents, one gets a sense of how computationally intensive they were: "*item* we have first [found] 118 mark [silver] from the *Nativity* of *Mary* to Christmas, of which 11 mark $\frac{1}{2}$ lot 1 $\frac{1}{2}$ quent to our merciful lord anno [14]70".[25] In each quarter, Römer would list the quantity of silver coming from each mining pit and calculate the tenth part, and then send this amount to the Saxon ruler as an in-kind tax. Römer also administrated the duke's monopoly on silver purchase on a commission basis: he was allowed to buy directly silver from the investing mining companies, at a fixed rate of 6 florins for a mark. He would then sell it on the behalf of the duke, to whom he then owed 1 $\frac{1}{4}$ fl. per mark. As Römer was able to sell silver at a much higher price through his trading offices in Frankfurt-am-Main, Venice, Augsburg, and Nuremberg, he could pocket the difference (Dietrich 1991, 41–46; Kahleyß 2013, 159).

4.3.1 The First Printed Practical Arithmetics: Ulrich Wagner and Johannes Widmann

The silver rush and the rise of Martin Römer both happened—it is important to emphasize—slightly before the first known German arithmetic was printed. This is important to consider in order to nuance the predominant analysis, according to which "the *Rechenbücher* emerged as a fusion of the oral experience knowledge with a written scholarly tradition" (Gärtner 2000, 281). The theme of a fusion of practical experience with the rational, mathematical principles has been subject of several studies lately, but it should be emphasized here that a written tradition of accounting and administration existed among practitioners within the mining administration (Smith 2010; Long 2015).

The *Bamberger Rechenbuch*, as it is now known, was published under various forms in the years 1482–1483 by the Nuremberg *Rechenmeister* Ulrich Wagner († 1490).[26] If we open this book, the close connection between the teaching of arithmetic it contains and existing practices seems to be clear. Mercantile operations and the

[25] (Hahn 1932, 36–37): "item zcum ersten 118 marck von Nativitatis Marie [8 septembre] biß uff Weinachten, davon 11 marck 13 1/2 lof 1 1/2 quent unsern gnedigen hern zcu zcehnden anno etc. 70°".

[26] About the xylographic *Bamberger Blockbuch* (ca.1471–1482), see (Vogel 1980).

thriving mining business form the bulk of this text: it contains sections on spices, precious metals, and exchanges, as well as metrological tables about weight and prices of silver (Kirnbauer 1960, 29).

The second German *Rechenbuch* was published by Johannes Widmann (ca. 1460–?). The *Behende vnd hubsche Rechenung auff allen kauffmanschafft* (Diligent and Fair Calculation of all Commercial Transactions) appeared in Leipzig in 1489, and five subsequent editions are known.[27] Widmann is known to have studied at the University of Leipzig, where he obtained his *magister artium* and offered the first course on algebra of the German-speaking countries. Interestingly, Widmann soon left Leipzig for the neighbouring Ore Mountains and settled in the booming mining cities of Annaberg—a mining city founded during the silver rush. His activities are unknown, but since his name is not found among the teachers of the Latin school, it can be reasonably assumed that he used his mathematical skills in mining-related activities.

There are obvious similarities between Wagner and Widmann's texts, not only when presenting the four operations and computations with fractions, but also in the succession of chapters dealing with various goods such as pepper, ginger, saffron and nails (although Widmann's text is substantially longer and contains many more examples). A new feature in Widmann's text, especially interesting for us, is his presentation of a "rule of alloying" (*regula alligationis*):

> Rule of alloying
>
> You should thus practice this rule by setting first the number of the compound, then alloy there the smallest with the biggest, and the middle between them on the first number set. And the thing is a middle, that [you] alloy with the least and also the most and set the difference as the alloyed number and is correctly done.[28]

Surprisingly enough, the examples used by the author in the following folios are neither related to metal alloys nor to mining, but deal successively with grains, sheep, donkeys, oxen, wax, and bells. Each of them is mixed with similar objects of different quality, and a pseudo-alloying situation arises. Widmann finally turns to a chapter on "coins" (*Muncz*) where the same rules are applied (Fig. 4.2):

> Item what every mint master should know, if there are many coins in the assay of which he wishes to know how much the *mr* [Mark] will cost him. And how much the coin should be against the *lot*—and then how many for 1 *fl* [florins] and how high the grain should be, and the crucible assaying [Schickung des tigelß] that is [bringing] the coin in a lead alloy, and how to compute the piece in the crucible, and similar things.[29]

---

[27] (Gärtner 2000, 48–54). The subsequent editions appeared in southern cities: Pforzheim, Hagenau and Augsburg.

[28] (Gärtner 2000, 462): "Regula alligationis. Diese regel soltu also practiciren Secz zum ersten die zal der mischung darnach alligir albeg das kleynste mit dem grosten. vnd die mittel vntereynander auff die erste gesaczte zal. vnd ist sach das eyn mittel ist. das alligir mit dem mynsten vnd auch meysten- Und die differentiam secz albeg auff die aligirten zal vnd ist recht gemacht."

[29] (Gärtner 2000, 465):"item eynem itlichen muncz meyster ist not zu wissen bestant mancherley muncz in der prob also das er mag wissen wie hoch die mr hyn kumpt. vnd wie vil der muncz auff das lot geen sollen- vnd darnach wivil für 1 fl vnnd wie hoch das kornn besten sol vnd die Schickung des tigelß daz die muncz in eynem wesen pleyb. vnd wie man die stuck in den tigel rechnen sol vnd des Gleichen".

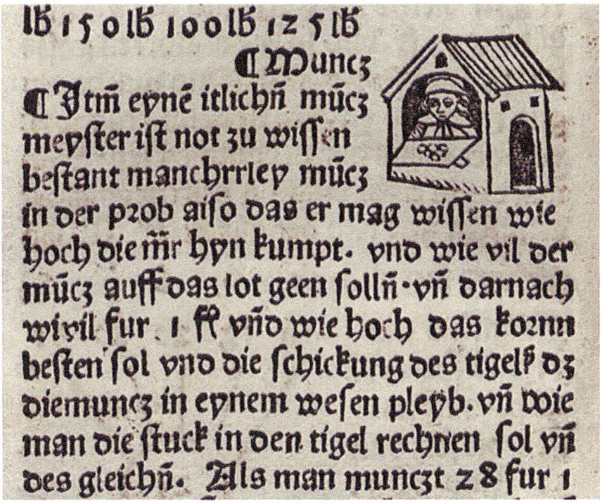
lb 1 5 o lb 100 lb 12 5 lb
Muncz
Item eyne itlichn mucz
meyster ist not zu wissen
bestant manchrrley mucz
in der prob also das er mag wissen wie
hoch die mir hyn kumpt. vnd wie vil der
mucz auff das lot geen solln · vn darnach
wivil fur 1 ꝶ vnd wie hoch das kornn
besten sol vnd die schickung des tigels dz
diemuncz in eynem wesen pleyb. vn wie
man die stuck in den tigel rechnen sol vn
des gleichn. Als man munczt z 8 fur 1

Fig. 4.2 Johannes Widmann, *Behende vnd hubsche Rechenung auff allen kauffmanschafft*, Leipzig, Kachelofen, 1489, §266 (Source: *München, Bayerische Staatsbibliothek—Inc.c.a. 82, CC BY-NC-SA 4.0)*

One finds in the Widmann text two kinds of problems related to alloying, with multiple numerical examples for each. The first one is usually named *Beschickung des Tiegels*, which can be translated as "crucible assaying" (see Fig. 4.2). Given a quantity of "gekörntes" [grained] silver, which means silver mixed with copper at a given proportion, one is asked how much pure silver should be added to heighten the proportion of silver up to a given threshold.[30] The second type of exercise is usually labelled "coin minting" (*Muntzschlag*) and amounts to conversions between weight and value units. Given one *mark* of pure silver, one is asked to produce a given number of coins of mixed silver with a given purity, and one asks the real silver value of each coin produced.[31]

Widmann's textbook is similar to the Italian tradition of abacus manuals inasmuch as some problems can be labelled "generic" or "imaginary". For instance, when Widmann writes "I have grain worth a *Scheffel* (or another measure, as you want)", it is clear that this grain example has a more generic tone—an impression confirmed by the complicated and hardly natural setting that he then uses.[32] Many alloying problems, however, seem to be of practical nature. They do not use abstract numbers, as pseudo-practical problems sometimes do, but concrete numbers with sound metrological units and realistic ranges. Moreover, the type of questions asked

[30] A summary explanation of this kind of problem can be found in (Ries and Deschauer 1992, 158–59)

[31] This section corresponds to the § 266–72 in the pagination of (Gärtner 2000). Further examples with silver and gold can be found in §273–92.

[32] "ich hab korn des gilt eyn scheffel (ader ander moß was du dan wilt)": this section corresponds to the § 263 in the pagination of (Gärtner 2000).

could indeed be useful to assayers: Widmann himself observes between two exercises that "all these things are necessary knowledge to any mint master".[33] We should be cautious, however, before reading this section as a complete, straightforward reflection of the practice of arithmetic as it existed in the Saxon mining administration. One should note that mining problems only make a minor part of Widmann's book, which has a very broad scope. The *Behende vnd hubsche Rechenung* begins with a classical *algorismus* part and ends with a geometry containing practical and pseudo-practical problems such as measuring the height of a tower using its shadow, a stick, or a cord.[34] Still, the ease with which Widmann plays with units and computation is striking. In this book – where incidentally the first known use of the + and-symbols are recorded – the author already uses them in "elaborated" cases such as the following minting problem:

> Item one wants to mint coin and wants to set the calculation [rechnung] to 1 *mr* of fine silver to $7\frac{1}{2}$ Rhine *florins* and wants that the coins have 7 *lot* of fine silver [for] 1 *mr* and wants to have $\frac{1}{4}$ for 1 *mr* coins as minting tax and the mining master also wants $\frac{1}{4}$ for 1 *mr* coins for his work and the costs that come with all things, and he wants that one coins 165 *pf* [Pfennige] for [each] Rhine *florin*. Now the question is, how many *pf* should be in each *mr* so that the coin has its equity, as should be. Do as such, the coin should have 7 *lot* of fine silver [for each] *mr* at $7\frac{1}{2}$ *fl*, which makes 7 *lot*, multiply with $7\frac{1}{2}$ *fl* and divide by 16 makes $3\frac{9}{32}$ *fl* to which you make $\frac{1}{2}$ *fl* [for] the mint master and the ruler, this makes $3\frac{25}{32}$ Rhine *fl*. - that you make to 1 *fl*. For 165 *pf* does *pf* $623\frac{29}{32}$ should to 1 *mr*, these you divide by 16 *lot* comes 39 *pf* $-\frac{3}{512}$ *pf* these should be for each lot and not more.[35]

Note here how clever the use of the minus sign is in the last operation, in which a mining master should compute how many silver *pfennig* should be in each *lot* used for mining Rhine florins. Instead of carrying out the complex division of 623 29/32 by 16, Widmann rounds up his divider to 624 and divides 624 by 16, immediately

[33] (Gärtner 2000, 465): "Und die ding alle seyn not zu wissen eynem itlichen munczmeyster."

[34] (Gärtner 2000, 508–9).

[35] (Gärtner 2000, 466): "Item eß ist eyn her der wil muncz schlahen vnd wil die rechnung seczen 1 mr feyn silberß auff $7\frac{1}{2}$ *fl* R*eynisch* vnd wil daz die muncz 1 mr sol haben 7 lot feyn silber vnd wil von eyner mr muncz haben $\frac{1}{4}$ fl *Reynisch* zcu schlagkschacz vnd der munczmeyster wyl auch haben $\frac{1}{4}$ von 1 mr muncz vor seyn arbeyt *vnd* kostung die er dar auff legt mit allen dingen, vnd der her wil daz man schlagt 165 pf für 1 fl R*eynisch* Nu ist die frag wye vil pf an 1 mr gen sol das die muncz yr. gerechtigkeit hab als ir gepurt. Machß al so die muncz sol haben 7 lot feyn silber 1 mr zu $7\frac{1}{2}$ fl was machen 7 lot die multiplicir mit $7\frac{1}{2}$ fl vnd teyls mit 16 kumpt $3\frac{9}{32}$ fl dar zu thu $\frac{1}{2}$ fl dem munczmeyster vnd herrnn daß macht $3\frac{25}{32}$ fl Reinisch· dy mach zu pf 1 fl für 165 pf thut pf $623\frac{29}{32}$ sol an 1 mr gen die teyl mit 16 lot so kumpt 39 pf$-\frac{3}{512}$ pf die sollen an 1 lot gen vnd nicht mer·".

obtaining 39, from which he subsequently subtracts what had previously been rounded up – here $\frac{3}{32}$ divided by 16 – obtaining his "$-\frac{3}{512}$".

4.3.2 *Adam Ries: The Mining Examples of a Teacher*

Adam Ries (1492–1559) is a household name in the German-speaking area, but as many Renaissance *Rechenmeister*, he is barely known outside this area and in some cases not even considered a mathematician.[36] Born in Staffelstein, Ries is known to have visited Zwickau, a rear base of Ore Mountains capitalism, in his youth, as his brother was studying in the local Latin school. He later paid a first visit to Annaberg in 1515, which was at the time the centre of Saxon ore mining. Once again, the contacts between the world of the arithmetic masters and the world of mining seem straightforward. Once again, however, we should remain cautious: Ries did not go to discover or work in the mining administration, but visited the local Latin school. He was invited by its rector Andreas Weidner who, like Ries, was born in Staffelstein (Schellhas 1977, 5).

Adam Ries later opened a merchant school in Erfurt, a medieval university town, in 1518. Just as Johannes Widmann had done a few years before, he was soon attracted by the fantastic wealth of mining cities and moved to the Ore Mountains. This is where the second edition of his famous textbook appeared—indicating on the cover that Ries was "now in Saint Annaberg" (Ries 1525, title page). This work was an exceptional publishing success with more than 120 editions, contributing significantly to the diffusion of Indo-Arabic numerals in the German-speaking area. While Ries' book is significantly shorter that Widmann's, with worse illustrations and altogether a more superficial approach, it has been convincingly argued that Ries achieved success by striking an excellent balance between readability and usefulness (Deschauer 2003).

Although he is today mostly known as a teacher and *Rechenmeister*, we actually know very little about his private arithmetic school.[37] We have much more information about his well-documented work for the mining administration. However, historians have rarely reflected about the interaction between his teaching activities and his professional use of mathematics.[38]

Ries' second *Rechenbuch* contains three sections in which mathematics and mining interest are mixed: "Reckoning with silver and gold", "Crucible assaying", and "Coin Minting". Things get especially interesting when we realize that all three are notably absent from the first *Rechenbuch* of 1518.[39] It is thus clear that in order to

---

[36] About Adam Ries and early Renaissance Rechenmeister, see (Wußing 2013; Folkerts 2011)

[37] An interesting reference is (Rochhaus 1996).

[38] The reference work on the topic is still (Schellhas 1977) which also includes a fac-simile of Ries' own accounting sheets. See also. (Kirnbauer 1960)

[39] "Silber und golt rechenung", "Schickung des tigels" and "Vom Muntzschlagk", Fiii sqq.

adapt to his new professional environment, or to promote the usefulness of his textbooks, Ries was led to include new problems and methods related to minting and mining. However, his new textbook was published after his first visit to the region, but before he settled there and became involved in the mining administrations. We should thus ask where his knowledge came from. It is clear that Ries had studied Widmann's book, and possibly other works deriving from it, but he could also have directly discussed with assayers and mining officials—the most likely being a combination of it all.[40]

In order to understand the nature of arithmetical knowledge presented and the interplay between theory and professional practice, let us consider one example:[41]

5		2
9		16
	11	

Item one has grained silver. If a mark has 9 lot [and] he wants to have it at 11 lot, how much fine silver should he add to one mark? Do this and set as stands here first how much the silver has, right after how much the added silver has and thirdly set just under how you want to have it, as stands here, then take the smaller from the middle one as 9 from 11 gives 2, this you write over 16 and take then 11 from 16 gives 5, this you set over 9 and continue with the *regel de tri* [which] says 5 lots need 2 lot [of] fine [silver] how much does a mark need *facit* 6 lot 1 quent 2 pfen[ig] of weight and 4/5 of a heller.

As in the examples taken from Widmann's book, Ries uses concrete numbers and metrological units. It should further be noted that the metrological units used throughout the examples are correct. This holds both within one given system (1 mark silver = 16 lot, 1 lot = 4 quenten, 1 quenten = 4 pfennig, and 1 pfenning = 2 heller) and for the correspondence between systems of prices and weights. When we read in Adam Ries's first example that "fine silver costs 8 florins a mark", this is consistent with contemporary sources: we have seen above the computations of Martin Römer, buying pure silver at 6 florins a mark, further paying $1\ \frac{1}{4}$ fl. to the

[40] The influence of Widmann seems hardly deniable on this specific point. Adam Ries writes in another instance that he has "very carefully read" Widmann's textbook (*mit gantzem vleyß gelesenn*), and deals with the same mining problems in a very similar order—although his text is shorter, better structured, and uses its own numerical examples. For a larger comparison between the two authors, see (Gärtner 2000, 217–19.

[41] (Ries 1525, §165). See a modern German transcription in (Deschauer 1992, 73):"Item einer hat gekornt Silber. Heldt ein margk 9 Lot wils haben zu 11 loten wievil sol er fein silber eyner margk zcu setzen thu ym unnd setz wie hie stet zum ersten wievil das silber helt darnach gleich wie vil das helt von dem man zu setzen viel und zum dritten setz gleych darunder wie man es haben wil / wie hie stet / Darnach nym das kleyner vom mittelsten als 9 von 11 pleyben 2 die schreyb uber 16 und nym als dan 11 von 16 pleyben 5 die setz uber 9 unnd machs fort nach der regel de tri sprich 5 lot bedorffen 2 lot fein wieviel bedorff ein margk facit 6 lott ein quentenn 2 pfen: gewicht und vier funffteyl eynes helles gewicht."

Saxon ruler and making a profit for himself (the final price must thus have been north of $7\frac{1}{4}$ florins).[42]

It might come as no surprise that Ries, a professional teacher now living in a mining city, would teach and present concrete situations in his textbook. After all, he was himself a mining official carrying out precisely such operations. This is nevertheless worth mentioning, since there existed at the time pseudo-practical textbooks in which examples were either computed with abstract numbers (without metrological units), or with unrealistic numbers chosen in order to ease computations. Here, on the contrary, the examples presented by Ries seem to have been directly relevant for occupational practices of the time.

4.3.3 Adam Ries: The Mathematics of a Mining Official

If we sum up our previous findings, we see that from the second half of the fifteenth century onward, mining officials and investors possessed and used efficient arithmetic methods in the minting and assaying businesses. These regions even attracted scholars such as Johannes Widmann, whose 1489 textbook presented realistic mining problems. Adam Ries followed his path and had obtained, even before settling for good in Annaberg in 1523, a certain knowledge of the arithmetic involved in mining and commercial operations—either directly from Widmann, or from another representative of the Rechenmeister tradition. In the mining city of Annaberg, Ries then quickly opened a reckoning school (*Rechenschule*) in his private house, his being only one of several schools in this thriving mining city. He became a citizen of Annaberg, acquiring several domains and houses. After his death, most of his children would pursue his arithmetic activities in various settings of the Ore Mountains.[43]

After explaining that his textbooks included real-life examples and were possibly useful for mining officials and essayer, let us switch the perspective and see how Ries, in his multifaceted activities as mining official, relied on practical arithmetic. Adam Ries indeed became a mining official shortly after arriving in Annaberg: in fact, both activities seem to be closely related.[44] In 1524 he became *Rezeßschreiber* in the local mining office. Mining was at the time made of private enterprises, based on a shareholder system (the so-called *Kuxensystem*) and regulated by the Duke George of Saxony. Adam Ries had thus to check the balance sheets sent quarterly by each mine supervisor, correct them, and forward them to the central mining office.[45]

[42] "fein silber kost ein margk 8 flo" (Ries 1525, §155).

[43] About Adam Ries' life in Annaberg, see (Wußing 2013, 22–32). About Adam Ries' descendants, see (Rüdiger 2017).

[44] About the interplay between mathematical skills and practical uses in the Ries family, see (Morel 2022).

[45] About Adam Ries activities as mining official, see (Schellhas 1977, 9–29).

His daily work as *Rezeßschreiber* thus involved extensive calculations about silver quantities—tasks very similar to the examples presented in his textbook.

Adam Ries must have given satisfaction to the higher mining administration, for in the following years he was appointed as *Gegenschreiber* and *Zehnter* both in Annaberg and in a neighbouring mining district. A *Gegenschreiber* recorded in a specific book the transactions of mining shares between investors; he further supervised the distribution of corresponding profits (or collecting compensations for a pit's losses) at the end of each quarter, dividing them according to the number of shares owned by the various investors. As a *Zehnter*, he was in charge of collecting the duke's tenth tax on all mines and as such was one of the most important officials (as explained above, this is precisely what Martin Römer was doing two generations before). Ries was also supposed to control the smelting processes and weight the pure silver at different stages (Schellhas 1977, 12–16; Laube 1974, 58).

These various positions show how Adam Ries was, for several decades, at the exact position where arithmetic was implemented in mining. Raw ore was extracted, then smelted and refined to give pure silver, as Georgius Agricola would later explain and beautifully depict in his *De re metallica* (1556). However, mathematics was just as necessary as the mining and smelting crafts for this process. In order to prevent tax avoidance and fraud, taxes and dividends could only be paid using coined silver.[46] Since the silver coins included a fluctuating proportion of pure silver, determined by the local ruler and implemented by the local mints, this was exactly where the various arithmetical methods described above—the *crucible assaying* (Beschickung des Tiegels) and the *coin minting* (Müntzschlag)—were required. Rulers often used their prerogatives to buy directly from the mine a large proportion of the pure silver at a preferable price— or farmed this right to investors such as Martin Römer—and records show precisely how Adam Ries realized the computations.[47]

Summing up this too long a discussion, we can gather a few insights about the relation between the content of Ries' arithmetical textbooks and his professional activity as mining official. His reckoning school seems to have been very successful, to the point it was mentioned in the town's chronicles (Rochhaus 1996, 97). Moreover, his long career in the mining administration means both that he was a skilled official, and that there was a shortage of qualified people. This last point should be emphasized: despite the availability of practical arithmetic textbooks, there were so few people skilled enough to understand them and bridge the gap to real-life practices, that mining officials often combined many positions. Adam Ries, for instance, was not only teacher, *Rezeßschreiber, Gegenschreiber* and *Zehnter*, but also acted as local agent for rich foreigners willing to invest in the Saxon mines—a position which undoubtedly involved lots of calculations (Laube 1974, 108).

[46] See for instance the mining law of Saxony, 1554, article 8.

[47] (Schellhas 1977, 20): it can be instructing to compare the real price used by Adam Ries (7 florins and 17 Groschen for one *Mark* of pure silver) with the examples of prices he mentioned in his *Rechenbuch* (between 8 and 8 ½ florins). This shows how realistic Adam Ries' examples were (and how profitable the prerogative of first buyer were).

Indeed, reckoning masters in the early sixteenth century seem to have often mixed their teaching activities with practical considerations. In a manuscript written before his arrival in Annaberg and entitled *Beschickung des Tiegels*, Adam Ries dealt in detail with assaying computations.[48] Many textbooks' authors also taught, practiced accounting, and possessed an extensive knowledge of various arts and crafts. This was clearly the case with Adam Ries. This does not necessarily mean that the content of their textbooks directly reflects "genuine" practices, but these texts should nevertheless be taken seriously.

They are in many cases the work of people knowing what they write about, either from previous reading—as did Adam Ries before arriving in Annaberg—or from direct experience—as he later did as teacher and mining official. Similar observations could be made in other regions at the forefront of economic and technical development. Ad Meskens has for instance shown for the city of Antwerp how the Coignet family was simultaneously involved in wine gauging and tax collecting. They also had their arithmetic school and published books—not in Latin, but in French or Flemish vernacular, such as an *Arithmetic, containing several beautiful questions & demands, proper & useful to all who frequent the traffic of goods* (Meskens 2013, chap. 4–5; Mennher and Coignet 1573).[49]

The same argument can be made from the opposite perspective: early modern craftsmen had, in many cases, a thorough knowledge of the mathematical arts and methods useful for their activities. In Annaberg, Adam Ries became friend with the assayer and mint master Hans Conrad, who besides the arithmetic needed for his daily work, also practiced the *ars magna* (algebra). The categories of teacher and mathematical practitioners should rather be seen as labels that sometimes blur and can be combined than as rigid identities.

It might be necessary to add a few caveats to the previous discussion by underlining that the professional and teaching uses of mathematics did not necessarily perfectly fit together. They might be extremely close at times, but obeyed a different logic. As a mining official, Adam Ries followed a logic of administration, which for instance means that some of the documents he produced were not meant to be printed.[50] Similarly, he had to use roman numerals in some contexts (although calculations were always made using the modern Indo-Arabic numerals), simply because the mining laws stipulated, for fear of frauds, that some registers had to be written "using German numerals." (Schellhas 1977, 20) Adam Ries is also exceptional for the breadth of his activities and the level of success he achieved. When inflation plagued the Ore Mountains, he computed specific numerical tables to deal with the

[48] This manuscript is now in the SLUB Dresden Mscr.Dresd.R.284. It can be speculated whether the purpose was really practical, or if Ries was displaying his skill to lure a prospective patron.

[49] Remarkably, Michel Coignet's textbook is an update of Valentin Mennher's, who moved to Antwerp from southern Germany (Swabia).

[50] A nice example is a detailed instruction on mining calculations, handwritten in 1554 and now conserved in the SLUB Dresden Mscr.Dresd.K.349. The document does not contain secrets methods or unknown formulas, but is rather of collection of templates for noting mining calculations down.

dearness of bread; when the Saxon ruler decided to debase his currency, Ries was enrolled to write a supporting pamphlet, the *Münz-Bedenken* (Gebhardt 2006).

It is also clear that professional interest in the arithmetic of mining intersected with a more general curiosity and culture of mathematical questions. In his third *Rechenbuch*, Adam Ries mentions an example presented to him in 1509 by "Thomas Meiner, now alderman in Annaberg", where he was also mining tax collector. The riddle, however, is not related to mining but to a "company" problem in which three persons pool their money to buy Ginger, Pepper, and Saffron, and wish to divide it fairly.[51] Similarly, we have seen that Adam Ries discussed the art of algebra with the smelter Hans Conrad, although it was not strictly necessary to solve the concrete problems they dealt with. In the Ore Mountains of Saxony as in other mining regions, a culture of accuracy developed and greatly heightened the prestige of geometry and arithmetic (Morel 2022, chap. 3).

As in Italy a century before, practical arithmetic manuals provided the mathematical foundation to this culture. It was not necessary for these texts to faithfully mirror the socio-economic practices that surrounded them, but rather to transmit mathematical skills and methods that could be applied to solve a variety of problems. These broadly applicable skills were particularly needed in a context of deep socio-economic change such as the cities of the Ore Mountains, in which practical mathematicians managed to cover new social positions. Finally, as was the case for the Italian abacus masters, whose manuals were used in a school context, Ries' *Rechenbuch* was written as a demonstration of his teaching skills, as the last paragraph explicitly states:

> [I] also want to please all those enthusiasts about calculations [liebhaber der rechnung] with this short booklet, and friendly ask them to receive the present [book] benevolently. If something is mistaken, or not fully described or rightly presented, I will try my best to show anyone, to the best of my means, and also at another time to faithfully instruct him in gauging [of barrels], the rules of algebra and bookkeeping.[52]

These dense and fluid exchanges, however, only lasted for a few generations. When the cossist and scholar Christoff Rudolff (1499–1545) published his famous arithmetic textbook in 1526, he prominently indicated on the title page that it contained "the computation of gold and silver, the crucible assaying and what belongs to the mint master".[53] A new edition in 1540 already dismissed this part from its title page, although the content related to mining operations was still present.[54] From then on,

[51] (Ries 1550, 162v–63r):"Thomas Meiner itzt des Raths auff S. Annaberg".

[52] "Wil allso mit diesem büchlein kürtz begrieffenn alle liebhaber der rechnung vorehret haben Bit die selbigen gar freuntlichen gegenwertiges gütlich anzunehmen / ob yrgents was vorsehenn / ader nicht gantz grüntlich beschrieben willigklich recht zuwertigen / wil ich vmb eynen yeden meynes vormügens gefliessen sein zuvordienen und zu eyner andern zeyt ym das Visirn: die regelen Algebre und das Buchalten treulich mit zu teylen geneygt seyn."

[53] (Rudolff 1526): „silber und goldt rechnung, was zu schickhung des tegels, was aynem muntzmaister- rechnung belangend zuhorig".

[54] In the 1540 version, there is still a chapter containing "Exempel von Bergwerck Silber und Goldt rechnung", "Regula alligationis" and "Müntzschlag". New editions of Rudolff in 1566 and 1588 seem to be—on these subjects—exact reprinting without improvements.

the practical arithmetic literature of the *Rechenbücher* would offer largely fossilized versions of problems that were found in the very first German writings of the late fifteenth century. The *regula alligationis* was still presented as a typical example from the Italian practice but the term "crucible assaying" (*Beschickung des Tiegels*) largely disappeared. Without conclusive evidence about the explanation of this gradual stagnation, one can only speculate: as the economic importance of mining towns declined, practical arithmetics were largely printed in commercial cities, or maybe the existing examples were considered as satisfying for the current needs.

There seems to have been a window of opportunity of a few generations, during which intense exchanges took place. The tremendous economic growth in mining regions, combined with new technical or economic problems and a lack of skilled professionals fuelled the circulation of men and useful knowledge.[55] The need to explain and concretely articulate what computation meant and how minting worked was crucial in regions where highly numerate officials lived in contact with workers who sometimes only knew of tally sticks or *abaci* (*Kerbhölzer* and *Rechenbrett*).[56] In the last decades of the fifteenth century as in the first decades of the following one, the printed textbook thus seems to have pretty closely reflected the nature of problems that were indeed used in minting and mining. As the mining administration became more structured, and the mining business increasingly found a stable model, mathematical experimentation slowed down, reflecting a more institutionalised socio-economic context.[57]

4.4 Conclusion

Both fifteenth-century Pisa and Annaberg in the Ore Mountains are good examples to show the complex nature of the development of practical mathematical knowledge. According to the subscientific model, as presented for instance by Jens Hoyrup, there would be a sharp distinction between a scientific knowledge, "pursued systematically and for its own sake" and a subscientific one, "acquired and transmitted in view of its applicability" (Høyrup 1994, 25). Looking at concrete examples, the situation often appears much more complicated, both because no clear distinction exists between scholars and practitioners, and because intermittent communication was possible under certain conditions. *Abaco* teachers might be interested both in theoretical considerations and in training merchants. For example, Maestro Benedetto, one of the most important Florentine abacus masters of the

[55] On the concept of „useful knowledge" in the history of science, and especially on the "demarcation between professional expertise and useful knowledge", see (Morel et al. 2016).

[56] About the social acceptance of measurement and the issue of numeracy, see (Schaffer 2015; Felten 2022, 44–57).

[57] Interestingly, while practical arithmetic gradually focused on more classical problems of weight and measures, sections on "crucible assaying" and the related computations remained in books on mining sciences. See for instance (Ercker 1574, xiii–xv).

fifteenth century, wrote both an extensive treatise of abacus mathematics that had an almost encyclopaedic character and a nimbler abacus manual that could appeal a readership of practitioners (Arrighi 1965; Ulivi 2002; Høyrup 2024, 306–345). Adam Ries spent years working on his masterpiece, the unpublished *Coß*, while working as reckoning master for the mining administration. These contacts seem to depend heavily on local circumstances, degree of numeracy, and the social role played by mathematical sciences at a given time.

The previous analysis suggests that the most productive periods in the tradition of practical arithmetic were moments of relatively loose forms of institutionalisation: windows of opportunities that allowed new intersections of different bodies—both of knowledge and of practitioners. In these moments of social and epistemological transformations, mathematical procedures were adapted to emerging needs and the identification of new solutions was made possible. Educational texts of practical arithmetic can be read as snapshots of this broader process. In order to understand these texts not as immediate and complete descriptions of contemporary practices, but as the timely written record of fluid and developing practices, we have proposed the concept of "crystallization". If we understand texts of practical mathematics as the outcome of these tensions—both intertextual and extratextual—we can account for the varying degrees of proximity to contemporary practices that can be found in them. In these texts, examples drawn from long mathematical traditions could coexist with fictitious examples and with problems stemming directly from contemporary practices.

Like in crystal growth new particles accumulate incrementally around a stable nucleus, in a similar way texts of practical arithmetic include formulations of mathematical knowledge that were codified at different moments of time. A few of these codifications may belong to some of the oldest codifications of mathematical knowledge, while others may have just crystallized out of the liquid process of mathematical experimentation aimed at addressing open problems and emergent needs. In this metaphor, the letter of previous mathematical codification served as the nucleus around which new codifications crystallized, sometimes continuing previous structures, sometimes diverging from them to build in new directions. As these were practical texts, previous codifications of practical knowledge were reproduced depending on their relevance to local needs. Since texts of practical arithmetic were often used in a learning environment, these needs mostly came from educational practices.

Rather than reproducing faithfully their contemporary socio-economic context, instructional texts needed to keep a written record and to transmit a set of lived mathematical skills that could be adapted and applied to the multifaceted contingencies of the world of practice (Smith 2022). These texts did not transmit a general mathematical theory, but long lists of problems in which mathematics was used as a problem-solving tool. At the same time, the authors of these texts were aware that their context-specific rules could have been used to address problems of different kinds. Generality was not achieved through deductive demonstration, but rather through exemplification and analogization (Daston 2022, 58). The sections of miscellaneous problems—which are common in both the Italian and in the German tradition of these texts—could be used to train this analogical skill: to teach the reader/pupil to see analogies and common patterns in the ever-changing variety of the world of practice, and to identify the most appropriate mathematical method to engage with this world. While these texts did not build a systematic form of knowledge, they contributed to spreading a mathematical attitude, a practical knowledge that gestured towards a peculiar kind of generalisation.

Bibliography

Arrighi, Gino. 1965. Il codice L.IV.21 della Biblioteca degl'Intronati di Siena e la "bottega dell'abaco di Santa Trinita di Firenze". *Physis* VII:369–400.

Arrighi, Gino. 1966. Un "programma" di didattica di matematica nella prima metà del quattrocento (dal codice 2186 della Biblioteca Riccardiana di Firenze). *Atti e memorie dell'accademia Petrarca di lettere, arti e scienze di Arezzo* 38:117–128.

Black, Robert. 2007. *Education and Society in Florentine Tuscany*, Education and Society in the Middle Ages and Renaissance, v. 29. Leiden; Boston: Brill.

Bocchi, Andrea. 2017. *Lo livero de l'abbecho*. Pisa: ETS.

Braunstein, Philippe. 2016. *Les allemands à Venise (1380–1520)*. École française de Rome.

Castellani, Arrigo Ettore. 1958. Frammenti d'un libro di conti di banchieri fiorentini del 1211. *Studi di filologia italiana* 16:19–95.

Danna, Raffaele. 2021. Figuring Out: The Spread of Hindu-Arabic Numerals in the European Tradition of Practical Mathematics (13th–16th Centuries). *Nuncius* 1:1–44. https://doi.org/10.1163/18253911-bja10004.

Danna, Raffaele. 2026. *The Craft of Indo-Arabic Numerals: How Practical Arithmetic Shaped Commerce and Mathematics in Western Europe, 1200–1600*, Cambridge (Mass.), London: Harvard University Press, ISBN: 978-0-674-27933-9.

Danna, Raffaele and Locatelli, Stefano. 2026. Dalla teoria alla prassi della matematica: libri d'abaco e pratiche di mercatura nel tardo medioevo. In Bassani, Alessandra, Gazzini, Marina, and Olivieri, Antonio (eds), "Rationem reddere. Diritto, assistenza, cultura tecnica e scientifica (medioevo e prima età moderna)", *Quaderni degli Studi di Storia Medievale e di Diplomatica* 10: 199–223. https://riviste.unimi.it/index.php/SSMD/article/view/31326.

Daston, Lorraine. 2022. *Rules: A Short History of What We Live By*, Lawrence Stone Lectures. Princeton: Princeton University Press.

De Munck, Bert, Steven L. Kaplan, and Hugo Soly, eds. 2007. *Learning on the Shop Floor: Historical Perspectives on Apprenticeship*, International Studies in Social History. Vol. 12. New York: Berghahn Books. https://cam.ldls.org.uk/vdc_100132037075.0x000001.

De Roover, Raymond. 1953. The Commercial Revolution of the Thirteenth Century. In *Enterprise and Secular Change: Readings in Economic History*, American Economic Association Series, ed. Frederic Chapin Lane and Jelle C. Riemersma, 80–85. London: Allen and Unwin.

De Roover, Raymond. 1956. The Development of Accounting Prior to Luca de Pacioli According to the Account Books of Medieval Merchants. In *Studies in the History of Accounting*, ed. Basil S. Yamey and Ananias Charles Littleton. London: Sweet & Maxwell.

Denzel, Markus A. 2002. Die Bedeutung der Rechenmeister für die Professionalisierung in der oberdeutschen Kaufmannschaft des 15./16. Jahrhunderts. In *Verfasser und Herausgeber mathematischer Texte der frühen Neuzeit*, Schriften des Adam-Ries-Bundes Annaberg-Buchholz, 23–30. Freiberg: Medienzentrum der TU Bergakademie Freiberg.

Deschauer, Stefan. 1992. *Das zweite Rechenbuch von Adam Ries: eine moderne Textfassung mit Kommentar und metrologischem Anhang und einer Einführung in Leben und Werk des Rechenmeisters*. Braunschweig: Vieweg.

Deschauer, Stefan. 2003. Das 2. Rechenbuch von Adam Ries ein Bestseller für das Wirtschaftsleben des 16. Jahrhunderts. In *Gemeinnützige Mathematik: Adam Ries und seine Folgen*, 9–32. Erfurt: Verlag der Akademie gemeinnütziger Wissenschaften.

Dietrich, Richard. 1991. *Untersuchungen zum Frühkapitalismus im mitteldeutschen Erzbergbau und Metallhandel*. Hildesheim: Olms.

Ercker, Lazarus. 1574. *Beschreibung: Allerfürnemisten Mineralischen Ertzt, vnnd Berckwercksarten: wie dieselbigen, vnnd eine jede in sonderheit, jrer natur vnd eigenschafft nach, auff alle Metaln Probirt, vnd im kleinem fewer sollen versucht werden, mit erklerung etlicher fürnehmen nützlichen Schmelzwerckem im grossen fewer, auch schaidung Goldt, Silber, vnnd andere Metalln ... vnd was denen allen anhengig in fünff Bücher verfast, Dergleichen zuuorn niemals in Druck kommen ... mit schönen Figuren vn[d] abriß der Instrument, trewlich vn[d] fleissig an Tag geben*. Schwartz.

Ermisch, Hubert. 1887. *Das sächsische Bergrecht des Mittelalters*. Leipzig: Giesecke & Devrient.

Felten, Sebastian. 2022. *Money in the Dutch Republic: Everyday Practice and Circuits of Exchange*. Cambridge: Cambridge University Press. https://www.cambridge.org/core/books/money-in-the-dutch-republic/1597F40C3D62B07D897266C3E22C9E26.

Field, Judith Veronica. 2016. The Unhelpful Notion of "Renaissance Man". *Interdisciplinary Science Reviews* 41 (2–3): 188–201. https://doi.org/10.1080/03080188.2016.1223585.

Folkerts, Menso. 2001. Early Texts on Hindu-Arabic Calculation. *Science in Context* 14 (1–2): 13–38. https://doi.org/10.1017/S0269889701000023.

Folkerts, Menso. 2011. Die Mathematik im sächsisch-thüringischen Raum im 15. und 16. Jahrhundert. In *Kaufmanns—Rechenbücher und mathematische Schriften der frühen Neuzeit*, Adam-Ries-Bund, vol. 22, 1–22.

Folkerts, Menso, and Paul Kunitzsch. 1997. *Die älteste lateinische Schrift über das indische Rechnen nach al-Ḫwārizmī*. Abhandlungen (Bayerische Akademie der Wissenschaften. Philosophisch-Historische Klasse); n.F., Heft 113. Verlag der Bayerischen Akademie der Wissenschaften.

Franci, Raffaella. 2003. Leonardo Pisano e la trattatistica dell'abaco in Italia nei secoli XIV e XV. *Bollettino di storia delle scienze matematiche* 23 (2): 33–54.

Gärtner, Barbara. 2000. *Johannes Widmanns 'Behende und hubsche Rechenung': Die Textsorte Rechenbuch in der Frühen Neuzeit*. Tübingen: Max Niemeyer. https://doi.org/10.1515/9783110919967.

Gebhardt, Rainer, ed. 2006. *Zur Wirkungsgeschichte der Brotordnung von Adam Ries*, Schriften Des Adam-Ries-Bundes Annaberg-Buchholz. Vol. 18. Freiberg: Medienzentrum der TU Bergakademie Freiberg.

Giusti, Enrico, and Paolo D'Alessandro, eds. 2020. *Leonardi Bigolli Pisani vulgo Fibonacci Liber Abbaci*, Biblioteca di «Nuncius». Firenze: Olschki.

Goldthwaite, Richard A. 2009. *The Economy of Renaissance Florence*. Baltimore, Md: Johns Hopkins University Press.

Goldthwaite, Richard A. 2015. The Practice and Culture of Accounting in Renaissance Florence. *Enterprise & Society* 16 (3): 611–647. https://doi.org/10.1017/eso.2015.17.

Grendler, Paul F. 1989. *Schooling in Renaissance Italy: Literacy and Learning, 1300–1600*, The Johns Hopkins University Studies in Historical and Political Science. Series 107 1. Baltimore: Johns Hopkins University Press.

Hahn, Karl. 1932. Die ältesten Schneeberger Zehntrechnungen. *Neues Archiv für Sächsische Geschichte und Altertumskunde* 53:35–50.

Heeffer, Albrecht. 2011. On the Curious Historical Coincidence of Algebra and Double-Entry Bookkeeping. In *Foundations of the Formal Sciences VII: Bringing Together Philosophy and Sociology of Science*, ed. Karen François, Benedikt Löwe, Thomas Müller, and Bart Van Kerkhove, 109–130. London: College Publications.

Herzog, Emil. 1865. Martin Römer: Ein biographischer Beitrag zur sächsischen Culturgeschichte. *Mittheilungen des königlich sächsischen Vereins für Erforschung und Erhaltung* 14:49–63.

Hoock, Jochen, and Pierre Jeannin, eds. 2001. *Ars mercatoria: eine analytische Bibliographie. Band 3, Analysen (1470–1700)*. Paderborn: Schöning.

Høyrup, Jens. 1994. *In Measure, Number, and Weight: Studies in Mathematics and Culture*. Albany: State University of New York Press.

Høyrup, Jens. 2008. *About the Italian Background to Rechenmeister Mathematics*. Preprint of the Max Planck Institute for the History of Science 349. https://doi.org/10.1007/978-3-030-19258-7_13.

Høyrup, Jens. 2024. *The World of the Abbaco: Abbacus Mathematics Analyzed and Situated Historically between Fibonacci and Stifel*. Cham: Springer.

Jenks, Stuart, and Michael North, eds. 1993. *Der Hansische Sonderweg? Beiträge zur Sozial—und Wirtschaftsgeschichte der Hanse*, Quellen und Darstellungen zur hansischen Geschichte, n. F., Bd. 39. Köln: Böhlau.

Kahleyß, Julia. 2013. Der wirtschaftliche Aufstieg des Martin Römer. Soziale Mobilität im westerzgebirgischen Bergbau des 15. Jahrhunderts. *Vierteljahrschrift für Sozial—und Wirtschaftsgeschichte* 100 (2): 154–177. https://doi.org/10.25162/vswg-2013-0005.

Karpinski, Louis C. 1921. Two Twelfth Century Algorisms. *Isis* 3 (3): 396–413. https://doi.org/10.1086/357950.

Kirnbauer, Franz. 1960. *Adam Riese und der Bergbau*, Leobener Grüne Hefte 49. Wien: Montan-Verlag.

Laube, Adolf. 1974. *Studien über den erzgebirgischen Silberbergbau von 1470 bis 1546*. Berlin: Akademie Verlag.

Long, Pamela O. 2015. Trading Zones in Early Modern Europe. *Isis* 106 (4): 840–847. https://doi.org/10.1086/684652.

Lopez, Robert Sabatino. 1956. Back to Gold, 1252. *Economic History Review* 9 (2): 219–240. https://doi.org/10.1111/j.1468-0289.1956.tb00658.x.

Maher, David W., and John F. Makowski. 2001. Literary Evidence for Roman Arithmetic with Fractions. *Classical Philology* 96 (4): 376–399. https://doi.org/10.1086/449557.

Mennher, Valentin, and Michel Coignet. 1573. *Livre d'arithmetique, contenant plusieurs belles questions & demandes, propres & utiles a tous ceux qui hantent la trafique de marchandise*. Anvers: Jean Waesberghe.

Meskens, Ad. 2013. *Practical Mathematics in a Commercial Metropolis Mathematical Life in Late 16th Century Antwerp*. Dordrecht: Springer. https://doi.org/10.1007/978-94-007-5721-9.

Morel, Thomas. 2022. *Underground Mathematics: Craft Culture and Knowledge Production in Early Modern Europe*. Cambridge: Cambridge University Press. https://doi.org/10.1017/9781009267274.

Morel, Thomas, Giuditta Parolini, and Cesare Pastorino, eds. 2016. *The Making of Useful Knowledge*. Preprint of the Max Planck Institute for the History of Science 481.

Murano, Giovanna. 2015. "Nella invenzione della quale molto pensai". Libri e autografi del matematico ed astronomo fiorentino Paolo dell'Abbaco († 1367). *Archives Internationales d'Histoire des Sciences* 65 (174): 55–75.

Nothaft, C. Philipp E. 2014. The Reception and Application of Arabic Science in Twelfth-Century Computistics: New Evidence from Bavaria. *Journal for the History of Astronomy* 45 (1): 35–60.

Ries, Adam. 1525. *Rechenung auff der linihen und federn in zal, mass und gewicht auff allerley handierung, gemacht in 1522. Itzt fleissig ubersehen, und zum letzten eine hübsche underrichtung angehengt*. Maler.

Ries, Adam. 1550. *Rechenung nach der lenge, auff den Linihen vnd Feder. Darzu forteil vnd behendigkeit durch die Proportiones, Practica genant, Mit gruentlichem vnterricht des visierens*. Leipzig: Jakob Bärwald.

Ries, Adam, and Stefan Deschauer. 1992. *Das zweite Rechenbuch von Adam Ries: eine moderne Textfassung mit Kommentar und metrologischem Anhang und einer Einführung in Leben und Werk des Rechenmeisters*. Braunschweig: Vieweg.

Rochhaus, Peter. 1996. Adam Ries und die Annaberger Rechenmeister zwischen 1500 und 1604. In *Rechenmeister und Cossisten der frühen Neuzeit*, Schriften des Adam-Ries-Bundes, ed. Rainer Gebhardt, vol. 7, 95–106.

Rowe, David E., and Joseph W. Dauben, eds. 2024. *A Cultural History of Mathematics*. 6. vols. London: Bloomsbury Publishing.

Rüdiger, Bernd. 2017. Adam Ries d. Ä. und dessen Söhne als Rechenmeister und Mathematiker sowie die Wandlungen im 16. Jahrhundert. In *Rechenmeister und Mathematiker der frühen Neuzeit*, Schriften des Adam-Ries-Bundes, vol. 25, 151–170.

Rudolff, Christoff. 1526. *Künstliche rechnung mit der ziffer und mit den zal pfennigen, daraus, nit allain alles so sich in gemainen kaufmans hendeln zuetregt, sunder auch was zu silber und goldt rechnung, was zu schickhung des tegels, was aynem muntzmaister—rechnung belangend zuhorig, beide durch die Regl de ref und die Welhisch practick außtzurichten.* Wien: Joannem Singriener.

Sangster, Alan. 2016. The Genesis of Double Entry Bookkeeping. *The Accounting Review* 91 (1): 299–315.

Schaffer, Simon. 2015. Ceremonies of Measurement. *Annales. Histoire, Sciences Sociales* 70, no. 2 (27 July): 409–35.

Schellhas, Walter. 1977. *Der Rechenmeister Adam Ries (1492 bis 1559) und der Bergbau*. Freiberg: Wissenschaftliches Informationszentrum der Bergakademie.

Schneider, Ivo. 1993. *Johannes Faulhaber 1580–1635. Rechenmeister in einer Welt des Umbruchs*, Vita Mathematica. Basel, Boston, Berlin: Birkhäuser. https://doi.org/10.1007/978-3-0348-7274-4.

Smith, Pamela H. 2010. Why Write a Book? From Lived Experience to the Written Word in Early Modern Europe. *Bulletin of the German Historical Institute* 47:25–50.

Smith, Pamela H. 2022. *From Lived Experience to the Written Word: Reconstructing Practical Knowledge in the Early Modern World.* Chicago: University of Chicago Press.

Spufford, Peter. 1988. *Money and Its Use in Medieval Europe*. Cambridge; New York: Cambridge University Press.

Stussi, Alfredo, ed. 1967. *Zibaldone da Canal: Manoscritto mercantile del sec. XIV*, Fonti per la storia di Venezia, Sez. 5: Fondi Vari. Venezia: Comitato per la pubblicazione delle fonti relative alla storia di Venezia.

Travaini, Lucia. 2000. Aree monetarie e organizzazione delle zecche nella Toscana dei secoli XII e XIII. In *L'attività creditizia nella Toscana comunale*, ed. Antonella Duccini and Giampaolo Francesconi, 25–42. Pistoia: Società pistoiese di storia patria.

Tropfke, Johannes. 1980. *Geschichte der Elementarmathematik*, Vol. I. Arithmetik und Algebra. Berlin, New York: Walter de Gruyter.

Ulivi, Elisabetta. 2002. Scuole e maestri d'abaco in Italia tra Medioevo e Rinascimento. In *Un ponte sul Mediterraneo: Leonardo Pisano, la scienza araba e la rinascita della matematica in Occidente*, ed. Enrico Giusti and Raffaella Petti, 121–159. Firenze: Polistampa.

Ulivi, Elisabetta. 2008. Scuole d'abaco e insegnamento della matematica. In *Il Rinascimento italiano e l'Europa*, ed. G. L. Fontana and Luca Molà, vol. 5, 403–420. Treviso-Vicenza: Angelo Colla Editore.

Van Egmond, Warren. 1976. *The Commercial Revolution and the Beginnings of Western Mathematics in Renaissance Florence, 1300–1500* (PhD Dissertation). University Microfilms International.

Van Egmond, Warren. 1977. New Light on Paolo Dell'Abbaco. *Annali Dell'Istituto e Museo Di Storia Della Scienza Di Firenze* 2 (2): 3–21.

Van Egmond, Warren. 1983. The Algebra of Master Dardi of Pisa. *Historia Mathematica* 10 (4): 399–421. https://doi.org/10.1016/0315-0860(83)90003-4.

Van Egmond, Warren. 2005. The Beginnings of European Mathematics in Renaissance Italy and its Spread to Germany. In *Arithmetische und algebraische Schriften der frühen Neuzeit*, Schriften des Adam-Ries-Bundes Annaberg-Buchholz, vol. 17, 1–22. Freiberg: Medienzentrum der TU Bergakademie.

Vogel, Kurt. 1980. *Das Bamberger Blockbuch: Inc. typ. Ic I 44 der Staatsbibliothek Bamberg: ein xylographisches Rechenbuch aus dem 15. Jahrhundert*. Saur.

Wußing, Hans. 2013. *Adam Ries*. Springer.

Yeldham, Florence A. 1927. Notation of Fractions in the Earlier Middle Ages. *Archeion* 8 (3): 313–329.

Chapter 5
Winding and Warping the Yarn Arithmetically in Early Modern Europe: Leibniz, Krüger and Practices of Measuring Threads During the Seventeenth Century

Michael Friedman

Abstract During the seventeenth century there was an increased interest in artisanal practices on behalf of European natural philosophers, which manifested itself also in examining textiles and textile-related practices mathematically. Were these reflections only an attempt to record artisanal knowledge, or rather to extract from it mathematical reasoning that the practitioners did not know how to formulate? The paper will examine the manuscripts of Daniel Krüger and Gottfried Wilhelm Leibniz, dealing with arithmetical calculations done in the course of yarn winding and warping. While these texts had different aims and their writings appeared against different social and economical backgrounds, both scholars aimed not only to show explicitly the arithmetical rules governing the practices of winding and spinning, but also to prompt the externalization of such artisanal knowledge.

Keywords Daniel Krüger · Gottfried Wilhelm Leibniz · Arithmetic of winding · Warping frame · Spinning bee · Measurement systems of threads

5.1 Introduction

During the seventeenth century, numerous European philosophers, mathematicians and natural philosophers expressed an interest in artisanal practices and household activities. During the early modern period, recognition that artisanal practices carried knowledge prompted attempts to find out whether practitioners and artisans employed geometric, arithmetical, or in general, mathematical procedures or

M. Friedman (✉)
Mathematical Institute, Bonn University, Bonn, Germany
e-mail: friedman@math.uni-bonn.de

Senthil Babu D., R. Wagner (eds.), *Normative Texts and Occupational Practices*, Boston Studies in the Philosophy and History of Science 350,
https://doi.org/10.1007/978-3-032-28419-8_5

reasoning. It also prompted attempts to reformulate such artisanal practices in a more formalized, mathematical way, leading, to cite Pamela O. Long, to several 'trading zones' between scholars and artisans,[1] zones which were also accompanied by various 'mathematical practitioners'.[2] While I will argue later that such a conception of 'trading zones' should be revised, one of these household domains which may have functioned as such a 'zone' consisted of textile-related activities: spinning, winding and unwinding the yarn, braiding, knitting and weaving, among other activities. Famous in this respect is Descartes' call in the tenth rule of his *Regulae ad directionem ingenii*, that one "must first tackle the simplest and least exalted arts, and especially those in which order prevails—such as those of the craftsmen, who weave fabrics and carpets, or of the women, who either embroider or intermix [interlace] threads in infinite ways to produce various weaves."[3] (Descartes 1985: 35) Bacon, suggesting to gather knowledge via experiments, underlines that these experiments should also take into account artisanal knowledge: the "particular arts to be preferred" in such a project of investigating nature are those that "exhibit, alter and prepare natural bodies and the material of things such as agriculture, cookery, chemistry" as well as "weaving, carpentry [or] architecture" (Bacon 1858: 363–4). Moreover, between 1620 and 1640, Joachim Jungius wrote a collection of notes titled *Texturæ Contemplatio*, dealing with textile practices, presenting also related arithmetical and geometrical reflections.[4] The arithmetical part presents few calculations done by weavers; Jungius describes how certain measurements and arithmetical calculations were made before starting to weave: after winding a group of threads on an instrument called warping frame with several sticks, also called a warping frame (see Fig. 5.1(1)), the weaver should first determine how long the threads will be, and then use them as the warp.

While Jungius' manuscript itself was never published, it is known that Gottfried Wilhelm Leibniz borrowed *Texturæ Contemplatio* in 1678, never returning it; Leibniz copied this manuscript, and in this copy one finds a few additional folios. One of these folios, titled "Leinweber Rechnung" ("Calculation of linen weaver"),[5] has numerous calculations, apparently of a scholar from Prussia. Such calculations, consisting mainly of conversions of various measurements, describe also how to wind the spun yarn with the warping frame, an instrument, already mentioned above, assisting in making skeins or hanks of yarn in a predetermined size. These calculations were not only common for the weavers, being a preliminary step before winding a group of threads on the warp frame, but were also noticed by other schol-

[1] On the encounters between artisans and scholars and the knowledge transfer between the two groups in the early modern period, see, among others: Smith (2004, 2022), Werrett (2019), de Munck (2019), Long (2011, 2015).

[2] See: Cormack et al. (2017).

[3] The translation was slightly modified by the author.

[4] For a detailed analysis of this manuscript, see: Friedman (2022, 2023a).

[5] See: Leibniz-Handschriften zur Technica, LH 38, fol. 35v. LH 38, or LH XXXVIII, which is found in the Gottfried Wilhelm Leibniz Bibliothek—Niedersächsische Landesbibliothek and will be abbreviated from now on as LH 38.

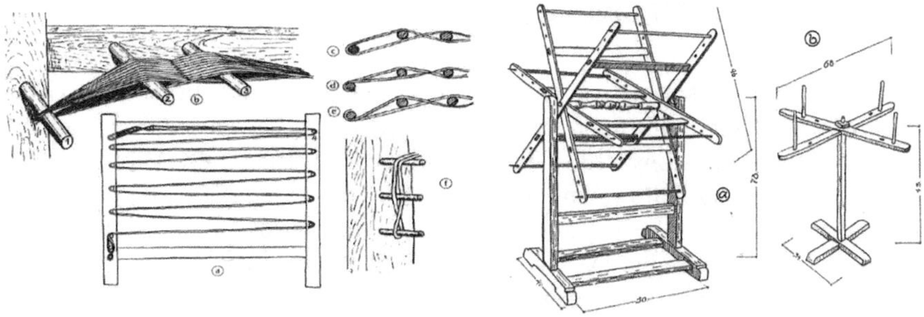

(1) (2)

Fig. 5.1 (1) A depiction of a warping frame, and how one winds up several threads on it (Bomann 1927: 253); (2) Depictions of a swift (**a**) and a vertical warp reel (**b**) (Bomann 1927: 251)

ars. Daniel Krüger's book *Schola textoria nova* from 1664, being a more elaborate text than the "Leinweber Rechnung", contains similar calculations.

The question arises, how Krüger's book as well as the additional folio in Leibniz' copy of *Texturæ Contemplatio* took into consideration the 'real world' arithmetical calculations that were employed in the community of (linen) weavers. Examinations of the various measures of length employed by textile workers, as will be conducted in this paper, make central a variety of questions concerning practical work with such various units. How did these texts encode and react to the practices of spinning, of winding and of unwinding of yarn, and to their social context? Do these texts present the spinners and weavers in a position to assert the autonomy and uniqueness of their knowledge, or rather do these texts appropriate such knowledge, making it more 'mathematical', but at the same time marginalize the role of the practitioners? More specifically, was Leibniz only collecting notes which recorded and commented on these material activities, or did he seek to extract or develop more profound mathematical knowledge from these activities? Comparing Krüger's more elaborate text to Leibniz's own reflections on linen spinning and weaving, how can one characterize these texts, when each approaches (the arithmetic of) linen and yarn spinning and winding practices differently?

As we will see, both Leibniz and Krüger complicate and undermine the separation of the categories of theoretical thinking vs. embodied practice, but for each of them, such complications occur on different epistemological, economic and social backgrounds. Nevertheless, both texts prompt also a tension between artisanal knowledge and its transfer into the calculative realm that encodes it. To see this, I begin in Sect. 5.2 with a short introduction on how spinning, winding and warping were done during the early modern period. I then discuss in Sect. 5.3 the "Leinweber Rechnung" and Leibniz' own reflections on linen spinning and weaving, moving on to Krüger's book in Sect. 5.4. Section 5.5 summarizes these findings and discusses these texts in the larger framework of codifications of artisanal knowledge in the early modern period, whereas Sect. 5.6 is an appendix containing a transcription of "Leinweber Rechnung".

5.2 Practices of Spinning, Winding and Warping

While, when discussing textile practices, one may think immediately on weaving with a treadle loom as the main activity, such practice was preceded by several preparatory stages that consisted of preparing the threads themselves; one of these stages was spinning, followed by winding. As John Styles (2020: 201) notes, spinning was not simply a "matter of sitting at a spinning wheel and turning raw cotton into yarn". After receiving the raw cotton, it had to be loosened out, cleaned, and carded into a form that "the spinner's fingers could draft, spun at a spinning wheel or with a hand spindle." It is only after this stage that the spun yarn could be prepared into hanks or skeins, using the warp reel, the warping frame or the warping mill (see Fig. 5.1(2)). Styles underlines that reeling could also "be a way of measuring and policing its [the yarn's] fineness, and hence its quality. If a reel was of a standard length or circumference, and a hank consisted of a standard number of turns, then the number of hanks per pound weight provided a measure of the yarn's fineness, termed its count." (Ibid.: 202) As we will see later when discussing Krüger's work, such standardization of measurement—or at least an agreement about the length of an "ell" and the circumference of the warp reel or warping frame—was indeed demanded both from the "Master" and the "Hausmutter".

The wrap reel, called in German "Haspel," is the tool used for reeling, and it has two main forms. One is with built spokes (see Fig. 5.1(2), on the left), which can be folded together and spread apart again, so that the yarn can be laid around and then pulled taut. The second is a smaller, usually vertical version of the first form, in which only two arms are configured as a cross balanced on top of a vertically positioned axis (see Fig. 5.1(2), on the right). Once reeling is done, the weavers wind several hanks, which consist of several threads. They do this either on the warp frame (see Fig. 5.1(1)) or on the warping mill, called "Scherrahmen" (see Fig. 5.2), in order to prepare the future warp or to sell such hanks to the weavers or merchants. The warp frame can be described as a wooden frame containing evenly spaced pegs on which the warp is wound in preparation for weaving. The pegs are spaced out in such a way that one should be able to measure the length of the future warp threads (if one follows the right path).[6] The weaver hence warps a constant number of threads back and forth on the frame or on the mill to obtain, after taking down the resulting taut threads, a hank of yarn. Each hank (in German, "der Gang" or "das Ging", in plural "Gänge") contains a predetermined number of threads—usually between 20 and 40.[7] While the length of the obtained threads should be eventually known after the predetermined number of windings, it is important to stress that the

[6] Jungius describes several numerical calculations done in the course of using the warp frame. See: Friedman (2023a: 263–4).

[7] Bomann (1927: 252–4) notes that each "Gang", obtained from winding on the frame, has 20 threads; also Krüger (1664: B, recto) notes that each "Gang" has 20 threads. However, in (Sennewald 1859), "Gang" consists of 40 threads. In the Grimm dictionary (Grimm: "Gang, m.", sect. III, 5, d), "Gang" is defined as having either 20 or 40 threads.

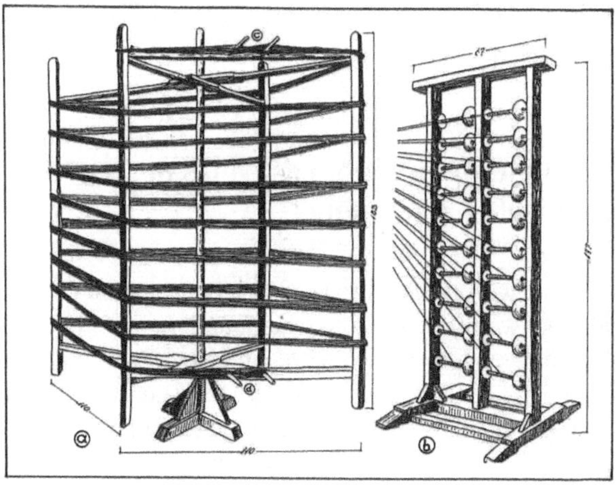

Fig. 5.2 A depiction of a rotating warping mill (left) and 20 coils of threads, which are coiled on the mill (Bomann 1927: 255)

number of "Gänge" being made in total does not depend on the length of the wished fabric, but on the width and fineness of the fabric (Bomann 1927: 254). As is however clear, the usage of such frames demands several calculations to determine how long the warp threads should be.

Warping with the warping frame was considerably facilitated by the so-called warping mill (see Fig. 5.2), which could now be rotated and hence save the time-consuming manual back-and-forth winding. Also here the number of threads for a "Gang" could change from 20 to 40. It is unclear when the rotating warping mill began to be employed extensively in the Holy Roman Empire, as both instruments were called "frame".[8] While neither Krüger nor Leibniz refer to the round form or to the rotation of such a possible warping mill, the anonymous *Gründliche Unterweisung vom Spinnen und Weben* from the beginning of the eighteenth century, which is a shortened version of Krüger's manuscript, already notes that the weaver's instrument is either the "frame" ("Rahmen") or the "Umläufer", meaning the rotating warping mill (Rüdiger 1740: 4).[9]

In this paper I concentrate on the stage of winding as done by the spinners: the preparation of coiled or wrapped units of yarn by using the wrap frame or the warping mill, instruments which both Leibniz and Krüger discuss in their texts. Such instruments, ensuring the yarn threads are coiled into hanks of a standard size, were hence usually themselves of a standard size. To ensure the required size, its revolutions or windings are counted as the yarn is wrapped around it. Hence, to reach the

[8] See also (Bohnsack 1981: 125–130) on silk mills, known also as (Luccheser) *Filatorium* (or *Seidenzwirnmühle*), a mechanisms with which numerous threads were twisted and wounded at the same time. Such a mechanism existed in Europe already starting the thirteenth century.

[9] Rosenfeld (1958: 11–12) notes that such an instrument already replaced the warping frame at the beginning of the eighteenth century.

desired size, one would have to know not only the circumference of the frame or the mill, but also the number of windings one makes.

5.3 Rules of Winding in Leibniz's Notes on Technique: The "Leinweber Rechnung"

I would like to begin with the first text that encodes a form of measurement and calculation with the warping frame, to be found in a folio titled "Leinweber Rechnung" in Leibniz's handwritings on technique.[10] This folio is interesting not only because it is one of the few pieces of evidence we have documenting the interest of seventeenth century natural philosophers in the arithmetical rules of winding, but also because it is found in a copy of one of Jungius' manuscripts. To make clear Leibniz's interest in such practices, I will start in Sect. 5.2.1 with explicating the socio-economic context of such an interest, moving on in Sect. 5.2.2 to analyze "Leinweber Rechnung".

5.3.1 Why Was Leibniz Interested in Spinning and Winding Practices?

It is well known that Leibniz highly appreciated Joachim Jungius (1587–1657), who was one of the most esteemed natural philosophers and logicians during the first half of the seventeenth century, known especially due to his *Logica Hamburgensis*. Leibniz made considerable efforts to examine Jungius' Nachlass and to have several of his writings published, and was in contact with the first administrator of this Nachlass, Martin Fogel (1643–1675). Indeed, in 1678, having visited Fogel's library three years after Fogel's death in 1675, Leibniz also purchased this library, and at the same time borrowed 86 sets of manuscript papers written (or copied) by Fogel, manuscripts which Leibniz never returned.[11] One of these manuscripts was a copy of a set of notes by Jungius titled *Texturæ Contemplatio*, a copy made by Jungius' colleagues and students of an unpublished collection of notes on textiles, weaving and knitting practices, written by Jungius between the early 1620s and the late 1640s.[12] Jungius himself, while elaborating in *Texturæ Contemplatio* on several textile-related artisanal practices besides weaving and knitting, hardly mentions spinning or winding. Folio 35r of *Texturæ Contemplatio* mentions briefly "the whorl of a spindle". Moreover, at fol. 19r, Jungius mentions Sebastian Münster (1488–1552)

[10] Note that it is not clear whether Leibniz was referring here to the warping frame or to the warping mill. I will use hence the term "warping frame" to denote the instrument Leibniz was referring to.

[11] On Fogel and his own Nachlass, see: Marten and Piepenbring-Thomas (2015).

[12] The transcription and translation of this manuscript is to be found in: Friedman (2023a).

Fig. 5.3 Drawing of a family of Suebi who spin, from Sebastian Münster's *Cosmographia*. The child in the middle is depicted as the one who winds the yarn on the warp reel (Münster 1550: 581; Bayerische Staatsbibliothek, Res/2 Geo.u. 51; CC BY-NC-SA 4.0)

and cites from his book *Cosmographia. Beschreibung aller Lender durch Sebastianum Munsterum* of 1544. While Jungius cites partially from page 581 of Münster's book, in the book itself, on this page, one finds a depiction of a family of Suebi who spin and wind (see Fig. 5.3); however, Jungius does not mention this drawing. Nevertheless, as mentioned above, he does depict and describe in *Texturæ Contemplatio* the warping frame, and he presents a few arithmetical calculations related to the length of the resulting warp.

Leibniz made a copy of *Texturæ Contemplatio*, probably around 1678–9, since it seems that he initially intended to return the 86 bundles back to Fogel's family, though this never happened. The copy is to be found in the manuscript bundle LH 38 (*Leibniz-Handschriften zur technica*), fols. 26r-40r. While recent works have already analyzed the content of *Texturæ Contemplatio* as well as Leibniz's copy (Friedman 2023a), one of the more relevant folios for the current paper is to be found in Leibniz' copy and not in the original manuscript *Texturæ Contemplatio*. This folio: fol. 35v, presented in Fig. 5.4,[13] is titled "Leinweber Rechnung" ("The calculation of a linen weaver"), and was added to the copy of *Texturæ Contemplatio*, probably after the copy was already made. The main text on this folio and the accompanying arithmetical table are not written by Leibniz; numerous remarks are added in the margins, as they were in all of the folios in the copy of *Texturæ Contemplatio,* and these remarks were certainly read by Leibniz. A transcription of this folio is found in the appendix in Sect. 5.5 at the end of this paper.

Before analyzing this folio, its table and the comments, the question arises, why was Leibniz interested in "the calculation of a linen weaver"? As was noted above, while it is not known when exactly Leibniz copied *Texturæ Contemplatio* (or had it

[13] Folio 35r is empty.

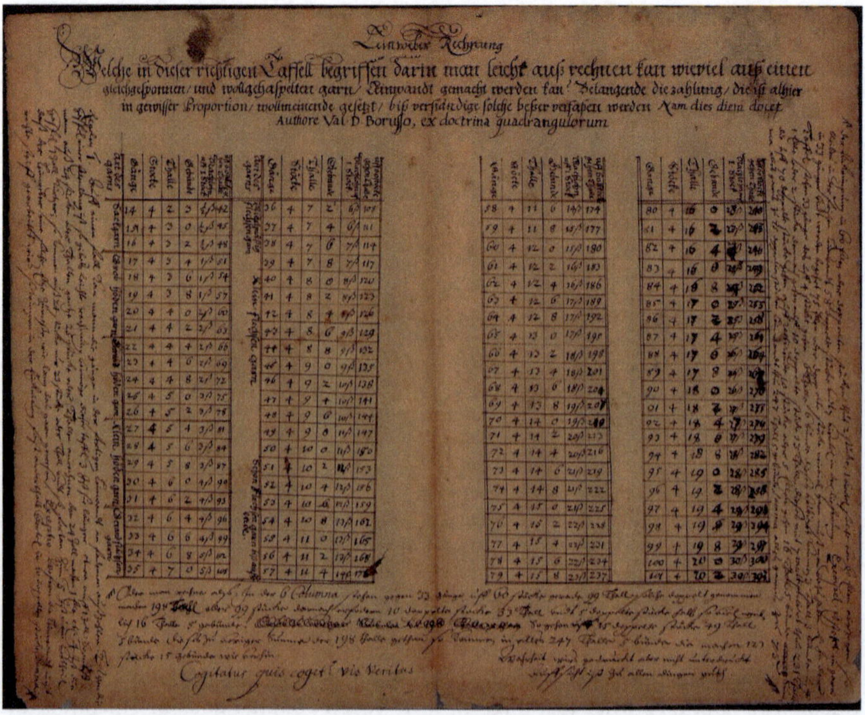

Fig. 5.4 The folio titled "Leinweber Rechnung" from (LH 38, fol. 35v). The notes in the margins contain, among others, instructions how to use the numerical table. © Gottfried Wilhelm Leibniz Bibliothek—Niedersächsische Landesbibliothek

copied, as several folios of this copy were not copied by Leibniz himself, but rather by someone else, with Leibniz later controlling the copies), one may assume that it was between approximately the second half of 1678 and 1679. While it may seem enticing to conjecture that the folio on "the calculation of a linen weaver" may have been added to Leibniz' copy of *Texturæ Contemplatio* during this period, other writings of Leibniz may indicate that it may have been added in a later period.

Indeed, around September 1688, right before *le Roi Soleil*, Louis XIV, published his *Mémoire de raisons* on 24 September 1688, which signaled the beginning of the nine years war, Leibniz reflected on how to counter Louis XIV's offensive foreign policy, by, for example, improving the finances of Emperor Leopold I.[14] Leibniz suggested, among other things, that Silesian handworkers could produce linen by themselves, instead of paying for and importing these products from Holland and

[14] A IV4: 90–99. From now on, when referring to Leibniz's edited works, I will use the letter 'A' followed by one Roman and one Arabic numeral (representing the series and volume number) to refer to the edition of Leibniz's collected works published in the Akademie der Wissenschaften edition of Leibniz's miscellaneous works: (Leibniz 1923–).

Genoa.[15] Shortly after the war erupted, in his "Denkschrift über den Leinenhandel" from the first half of October 1688, Leibniz suggests his plan for economic war against France: Spain should replace its linen imports from France with Silesian and Lower Saxonian products.[16]

Such economic suggestions were also proposed during the 1690s. During the first half of 1695, Leibniz was informed by Johann Daniel Crafft about several projects, among them one concerning linen drapery.[17] Moreover, in January 1695, Leibniz stressed the economic importance and advantages of linen production in Germany, and suggested the establishment of linen manufactories.[18] Leibniz underlined that one should develop the linen industry, which was all imported at the time. He stated that "Car si les Italiens ont des soyes, si les Anglois et les Espagnols ont leurs lames et draps; le lin et le chanvre est nostre soye [...]."[19] Spinning, and hence also the subsequent stage of winding the yarn, had to be done in the cities, but also in the countryside, where the spinners and weavers also work[20]; selling finished products (cloths, ribbons, etc.) would be more financially advantageous than selling yarn. Indeed, Leibniz, emphasizing the economic advantages that the state would have, even said that it was "shameful and damaging that our people should be so duped by foreigners", as they "get the least pay even though they do the hardest part of the work".[21] Since the table in the folio "the calculation of a linen weaver" also deals with wages to be paid, hence showing the predicted economical costs, it may very well be that this folio was of particular interest to Leibniz during the periods mentioned above (1688 or 1695), and therefore was inserted into this bundle during these years.

[15] A IV4: 93: "Wofern auch die Schlesier eine gewiße Qvantität tauglicher linwand anstatt des baaren Geldes der jahrlichen 100.000 in Holland und nach Genua zu liefern übernehmen[,] gienge des wegen kein geld auß dem lande, und würden die leute desto mehr zum fleiß angereget, ihnen aber würde solches an denen landes anlagen obgekürzet, so ohne dem an 100,000 th. und drüber in solcher provinz verstarcket werden können, dafern man sie den leuten nicht an baaren gelde sondern in natura abfordert, daher dann lezlichen, und in fine finali dem Kayser diese Million gar nichts kosten würde."

[16] A IV4: 101: "Damit man aber beßer heraus gehe, so ist es mit einem Worth der Linnenhandel welcher bey Spanien unentbehrlich, und von Franckreich auff Teutschland, sonderlich aber auff Schlesien und NiederSachsen zu transferiren."

[17] A III6: 268 and 334. Crafft notes that it was Baron Schwartzenstein who was responsible on this project.

[18] A I11: 167–169.

[19] A I11: 168.

[20] This is the work of the "paysan, qui par consequent s'étend par tout le pays, tant pour filer, que pour mettre par aprés le fil en oeuvre." (Ibid.)

[21] Ibid: "[Ils] ont le moins de paye, quoyqu'ils fassent la partie la plus rude du travail, parce qu'ils travaillent plus tost de force que d'industrie [...] il est egalement honteux et dommageable, que nos gens sont ainsi la duppe des estrangers [...]."

5.3.2 On the "Calculation of a Linen Weaver"

Having seen the background for Leibniz's interest in linen and spinning from an economical point of view, the question arises whether the folio on "the calculation of a linen weaver" reflects such an interest. As we will see, the remarks added to the folio do not reflect economic issues, but attempt much more to articulate arithmetically and in a written form what may be termed the artisanal, implicit rules of knowledge related to winding.

Examining this folio more thoroughly (see Fig. 5.4), below the title "Leinweber Rechnung", the following statement is written down: "How to grasp in this correct table, from it one can easily calculate, how much canvas [*Leinwandt*] can be made from spun and reeled yarn? Concerning the payment, this is all set here in certain well-meaning proportion, until such advisors will better formulate [it]. Since the day teaches the day. Author Val. D. of Prussia, of the doctrine [instructions] of the quadrangle."[22] While only from the title one might assume that economic issues are (also) discussed in this folio, this would be the wrong impression. Below the above statement is a table of length measurements that seems only to attempt to answer the question concerning the quantity and the length of the bundles which can be made from a yarn that has been already spun and reeled with the warping frame. The table deals hence with the length measurements of the hanks of yarn, formed when using such a frame, as elaborated in Sect. 5.1. Moreover, it seems that the title and the table were written by the Prussian "Val. D.," who is described as being acquainted with the "doctrine of the quadrangle", which may refer either to the square form of the reel or the square form of the frame.

As can be seen in Fig. 5.4, the table is partitioned into four parts, according to the second column titled "Gänge" (hanks), when, to recall, each "Gang" contains a predetermined number of threads. If each "Gang" consists of 20 threads (for example, since such a number is not mentioned in the folio), then 33 "Gänge" will consist of 660 threads. The number of "Gänge" progresses from 14 to 101, whereby the table contains one row for every integer between the above values. The first part (located on the rightmost side) progresses from 14 to 35 "Gänge", the second from 36 to 57 "Gänge", the third from 58 to 79, and the fourth from 80 to 101. In addition, the leftmost column is titled "Art des Garns" ("type of yarn"). Below this column one finds the various types of yarn; to stress, the more warp threads the fabric has, the finer the yarn is. Five types are listed in the first part of the table: "Sackgarn" (for 14 to 16 "Gänge"), "Grob Höden garn" (for 17 to 21), "Gerand Höden garn" (for 22 to 26), "Klein Höden garn" (for 27 to 31) and "Gerand flächsen garn" (for 32 to 35). Additional three types are listed for the second part of the table: "Mittelmäßig

[22] LH 38, fol. 35v: "Welche in dieser richtigen Taffell begriffen darin man leicht auß rechnen kan wieviel auß einen // gleichgesponnen / und wollgehaspelten garn Leinwandt gemacht werden kan? Belangende die zahlung die ist alhier // in gewisser Proportion wollmeinende gesetzt biß verständige solche beßer verfaßen werden [.] Nam dies diem docet // Authore Val. D. Borusso, ex doctrina quadrangulorum." (Since the title itself contains the punctuation mark slash: / , it is retained in the above citation, whereas the symbol // denotes where a line ends)

Flechsengarn" (for 36 to 38), "Klein flechsen garn" (for 40 to 50), and "Schon Flechsen garn biß auff ende" (for "Gänge" from 51 onwards).

The titles of the table's other columns relate, except one, to the frame, which is also later termed a "Haspel." These titles are, from left to right, "Gänge," "Stöcke" (or "Stocke"), "Thalle," "Gebünde," "Machlohn uf 1 Stock," and "Uff 60 stöcke gehen Thalle." "Stöcke," "Thalle," and "Gebünde" are all terms related to the length of the resulting hank, obtained from the warping frame: "Stöcke" has always the value 4; "Thalle" (also named, though not in this text, "Toll") and "Gebünde" (also named "Gebinde") are quantities which depend also on the number of "Gänge", where 1 "Thalle" is equal to 10 "Gebünde". In the remarks found in the text, "Thalle" is referred to also as "stücke," and can therefore be translated as 'pieces'; "Gebünde" can be translated as 'bundles.' However, such terms in English are not textile specific, and certainly do not refer to length measurements of yarn, hence any such translation would be misleading. "Machlohn uf 1 Stock" probably refers to the wage to be paid for the linen warp threads obtained this way. Moreover, examining the arithmetical relations between the columns, one can note that by denoting by G the number of "Gänge", by T the number of "Thalle," and by Ge the number of "Gebünde," the following formulae can be deduced from the table: $T = \lfloor G / 5 \rfloor$ and $Ge = 2(G - 5T)$.[23]

The question obviously arises, what are these lengths which "Stöcke," "Thalle," and "Gebünde" designate? While for such measures these arithmetical relations are not specified in the folio, in what seems like an attempt to answer this question as well as the question posed directly below the title, several instructions and examples are added in the left, right, and bottom margins of the folio. On the left margin, for example, one finds five "rules" concerning the warp reel or frame, beginning as follows: "Rule 1: In one case you can count the *Gänge* in the finished canvas [*Leinwandt*] by [looking] at the threads. [Rule] 2: If the reel [*Haspel*] is one ell long, there is an easy calculation [of the length of the obtained fabric] by means of this table. [Rule] 3: If it is longer than about 1 ell [cubit], one must count 25 pieces [stücken] or *Thallen* [when winding] 24 pieces [stücken] or *Thallen*, for 24 inches [Zoll] make 1 cubit [...]."[24] The first rule notes that one can count the number of hanks (i.e. "Gänge") used for making the finished fabric by counting the number of threads. The following three rules (rules 2–4) instruct how to compute the length of the obtained fabric when the reel is either exactly a cubit long or one inch more or one inch less. The fifth rule underlines that it is advantageous for the linen weaver to measure into cubits to know the length of the yarn. The question obviously remains, what is the length of such measurements. Johann Friedrich Krüger (1830:

[23] The notation used here is of course a modern one. Recall that $\lfloor x \rfloor$ denotes the rounding down of x to the nearest positive number.

[24] LH 38, fol. 35v: "Regelen 1. Auff einen fall kan mann die gänge in der fertigen Leinwandt an fademen außzehlen. 2. Wen die / Haspel einer Ellenlang ist so gibets leichte rechnung vermöge dieser taffel [.] 3 Ist sie länger etwa auff 1 Zoll: So muß / mann auß 24 stücken oder Thallen gantze 25 stücke oder Thallen anrechnen den[n] 24 Zoll machen 1 Elle etc".

333) notes that in Prussia, such measurements were calculated when the circumference of the warp reel was 3½ ells,[25] but in other regions in the Holy Roman Empire, the standard circumference of the warp reel could be also 4 ells (ibid.: 332–3, under "Garnhandel"). However, not only are the measures themselves not specified in "Leinweber Rechnung" but also the variations in the measures are not even mentioned, probably since this text was written for (or by) regional weavers in Prussia, and hence the measures themselves were regarded as common knowledge.

Here a short detour is needed: it is important to note that such measures were very much local, changed from region to region, and varied also within the regions themselves. Moreover, such measures were not determined by the desired properties of the produced fabric, but by the production processes. This has prompted a variety of measures that were highly specialized to the textile industry (Kramper 2019: 39–40). In the transition from the seventeenth to the eighteenth century, there were in the Holy Roman Empire various attempts to reduce this variety of measures and introduce a more standardized measurement system, by, for example, imposing a unified measure for the reel or for the frame.[26] However, these attempts usually ended in a failure, due to several reasons. First, when the administration in Prussia determined the fixed measure in "Berlin ells", for example, this measure was not really known in other regions. Second, some spinners probably objected to such standardization, since a variability of measurements allowed them to increase profit by, for example, reducing the length of the prepared hanks while still selling them for the same price. Third, the state administration lacked the means to impose such standardization on all the weavers and spinners, since winding reels and frames were practically in every household, and the ability to control them was highly limited. Indeed, the size of the frame or the number of threads per hank varied from region to region, and, hence, in eighteenth century Prussia and Saxony there were dozens of various measures.

To return to "Leinweber Rechnung": on the right margin, and then on the bottom margin, the text continues with explanations how to use the table and calculate with it; here it is important to note that for the author of the text, the measurements ells and 'doppelte stöcke' are identical. The question that the comments aim to answer is: "There is a yarn in 33 *Gänge* and 75 ells or 'doppelte stöcke' are desired, how much yarn must one have in addition for this?"[27]; this is done by looking at row number 33 in the table (which corresponds to 33 "Gänge"), which corresponds to "6 Thalle" and "6 Bünde" or to "99 Thall solche doppelt genommen".

The solution presents a series of calculations, some require implicit knowledge about the relations between the measures 'Thalle', 'Bünde', 'Elle' and 'doppelte stöcke'. According to the table, for 33 "Gänge", 4 'stöcke', being 2 'doppelte stöcke'—which are equivalent to 2 ells—result in "6 Thalle" and "6 Bünde". Hence,

[25] Krüger (1830: 333): "Zu Königsberg in Preußen; das Stück von 2 Toll oder 20 Gebinde oder 800 Faden mit 3½ Elle Haspellänge."

[26] I follow here (Kramper 2019: 89–90, 469–70).

[27] LH 38, fol. 35v: "Es stehet // ein garn in 33 gängen undt werden begehrt 75 Ellen oder doppelte stöcke wieviel garn muß man dazu haben".

multiplying in 5, 10 'doppelte stöcke' correspond to 33 'Thalle', and hence 75 'doppelte stöcke' correspond to 247 'Thalle' and 5 'Bünde'. The employed rule is that the half of a 'Thall' is 5 'Bünde' (or a 'Thall' is 10 'Bünde')—a rule which is not written down explicitly in the folio. Here there is hence an implicit use of the Rule of Three, but also of implicit knowledge of the relations between the measures of 'Thall' and 'Bünde'. One may assume that these rules and arithmetical calculations were well known and used by spinners and weavers, but the question obviously arises, who actually wrote down this numerical table, and were these implicit rules done systematically and formulated by spinners, weavers or administrators in the framework of an arithmetical theory of spinning related activities. That is, one may ask, who could have recorded these problems in writing, since it is not clear whether the practitioners themselves wrote this table.

As noted, Leibniz's interest in such household activities and in the textile-artisanal, practical activities was also manifest in him copying Jungius' *Texturæ Contemplatio*. The fact that Leibniz inserted this folio into his own copy shows Leibniz's interest in mechanical aids such as the warp reel and the warping frame. But why did the writer of this folio think one had to explicitly formulate and write down such rules? One possible answer would be that this table shows the connections of spinning and winding activities to arithmetic, as if the arithmetical rules and examples (how to determine the quantities derived from the number of "Gänge") must be stated explicitly. This points to Leibniz's own general approach to artisanal knowledge: it is based on certain logical or mathematical principles, but these must be explicitly formulated, though not necessarily by the artisans. Indeed, the tension, between how artisanal knowledge is embodied and practiced and how it may be presented or formulated as an ordered arithmetical knowledge is noted by Leibniz in his "Meditationes de cognitione, veritate, et ideis" from 1684. In this text, Leibniz underlines that painters and other artists ("artifices") know immediately when a work of art was manufactured well or badly, but when asked about the reasons for these judgements, they are unable to explain them.[28] While one may argue that this claim refers only to works of art, it should be also noted that similar views regarding artisanal knowledge were also expressed by Leibniz (Beirach and Friedman 2025), and that he may have considered such practices "as decidedly inferior to theoretical knowledge as a way of knowing", calling for artisanal knowledge "to be explicitly stated" (Jones 2016: 82).

These reflections show an ambivalent relation concerning artisanal knowledge: while it is not clear if one finds here a documentation done by practitioners, which record their own knowledge (as an attempt to write their own rules, or to bring into a written form their own way of calculating), one finds here a possible codification and arithmetization of artisanal practices. It could be that the existence of a table of such elaboration, with data and its attendant rules, suggests that it can only arise out of a well-established knowledge of an arithmetical theory of winding. So who were

[28] A VI4: 586: "Similiter videmus pictores aliosque artifices probe cognoscere, quid recte, quid vitiose factum sit, at judicii sui rationem reddere saepe non posse, et quaerenti dicere, se in re quae displicet, desiderare nescio quid."

these spinners who were in fact capable of formal arithmetization? The origins (i.e. provenance and authorship, among others) of such artisanal accounts, as is in our case, are almost impossible to trace, and the questions—whether this was indeed the method of writing down arithmetical tables used by weavers, spinners, or administrators, and if indeed the practitioners used such arithmetical tables, prepared before spinning and winding took place—remain open. However, as the rules of "Leinweber Rechnung" make explicit, one should at least know first of all the standard measure of an ell. That these measures changed from region to region was well known,[29] and such a variation shows that if one may assume that the author of such table knew of such variations, then he was certainly not a simple spinner. These questions become clearer and more pertinent in the following section, which presents the earlier and more elaborate work of Daniel Krüger on the same subject.

5.4 Krüger's Winding Practices: From What "verkleinern und verlåstern lassen" to Their Arithmetization

Arithmetical interests in spinning and winding were not unique to Leibniz and were already put in writing before Leibniz copied Jungius' *Texturæ Contemplatio*, inserting into his copy the folio discussed above. Daniel Krüger's *Schola textoria nova* from 1664, being a much more elaborate book on spinning and winding, describes similar calculations more thoroughly; the full title of his book refers explicitly to the "Haspel", i.e. to the warp reel: *Schola textoria nova oder neue Spinn- und Weberschule, darin gelehrt wird, wie viel Garn allerley Gattung, feines und grobes, auf jede Elle und folgens auf jedes Tuch, Recken, Schocken etc. nach richtigem Haspel zu geben.*[30] Krüger (1664: Aii, verso) notes in the introduction that while this craft is sometimes "diminished and blasphemed" ("verkleinern und verlåstern lassen") by "quite a few unjust masters" ("etliche ungerechte Meister"), it is highly useful and praiseworthy. With his book, which describes the practice of spinning and winding, Krüger wishes to appease the tensions and the "atrocious insults" ("greuliche Diffamationes") between the masters and the spinning women (named "Frauenzimmer"). As in Sect. 5.2, before turning to examine whether Krüger discusses the mathematization of winding practices (in Sect. 5.3.2), I will start in Sect. 5.3.1 by unfolding the social background of Krüger's manuscript. Section 5.3.3 will then move on to inspect the later editions of this manuscript and their own social and economic context.

[29] As was mentioned above (see: Krüger 1830: 332–3); see also (Wagner, Strackerjan 1855: vi), who note that the textile length measure "Schock" of 60 ells in Leipzig is equal to 59 ells in Augsburg, 61 in Baden, 41 in Bayern, 52 in Berlin etc.

[30] "Schola textoria nova, or New School of Spinning and Weaving, where one is taught how to wind the correct amount of yarn of all kinds—fine and coarse—onto each ell and, consequently, onto each piece of cloth [*Tuch*], *Recken*, *Schocken*, etc., using the proper reel." The terms *Recken* and *Schocken* designate textile units of measurement. See below for a discussion on such measurements. See also footnote 29 above.

5.4.1 Spinning and Sexuality in the Seventeenth Century

Before turning to Krüger's mathematical and arithmetical analysis, one question arises concerning these insults: to what was Krüger actually referring, when talking about the defamations directed towards the spinners? Did his introduction have only a rhetorical character, introduced only to emphasize that in general artisanal practices and household activities also carry a specific type of knowledge—i.e. embodied knowledge—which should not be underestimated by the "masters", or did rather Krüger refer to defamations directed specifically towards spinners? Were the "atrocious insults" expressed due to the financial advantages that the spinners might have gained from the variations in measurements that they were perhaps introducing, as noted above? The reason however might be different. Looking more closely at the activity of spinners and the *Spinnstuben* and *Rockenstuben* in Germany—i.e. the spinning bees, where spinning and winding of the yarn were taking place from the sixteenth until the eighteenth century—one can note that these places "throughout Germany were officially viewed as centers for scandalous socializing and were forbidden under penalty", as Alison G. Stewart (2003: 129) notes; the drawing of Sebald Beham from 1524 of a spinning bee illustrates a scene of brothel-like activities in a spinning bee, which Stewart analyzes in her work (see Fig. 5.5). This however was not always the case: Lieselotte E. Saurma-Jeltsch (2002: 298–307) underlines that during the late middle ages until the beginning of the sixteenth century, images of spinning women served to present a female ideal of virtue. Accordingly, such images were intended to demonstrate female role models and their theological connotations with reference to the iconography of Eve and Mary

Fig. 5.5 Sebald Beham's depiction of a spinning Bee, woodcut, ca. 1524, Ashmolean Museum of Art and Archaeology, Oxford

Fig. 5.6 Knitting, spinning and winding women at Jacob von der Heyden's "KunckelBrief oder SpinnStuben" (ca. 1620). The activity of winding with a square-shaped warp reel is depicted in the right corner. © Kunstsammlungen der Veste Coburg, Inventar Nr. XIII, 441, 8; Neg. Nr. 8685

spinning; in addition, spinning was also recommended to noble women as a contemplative activity. This image of the noble, unreachable spinning woman begins to change during the sixteenth century, when the spinning bees were considered as 'scandalous' meeting places between men and women.

As Craig Koslofsky describes, the spinning bees in the early modern period were of varied forms: up to twelve women would meet to spin wool or flax, knit, or sew by candlelight. In Fig. 5.6, which presents a more 'orderly' spinning bee, one can note the warp reel, drawn on the right side of the engraving. But while one may assume that the main activity done in the spinning bees was textile related, Koslofsky underlines that they were also a "vital place for the nocturnal courtship customs of early modern rural youth. […] This—in the authorities' view—is where the trouble began" (Koslofsky 2011: 203–4). Indeed, pastors and administrators—both outsiders to the village—denounced the spinning bees not only on "moral grounds, but also because the gatherings sustained local nocturnal" cultures (ibid.: 208). French and German authorities considered the spinning bee as a place where "intolerable corruption and detestable, hideous debauchery" take place, as a place "in which all kinds of immorality and fornication are carried on."[31] (Ibid.: 220–1) Nancy Locklin (2007: 132), analyzing women's work and activities in Brittany in the seventeenth and eighteenth century, stresses that these women's gathering were considered as having "demonic implications", causing, so the catholic synod of Saint-Malo, the "fall of man". However, from the second half of the eighteenth century, such gatherings were not any more seen as evil, but rather as a sign of "ignorant disrespect of the holy church, best met with improved education". As Koslofsky summarizes, beginning in the early seventeenth century, numerous laws, ordinances, and statutes

[31] For Spinning bees in early modern Germany, see also: Medick (1984).

were enacted to restrict or even prohibit the spinning bees, and this legislation "remained in place, relatively unchanged, for 150 years or more" (Koslofsky 2011: 222). But despite this prohibition, enforcement was consistently ineffective.

5.4.2 *Krüger's Arithmetic and the Mathematics of Winding*

Having seen the social background of the spinning bees, one may assume that Krüger goes against the characterization of the spinning bee as a place where there is not only no attempt at standardization of measures, but also demonic and sexual activities. By explaining the practice, Krüger aims to show that such defamations are not only made without any justification, but moreover will not persist, once the master understands that spinning and winding are "founded [on] *Arithmeticam*", i.e. on arithmetical rules (Krüger 1664: Aiii, verso). Krüger explains that the entire activity is based on three aspects (ibid.: B: "Die ganze Sache beruhet auff dreyen Stücken"): 1. On the weaver of the yarn and his instruments ("Werckzeuge"), among them the "frame" ("Rahm") and the measurement of ell ("Ellen"),[32] 2. On the spinning and winding housewife and her instruments (being the reel, that is, the "Haspel"), as well as the yarn and the ell, and 3. On the explanation and application of these instruments. When describing the warping frame, Krüger notes that every "Gang" consists of 20 threads (ibid.: B, recto). He stresses that having the correct circumference of the warping frame (being 3 ells) is "fundamental", and having a frame with the correct measures does in fact make various calculations redundant: "he [the yarn weaver] first invented it [as he] had to have a certain *fundamentum* before him; otherwise he would not have solved this difficult work for eternity, unless he was a good mathematician [*Mathematicus*]."[33] (Ibid.: C ij, recto) To stress that this knowledge is first-hand knowledge, Krüger then immediately notes that "so I also have a proper frame, three ells wide, in my house"[34] (ibid.). This length is then noted once more, when describing the various applications, Krüger also stresses that one should multiply ("multiplicando")[35] the length of the frame in order to obtain to correct length of the hank.

The main part of the book, after explaining the various spinning and warping practices, consists of numerical tables concerning the winding with the warp frame. The tables are similar to the one presented in the folio from Leibniz' LH 38, but are more elaborate (see Fig. 5.7). Indeed, Krüger underlines that the "tables serve the

[32] Rosenfeld (1958: 11) notes that it is not clear whether Krüger referred to the rotating warping mill or to the warping frame, indicaitng that he might not have even known the warping mill.

[33] "Dann der [der Garnweber] es erstlich erfunden / hat ein gewisses *fundamentum* vor sich haben müssen / dann sonsten haette er in Ewigkeit dieses schwere Werck nicht heben werden / sondern er ist ein guter *Mathematicus* gewesen."

[34] "So habe ich auch in meinem Hause einen richtigen Rahmen drey Ellen breit".

[35] Neither this page nor its verso side is numerated. It appears on the verso side of the page after the page B 3.

Die erste Tabell nach den Stücken.

12. Gänge.			13. Gänge.			14. Gäng.			15. Gänge.		
E.	st.	F.	E.	st.	F.	E.	st.	F.	E.	st.	F.
1		6	1		6½	1		7	1		7½
2		12	2		13	2		14	2		15
3		18	3		19½	3	1	1	3	1	2½
4	1	4	4	1	6	4	1	8	4	1	10
5	1	10	5	1	12½	5	1	15	5	1	17½
6	1	16	6	1	19	6	2	2	6	2	5
7	2	2	7	2	5½	7	2	9	7	2	12½
8	2	8	8	2	12	8	2	16	8	3	
9	2	14	9	2	18½	9	3	3	9	3	7½
10	3		10	3	5	10	3	10	10	3	15
11	3	6	11	3	11½	11	3	17	11	4	2½
12	3	12	12	3	18	12	4	4	12	4	10
13	3	18	13	4	4½	13	4	11	13	4	17½
14	4	4	14	4	11	14	4	18	14	5	5
15	4	10	15	4	17½	15	5	5	15	5	12½
16	4	6	16	5	4	16	5	12	16	6	
17	5	2	17	5	10½	17	5	19	17	6	7½
18	5	8	18	5	17	18	6	6	18	6	15
19	5	14	19	6	3½	19	6	13	19	7	2½
20	6		20	6	10	20	7		20	7	10

D

Fig. 5.7 Krüger's first conversion table, from ells to "stücke" and "Fitzen", according to the number of "Gänge" and then according to the ells, in: Krüger (1664: D, recto). The table appears in a section titled: "Folgen die Tabellen. Die erste nach dem ganzen oder vollem Haspel / wann gehaspelt 40. Fädem und 20. Fitzen"

purpose that one can easily see *without further reflection and calculation* how much yarn after the winding is obtained for cloths [or] rags […]."[36] (Ibid.: C ii j, recto) To the question, "how to find and organize oneself in [with] the tables"[37] (ibid.: C iii, recto), Krüger stresses that there are two types of tables. These two types are tables of conversion from ells to other measurements; as various spinners and weavers worked with various, different measures, which also stemmed from the fact that one worked with different warp reels, Krüger apparently aims to examine a variety of such measures, hence showing awareness that the arithmetical operations being done in practice were more complicated than one would have assumed, since some

[36] "[…] die Tabellen darzu dienen, daß man geschwinde ohne weiteres Nachdencken und Rechnen könne daraus sehen, wie viel Garnes nach den Gangen auf jede Ellen, und dann folgends zu Tüchern, Regken […]" (cursive by M.F.)

[37] "wie wird man sich in den Tabellen finden und richten?"

units of measurement could not be added with others without knowing in advance their relations. Such plurality of measurements and conversions do not appear in Leibniz's folio.

With the first type of table one converts, according to how many "Gänge" are wound, ells to "stücke" and "Fitzen"; the second type of table converts, again according to how many "Gänge" are wound, ells to "Zaspeln" and to "Fitzen". "Stücke", "Fitzen" and "Zaspeln" (also known as "Zaspel") are all units of measurement of length,[38] but as noted above, also these measures changed regionally,[39] a fact not mentioned by Krüger, maybe since it was obvious. Hence, for the first type of the table, one has conversion tables for 12, 13, 14… to 22, and then for 24, 26, 28, … to 60 "Gänge", while ells progress from 1 to 20. For the second type of the table, one has conversion tables for 12, 13, 14… to 43 "Gänge", while again ells progress from 1 to 20. Beside the conversion of measurements, Krüger gives a few examples of how to practically use the tables. One of them is the following (ibid.: C iii, recto-verso): given 15 "Gänge", the "Hausmutter" wants to prepare hanks in length of 66 "Zaspeln". What is the length in ells of the yarn to be prepared? Krüger instructs the housewife to search in the second type of table, in the corresponding table for 15 "Gänge". There one notes that 15 "Zaspeln" correspond to 20 ells. Hence, Krüger says, "she calculates" (ibid.: C iii, verso), reaching the conclusion that 60 "Zaspeln" correspond to 80 ells. For the remaining 6 "Zaspeln" Krüger instructs the housewife to search again in the same table for 6 "Zaspeln", which correspond to 8 ells. Hence one obtains that 88 ells are equivalent to 66 "Zaspeln".

Such examples explicate Krüger's approach, that the tables can be employed "without further reflection and calculation". Here one can note an affinity to how Leibniz considered spinning and winding in particular and artisanal knowledge in general: the rules and tables must be stated explicitly, in order to save time and in order to *not calculate*. In addition, one should recall that Leibniz compares another textile machine—the stocking frame[40]—to his own calculating machine, his "arithmetical instrument, which carries over all the work of the mind to wheels".[41] Moreover, in 1685, Leibniz notes regarding his own calculating machine, in a similar manner to how Krüger describes the arithmetical tables, that "it is unworthy of excellent men to lose hours like slaves in the labor of calculation, which could be safely relegated to anyone else if the machine were used."[42]

[38] In the Grimm dictionary (Grimm: "FITZE, f."), "Fitze" has several definitions; one, from 1821, defines "Fitze" as a bundle of 40 threads with the length of 4 ells; the other, from 1741, defines "Fitze" only as a bundle of threads. "Zaspel" has also various definitions: either 400 threads of 4 ells length; or other quantities of threads, e.g. 240, 450 or 480.

[39] For example, according to (Krüger 1830: 86, 94) "Fitze" in Berlin and Preußen is equivalent to 140 Berliner ells, whereas in Mark Brandenburg and Pommern it is equivalent to a measure between 120 and 160 ells. See also footnote 29 above, concerning the measurement "Schock".

[40] The stocking frame, a machine which mechanized knitting, was invented by William Lee in 1589.

[41] A VII6: 488. See also: Friedman (2023b).

[42] Cited from: Smith (1959: 181).

That is, while both scholars saw the practice of winding of the yarn with the warp frame as praiseworthy, and while both note that such a practice is based on arithmetic laws—and indeed, Krüger emphasizes several times the various calculations done by the housewives—at least according to Krüger's view, once it is made formal and presented as theoretical knowledge, one should not reflect on it anymore. In this sense, the arithmetical tables serve as a paper instrument which externalize the artisanal, embodied knowledge, and present the practitioner's implicit calculations as dispensable. However, it is clear that the aim of Krüger was different than that of Leibniz, as Krüger was probably aware of the reputation which the spinning bees had during the early modern period; such an awareness is not to be found in Leibniz's writings, as Leibniz was much more concentrating on the economic aspects of spinning as well as those related to commerce.

5.4.3 The Later Editions of Krüger's Book

The emphasis Leibniz puts on trade, commerce and the importance of establishing spinning centers in the Holy Roman Empire—in short, the emphasis on the economic aspects—is also to be noted in the second edition of Krüger's manuscript from 1688; while the main text remained unchanged, three appendices were added. The first one is an edict titled "Von Leinwebern" of Augustus the Elder, Duke of Brunswick and Lüneburg (1568–1636) from 1645, listing several ordinances concerning the criteria according to which the linen weavers will receive their salary (Krüger 1688: F, recto–F ii, verso). The second is an edict of George Wilhelm, Duke of Brunswick and Lüneburg (1624–1705) from 13 February, 1683. The edict, titled "Edict wegen des Kauffgarns", underlines that the sold spun yarn should have the "appropriate quantity of threads"[43] (ibid.: F ii, verso) and that the "correct quantity and measure" should be kept (ibid.: F iii, recto)[44]; otherwise, when the warp reel does not have the correct measures, the merchants will claim that they are deceived. The edict calls to keep these measures for the prepared products, so that everyone who wants to spin will get his proper salary. The third appendix, to be found on the last page of the book (ibid., not numerated), is an excerpt from the first edition of Johann Joachim Becher's *Politische Discurs* from 1668; Chap. VII in Becher's book, titled "Von dem Leinen-Handel", deals with the economic aspects of commerce with linen, and is cited in its entirety in this excerpt. In contrast to the edict from 1683, spinning and winding are hardly mentioned. Becher only notes the following: "Concerning the linen trade, it is to be known that it is not only a sincere trade [...] but this trade consists of six kinds [...] for the third kind, there is a special activity with spindles or spun yarns, which are sold by the thousands, transported to

[43] "[...] die gebührend Anzahl Fadem".

[44] "richtige Zahl und Masse".

other places and woven there."[45] (Becher 1668: 46–47) Any mention of the correct measures is not to be found in Becher's account.

The second edition of Krüger's manuscript had also several editions and reprints: in 1706, it was included in *Arcana Et Curiositates Oeconomicæ* (Meier 1706: 143–204). It was also republished in 1733, in the tenth volume of *Oeconomische Fama* (Krüger 1733); in this volume, the appendices of the second edition were not published.[46] The tables themselves were also reprinted, with a short introduction, in 1734, in an appendix to "Corpus Constitutionum Marchicarum"; the appendix is titled "General Privilegium und Gülde Brief des Garn-Weber-Gewercks / in der Chur und Marck Brandenburg dies und jenseits der Oder und Elbe insonderheit des Garn Weber Gewercks in Berlin / De Dato den 14 April 1734",[47] and the tables are to be found in (General-Privilegium 1734: 13–28). The appendix however first stresses in its introduction to the tables that having a proper reel is the "Fundament", and that its length should be "4 Berlinische Ellen" (ibid.: 12). Moreover, the anonymous *Gründliche Unterweisung vom Spinnen und Weben* (Rüdiger 1740) from the beginning of the eighteenth century is a shortened version of Krüger's manuscript, but without mentioning Krüger's name. As noted above, the weaver's instrument is noted to be either the "frame" ("Rahmen") or the "Umläufer", meaning the warping mill (ibid.: 4); also in this manuscript, the tables are reprinted.

Such new editions, reprints and summaries obviously prompted the dissemination of Krüger's work, sometimes without mentioning his name, hence presenting his tables as a means to make the practice of such calculations public and not associated to one single person, and in a sense, independent of the region where it was published. Some republications—as well as the various edicts and the privileges—also acted as though they were dealing only with regional artisanal practice (e.g. in Prussia or in Brandenburg); this is seen also due to the fact that the various measures changed from region to region. This is in contrast to the first edition, which may be seen as an attempt to emphasize that spinning and winding may be 'arithmetized', or, if one examines this edition in the framework of the various encounters of natural philosophers and artisans, as an attempt to extract arithmetic rationality from such artisanal, household works.

What is unique to the practice of winding yarn was that it was already arithmetized, even if implicitly; not only did the spinners have to count how many windings were made in order to later measure the spun yarn and to produce bundles in the desired size, but also the various systems of measurement units formed and forced a unique arithmetical thinking. However, one may read Krüger's introduction not only as an attempt to write down the implicit arithmetical rules of winding and warping; by noticing the social background of the spinning bees, he may have

[45] "Den Leinen-Handel belangend / so ist zu wissen / daß selbiger nicht allein ein herzlicher Handel / […] es bestehet aber dieser Handel in sechserley Arthen […] Drittens ist eine absonderliche Handlung mit Spindel oder gesponnen Garen / welches bey tausenden verkaufft / in andere Oerther gefuͤhrt / und allda verwebt wird".

[46] On the *Oeconomische Fama*, see: Kohfeldt (2010: 54–59).

[47] "General Privilege and Guild Charter of the Yarn Weavers' Guild / in the Duchy and Margraviate of Brandenburg on both sides of the Oder and Elbe Rivers, specifically the Yarn Weavers' Guild in Berlin / Dated April 14, 1734".

presented arithmetic as what may dissolve and free this embodied knowledge from its sexual connotations. These connotations or the mentioning of such "insults" are slowly disappearing from the later editions and reprints of (parts of) Krüger's manuscript.

5.5 Conclusion: Who Considers Winding and Warping as a Mathematical Practice?

Numerical tables were written by scholars, mathematicians and natural philosophers for describing the practitioners' and artisans' work, aiming, for example—when these practitioners indeed read these texts—to aid these practices to be done quickly and correctly. These were not only to be found in the texts discussed above; Leonard Digges's book *Tectonicon* from 1556 also contained numerical tables, which should be used by 'mathematical practitioners', especially by land surveyors.[48] However, one may ask whether the textile-related conversion tables discussed above were abundant, a question which remains open. Nevertheless, the fact that Krüger's tables were published and republished numerous times shows a need—at least on behalf of the administrators and natural philosophers—to make such systems of conversion public, even if at that time there was no standardization of textile-related measures.

Whether European spinners can be indeed called 'mathematical practitioners', in the sense of "men [and in our case, women] who deployed mathematical concepts for practical ends" (Cormack 2017: 3), is a question for which (for now) a satisfying answer cannot be given. First of all, what we do not possess are the calculations made by the actors themselves, i.e. the spinners and weavers. Did they, for example, learn or formulate by themselves the Rule of Three? Or whether these arithmetical rules, when these were used by the practitioners, were more implicit or based on rote learning of several measures? Due to the lack of written sources by the practitioners, it is hard to deduce the existence of 'arithmetical' "artisanal epistemology" for the spinning and winding women, which is based on an "ability of craftspeople to produce material things [which] rests upon experientially derived systems of knowledge that are simultaneously practical and conceptual and can be employed rigorously and methodically to generate new knowledge." (Smith 2022: 232) Second, what we do have are the calculations and reflections written or added by Krüger and Leibniz, i.e. texts which are external to the practitioners. Krüger emphasizes that arithmetical knowledge is indeed to be discovered in the practices he describes. Spinning and winding, he underlines, are "founded [on] *Arithmeticam*", i.e. on arithmetical rules, but he does not note that such rules may change once the measures themselves vary. The text in Leibniz's notes on technique, "Leinweber Rechnung", also contains arithmetical examples, which can only be understood once one works with the arithmetical table and has implicit knowledge about certain

[48] There are several accounts on role and usage of mathematical tables; see for example (Campbell-Kelly et al. 2003; Tobies and Tournès 2011).

measures. This means that both Krüger and Leibniz saw the practice of winding at least as embodiments of arithmetical knowledge, but it is not clear whether one can really speak about 'trading zones', in the sense that the practitioners of spinning and winding at the spinning bee in rural areas in Germany had access, for example, to the various copies of Krüger's book, hence prompting a 'trade' of knowledge.

While Krüger emphasizes arithmetic at the basis of the practice of winding, he also notes that the externalization of this arithmetic in the form of a table renders unnecessary the various calculations probably performed by the spinners. One may perhaps offer to consider two types of arithmetization: on the one hand, the arithmetical calculations made by the spinning, winding and weaving persons, converting and adding measurements; on the other hand, the 'external' texts and tables, external in the sense they were imposing an 'external' (system and form of presentation of) knowledge. This knowledge was presented by philosophers and administrators as evidence that, while there was mathematical knowledge involved in artisanal practices, such knowledge was understood as inferior to the philosopher's explication.

In this respect, Krüger is not alone when thinking about the relations between external codification of knowledge and textile artisanal practices: the weaver Xaver Friepes notes in 1799, in his book titled *Allgemeines Schnier Buch*, that weaving notation is for weavers "who cannot think, who do not want to think, or do not have time to think".[49] This statement, which implies that weavers can relegate their thinking to certain systems of notation, also sarcastically suggests that there are weavers who do not have time to think, as if they only follow various notation books. In this sense, the tables in the "Leinweber Rechnung" and those in Krüger's book function as 'forgetting machines': along notebooks, commonplace-books, card indexes, filing cabinets and libraries, all of these, as part of the rising European knowledge management during the early modern period, "enable users to remember substantially more than before because these media allow users to forget substantially more than before" (Cevolini 2016: 19). However, whether spinning and winding women in the early modern period indeed used such tables and hence forgot how to calculate, is another question. Krüger explicitly and Leibniz implicitly call for the externalization and writing down of arithmetical knowledge, hence to its forgetting, but with this call they also marginalize the artisanal epistemology of spinning, winding and warping.

Acknowledgements This research was supported by the Israeli Science Foundation (grant no. 461/21). I thank warmly Ellen Harlizius-Klück and Charlotte Wahl for conversations on Krüger, Leibniz and the various measurements of weavers. I also thank the reviewers of this paper for helpful comments and Gudrun Bernhardt for her help with the transcription of fol. 35v from Leibniz' LH 38.

[49] Cited from: Kreindl (1935: 79): "[…] ein blosser Liebhaber verschiedene Bilder zu denken und zu zeichnen, damit sich andere liebhaber dieser arbeit, auch selbst die Webermaister, welche nicht denken können, nicht denken wollen, oder zum denken nicht Zeit haben, hieraus gefällige Bilder erwählen können […]."

5.6 Appendix: Transcription of “Leinweber Rechnung” (LH 38, fol. 35v)

The following is a transcription of fol. 35v from Leibniz’ LH 38, titled. “Leinweber Rechnung” (see Fig. 5.4). Since the text itself contains the punctuation mark / , it is retained in the transcribed text, whereas the symbol // denotes where a line ends.

Leinweber Rechnung.

Welche in dieser richtigen Taffell begriffen darin man leicht auß rechnen kan wieviel auß einen // gleichgesponnen / und wollgehaspelten garn / Leinwandt gemacht werden kan? Belangende die zahlung / die ist alhier // in gewisser Proportion / wollmeinende gesetzt / biß verständige solche beßer verfaßen werden[.] Nam dies diem docet // Authore Val. D. Borusso, ex doctrina quadrangulorum.

[on the left margin:]

Regelen 1. Auff einen fall kan mann die gänge in der fertigen Leinwandt an fademen außzehlen. 2. Wen die // Haspel einer Ellenlang ist so gibets leichte rechnung vermöge dieser taffel 3 Ist sie länger etwa auff 1 Zoll: So muß // mann auß 24 stücken oder Thallen gantze 25 stücke oder Thallen anrechnen den[n] 24 Zoll machen 1 Elle etc. 4 Ist die // Haspel 1 Zoll kürtzer; so kommen auß 24 stücke nur 23 stücke oder Thall undt so fortan Zum 5 Ist es ein vortheil // daß der Leinweber eine solche Elle zumeße, wie lang das garn gewesen. Exceptio. Wofern die Leinwandt nicht // rechte steiff gearbeitet wird so krimpen in der Einknieung fast anderthalb viertel in 20 doppelten stöcken, Summa

[on the right margin:]

der Einkrimpung in 60 Ellen oder doppelten stöcken thut 2 stöcke, 1 viertel solche von 60 Ellen abgezogen so // bleiben in der frischen Leinwandt 58 doppelte stöcke undt 7 viertel in der lieferung. Exempel. Es stehet ein garn // in 33 gängen undt werden begehrt 75 Ellen oder doppelte stöcke wieviel garn muß man dazu haben. R. In der // Taffel stehen 33 gänge das uff 4 stöcke gehen. 6 Thalle 6 Bünde dieses halbiret kommen 3 Thall 3 bünde uff // 1 Elle oder 2 stücke demnach gehen auff 10 doppelte stöcke 33 Thall, dieses mit 7 multiplicirt thun 231 Thalle // die uff 70 doppelte stöcke gehen auff restirende 10 stöcke oder 5 Ellen gehen 16 Thall 5 bünde Sum–//ma alles garns zu 75 doppelten stöcken Leinwandt thun 247 Thall 5 gebünde summa alles garns zu 75

[on the bottom margin:]

Oder man rechne also: In der 6 Columna stehen gegen 33 gänge, uff 60 stöcke gerade 99 Thall solche doppelt genommen // machen 198 Tall oder 99 stücke darnach erfo[r]dern 10 doppelte stocke 33 Thall undt 5 doppelte stöcke halb so viel nemb– // lich 16 Thalle 5 gebünde. So gehen uff 15 doppelte stöcke 49 Thall // 5 bünde dieses zu voriger Summe der 198 thalle gethan so kommen in alles 247 Thalle 5 bünde die machen 123 // stöcke 15 gebünde wie vorhin.

Cogitatus quis coget? Vis Veritas

Wahrheit wird gedruckt aber nicht unterdruckt
Auffsicht ist zu allen dingen guth.

Abbreviations

A	Leibniz, Gottfried Wilhelm (1923–). *Sämtliche Schriften und Briefe*. Darmstadt, Leipzig, and Berlin: Deutsche Akademie der Wissenschaften zu Berlin. When referring to Leibniz's edited works, I will use the letter 'A' followed by one Roman and one Arabic numeral, representing the series and volume number
Grimm	Deutsches Wörterbuch von Jacob Grimm und Wilhelm Grimm, digitalisierte Fassung im Wörterbuchnetz des Trier Center for Digital Humanities, Version 01/23, <https://www.woerterbuchnetz.de/DWB>, accessed on 02.04.2024.
LH 38	*Leibniz-Handschriften zur Technica*, Gottfried Wilhelm Leibniz Bibliothek—Niedersächsische Landesbibliothek.
General-Privilegium	1734. General Privilegium und Gülde Brief des Garn-Weber-Gewercks ..., Anhang No I vom Garnweber Handwergk, in: *Des Corporis Constitutionum Marchicarum, Fünffter Theil Von Policey- Hochzeit- Kindtauffen-Begräbniß- und Trauer- Kleider-...*

References

Bacon, Francis. 1858. The New Organon. In *The Works of Francis Bacon,* ed. James Spedding, Robert Leslie Ellis, and Douglas Denon Heath, 15 vols., 1857–1874, vol. VIII (translations of the philosophical works, vol. I), 59–350. London: Longman & Co.

Becher, Johann Joachim. 1668. *Politischer Discurs*. Frankfurt am Main: Zunners.

Beirach, Yoav, and Michael Friedman. 2025. "'Tirer d'eux leurs secrets': Leibniz on Artisanal Knowledge and 'Secret' Geometry", to be published in *NTM. Zeitschrift für Geschichte der Wissenschaften, Technik und Medizin*.

Bohnsack, Almut. 1981. *Spinnen und Weben: Entwicklung von Technik und Arbeit im Textilgewerbe*. Hamburg: Rowohlt.

Bomann, Wilhelm. 1927. *Bäuerliches Hauswesen und Tagewerk im alten Niedersachsen*. Hanover: Th. Schäfer.

Campbell-Kelly, Martin, Mary Croarken, Raymond Flood, and Eleanor Robson, eds. 2003. *The History of Mathematical Tables. From Sumer to Spreadsheets*. Oxford: University Press.

Cevolini, Alberto. 2016. Knowledge Management Evolution in Early Modern Europe: An Introduction. In *Forgetting Machines: Knowledge Management Evolution in Early Modern Europe*, ed. Alberto Cevolini, 1–33. Brill: Leiden.

Cormack, Lesley B. 2017. Introduction: Practical Mathematics, Practical Mathematicians, and the Case for Transforming the Study of Nature. In *Mathematical Practitioners and the Transformation of Natural Knowledge in Early Modern Europe*, ed. Lesley B. Cormack, Steven A. Walton, and John A. Schuster, 1–10. Cham: Springer.

Cormack, Lesley B., Steven A. Walton, and John A. Schuster, eds. 2017. *Mathematical Practitioners and the Transformation of Natural Knowledge in Early Modern Europe*. Cham: Springer.

Descartes, René. 1985. *The Philosophical Writings of Descartes*, vol. 1. Trans. John Cottingham, Robert Stoothoff, and Dugald Murdoch. Cambridge: Cambridge University Press.

Friedman, Michael. 2022. Joachim Jungius and the Transfer of Knowledge from Weaving to Mathematics in the 17th Century. *Cahiers François Viète* III-13:43–82.

Friedman, Michael. 2023a. *On Joachim Jungius' Texturæ Contemplatio. Texture, Weaving and Natural Philosophy in the 17th Century*. Cham, Springer.

Friedman, Michael. 2023b. Leibniz and the Stocking Frame: Computation, Weaving and Knitting in the 17th Century. *Minds & Machines*. https://doi.org/10.1007/s11023-023-09623-3.

Jones, Matthew L. 2016. *Reckoning with Matter: Calculating Machines, Innovation, and Thinking About Thinking from Pascal to Babbage*. Chicago, London: The University of Chicago Press.

Kohfeldt, Christian. 2010. Justus Christoph Dithmar: *Die Oeconomische Fama*. Der erste Versuch eine ökonomische Zeitschrift zu etablieren. In *Landschaften agrarisch-ökonomischen Wissens. Strategien innovativer Ressourcennutzung in Zeitschriften und Sozietäten des 18. Jahrhunderts*, ed. Marcus Popplow, 49–60. Münster: Waxmann.

Koslofsky, Craig. 2011. *Evening's Empire. A History of the Night in Early Modern Europe*. Cambridge: Cambridge University Press.

Kramper, Peter. 2019. *The Battle of the Standards. Messen, Zählen und Wiegen in Westeuropa 1660–1914*. Berlin/Boston: de Gruyter.

Kreindl, Fritz. 1935. Die Mühlviertler Bröselmaschine (Webmaschine). *Heimatgaue* 16:78–81.

Krüger, Daniel. 1664. *Schola textoria nova oder neue Spinn- und Weberschule* [...]. Frankfurt an der Oder: Erasmum Rösnern.

Krüger, Daniel. 1688. *Schola textoria nova oder neue Spinn- und Weberschule [...]*. 2nd ed. Frankfurt an der Oder: Erasmum Rösnern.

Krüger, Daniel. 1733. *Schola textoria nova oder neue Spinn- und Weberschule* [...]. In *Die Oeconomische Fama*, vol. 10, pp. 2–53.

Krüger, Johann Friedrich. 1830. *Vollständiges Handbuch der Münzen, Maße und Gewichte aller Länder der Erde*. Quedlinburg/Leipzig: Gottfried Basse.

Leibniz, Gottfried Wilhelm (1923). *Sämtliche Schriften und Briefe*. Darmstadt, Leipzig, and Berlin: Deutsche Akademie der Wissenschaften zu Berlin.

Locklin, Nancy. 2007. *Women's Work and Identity in Eighteenth-Century Brittany*. Burlington, VT: Ashgate.

Long, Pamela O. 2011. *Artisan/Practitioners and the Rise of the New Sciences, 1400–1600*. Corvallis: Oregon State University Press.

Long, Pamela O. 2015. Trading Zones in Early Modern Europe. *Isis* 106 (4): 840–847. https://doi.org/10.1086/684652.

Marten, Maria, and Carola Piepenbring-Thomas. 2015. *Fogels Ordnungen: Aus der Werkstatt des Hamburger Mediziners Martin Fogel (1634–1675)*. Frankfurt am Main: Klostermann.

Medick, Hans. 1984. Village Spinning Bees: Sexual Culture and Free Time Among Rural Youth in Early Modern Germany. In *Interest and Emotion: Essays on the Study of Family and Kinship*, ed. Hans Medick and David Warren Sabean, 317–339. Cambridge: Cambridge University Press.

Meier, David. 1706. *Arcana Et Curiositates Oeconomicæ. Das ist Curieuse Heimligkeiten Betreffend die Hauß-Haltung*.

de Munck, Bert. 2019. Artisans as Knowledge Workers: Craft and Creativity in a Long Term Perspective. *Geoforum* 99:227–237.

Münster, Sebastian. 1550. *Cosmographiae uniuersalis lib. VI*. Basel: Heinrich Petri.

Rosenfeld, Hans-Friedrich. 1958. *Wort- und Sachstudien. Untersuchungen zur Terminologie des Aufzugs, zu Webstuhl und Schermethode der germanischen Bronze- und Eisenzeit und zur Frauentracht der Bronzezeit sowie der Frage ihres Fortlebens in der Volkstracht*. Berlin: Akademie Verlag.

Rüdiger, Johann Andreas, ed. 1740. *Gründliche Unterweisung von Spinnen und Weben, wie viel das löbliche Frauenzimmer, von allerley Garn, auf jede Elle, nach einem richtigen Haspel dem*

Weber zu geben hat: Und wie viel Ellen und Schock der Weber von allerley Garn wieder erstatten muß. Berlin: Rüdiger.
Saurma-Jeltsch, Lieselotte E. 2002. Profan oder sakral? Zur Interpretation mittelalterlicher Wandmalerei im städtischen Kontext. In *Erscheinungsformen höfischer Kultur und ihre Träger im Mittelalter: Band I: Erscheinungsformen höfischer Kultur und ihre Träger im Mittelalter*, ed. Eckart Conrad Lutz, Johanna Thali, and René Wetzel, 283–328. Berlin, New York: Max Niemeyer Verlag.
Sennewald, Georg. 1859. *Lehr- und Muster-Buch der gesammten Linnen- u. Tischzeugweberei*. Weimar: Voigt.
Smith, D. E. (ed.). 1959. *A source book in Mathematics*, 1 vol. Dover Publications.
Smith, Pamela H. 2004. *The Body of the Artisan: Art and Experiment in the Scientific Revolution*. Chicago: University of Chicago Press.
Smith, Pamela H. 2022. *From Lived Experience to the Written Word: Reconstructing Practical Knowledge in the Early Modern World*. Chicago: University of Chicago Press.
Stewart, Alison G. 2003. Distaffs and Spindles: Sexual Misbehavior in Sebald Beham's Spinning Bee. In *Saints, Sinners, and Sisters: Gender and Northern Art in Medieval and Early Modern Europe*, ed. Jane L. Carroll and Alison G. Stewart, 127–154. Aldershot: Ashgate.
Styles, John. 2020. The Rise and Fall of the Spinning Jenny: Domestic Mechanisation in Eighteenth-Century Cotton Spinning. *Textile History* 51 (2): 195–236.
Tobies, Renate, and Dominique Tournès (ed.), 2011. *Mini-Workshop: History of Numerical and Graphical Tables*, Mathematisches Forschungsinstitut Oberwolfach, Report No. 12/2011. https://doi.org/10.4171/OWR/2011/12.
Wagner, Gustav, and Friedrich Anton Strackerjan. 1855. *Compendium der Münz-, Maass-, Gewichtsund Wechselcours-Verhältnisse*. Leipzig: Teubner.
Werrett, Simon. 2019. *Thrifty Science: Making the Most of Materials in the History of Experiment*. Chicago: The University of Chicago Press.

Chapter 6
Between Theory and Practice: Teodoro Bonati and the Teaching of Hydrostatics at the University of Ferrara

Maria Giulia Lugaresi

Abstract In 1771, the teaching of hydrostatics was included in the curriculum of the "Faculty of Artists" at the University of Ferrara, and Teodoro Bonati (1724–1820) was appointed as teacher. Before becoming a professor, Bonati had worked as hydraulic consultant for the city of Ferrara for more than twenty years, during which he performed many surveys on practical hydraulics on the territory of Ferrara, especially along the course of the river Reno in the Po Plain, in the area between the provinces of Bologna, Ferrara, and Ravenna. Bonati's fieldwork experience with water regulation was rewarded with an appointment as professor of hydrostatics; he also published several important works on water measurement. A large share of his scientific production remained unpublished (and is still preserved in Ferrara), but is equally important for sketching the biographical and professional life of this mathematician. In focusing especially on this double role of theoretician (professor) and practitioner (technical consultant), I intend to discuss Bonati's contribution to the science of water measurement and water management.

Keywords Science of waters · Teodoro Bonati · Chair of hydrostatics · University of Ferrara · Water measurement

6.1 Introduction

On January second, 1820, at the age of ninety-six, Teodoro Bonati died in Ferrara. An obituary was published in the *Rendiconti dell'Accademia Nazionale delle Scienze*:

M. G. Lugaresi (✉)
University of Ferrara, Ferrara, Italy
e-mail: lgrmgl@unife.it

Senthil Babu D., R. Wagner (eds.), *Normative Texts and Occupational Practices*, Boston Studies in the Philosophy and History of Science 350,
https://doi.org/10.1007/978-3-032-28419-8_6

> The Knight Bonati, in combining theoretical studies with a constant practice, was always useful to his homeland, protected it from many threatening dangers and promoted the most real advantages. […] The application of mathematics to hydraulic science was the art Professor Bonati most excelled in. The services he gave with his work were very celebrated, not only in Ferrara, but also in every part of Italy. He was recognised as the first hydraulic expert of his time.[1]

The scientific figure of Teodoro Bonati is representative of the Italian tradition of studies on the motion of waters between the second half of the 18th and the first two decades of the 19th century.[2] Bonati was born in 1724 in Bondeno, near Ferrara. In 1740 he moved to Ferrara to study medicine at the University, where he graduated in 1746. Among the disciplines included in the curriculum of medicine was mathematics, a discipline that, in turn, prescribed the reading of the first six books of the *Elements* by Euclid, the *Sphaera* by Sacrobosco, and the *Theorica planetarum*.[3] The study of mathematics in Ferrara developed in close connection with the study of hydraulics. Because of its peculiar geographical setting, throughout its history the city had always paid special attention to advanced scientific techniques in order to solve problems related to controlling waters, managing riverbanks, and surveying land for water control purposes.[4]

Bonati lived and worked as a personal doctor in the palace of the marquis Guido Bentivoglio and could also take advantage of the presence in Ferrara of the mathematician Gianfrancesco Malfatti (1731–1807).[5] Thanks to Romualdo Bertaglia

[1] "Il Cav. Bonati congiungendo agli studii teorici una costante pratica giovò sempre alla patria, la difese dai tanti pericoli che la minacciavano, e ne promosse i più reali vantaggi. […] La matematica applicata alla Scienza idraulica fu quella parte in cui più d'ogni altra versò il Professor Bonati, e insigni furono i servigi che rese con la sua penna, e più con l'opera sua non tanto a Ferrara, quanto all'Italia tutta, che il riconobbe per il primo Idraulico de' suoi tempi" (Lombardi 1823–1825, p. CLXXIII). Unless otherwise attributed, the translations from Italian and French are my own. A French biography of Bonati was written by Gaspard de Prony and can be found in Prony (1835). An in-depth biography of Bonati can be found in Borgato et al. (1992, pp. 1–34).

[2] Many works published in the last thirty years deal with the science of water in Italy. See Fiocca (1998) and Fiocca et al.(2003). On the theoretical contribution of Italian scientists to the science of waters see Maffioli (1994, 2010). A fundamental contribution to the study of the history of fluids mechanics was given in Truesdell (1954). Some years later, Roger Hahn provided an extensive summary of the development of hydrodynamics in the 18th century: Hahn (1965). On mathematical practitioners see Cormack et al. (2017).

[3] The *Sphaera* and its commentaries were read in all European universities up until the 17th century. Gerard of Cremona composed an original *Theorica planetarum* (12th century) to present the mathematical systems of the *Almagest* suitably simplified. The *Theoricae novae planetarum* (1472) by Georg Peurbach replaced Gerard of Cremona's *Theorica planetarum* as the standard introduction to Ptolemaic planetary astronomy.

[4] Since the 16th century, a lively scientific debate and many hydraulics projects addressed the issue of water regulation in the lower Po valley (Lugaresi 2021).

[5] Gianfrancesco Malfatti, born in Ala, near the city of Trento, studied in Bologna under the guidance of the Jesuit Vincenzo Riccati (1707–1775). Then he moved to Ferrara, accepting the appointment of personal librarian of the marquis Cristino Bevilacqua. In Ferrara, Malfatti could devote himself to scientific studies, but, because of the lack of local interlocutors, he began to correspond with other mathematicians all over Italy. The correspondences between Malfatti and Giordano

(1688–1763), a surveyor and hydraulic technician of Ferrara, Bonati oriented his studies towards applied mathematics. He began collaborating with Bertaglia in water surveying: thanks to his experience in the field, he learnt the rudiments of hydraulics. The fundamental principles of this discipline were two: velocity proportional to the square root of the height of the drop, and reciprocity of the cross-sections with their average velocities in the hypothesis of stable and continuous flow (Castelli's law). If these two principles were satisfactory in light of the theory, which did not take resistances into consideration, they were not satisfactory in practice. So, "we have to look for a third principle in favour of practice, i.e. the law of hydraulic resistances. As it was too difficult to find this law just with theoretical principles, it is necessary to turn to practical experiments. In past times, Newton, Daniel Bernoulli, and Poleni found something, but they couldn't develop it because they could do only few experiments and on a smaller scale, though with the utmost care". In short, testing the validity of this law would require a great number of experiments.[6]

6.2 Working and Learning in the Field: The Activity of Teodoro Bonati as Hydraulic Expert

Bonati worked with Bertaglia as hydraulic consultant for the city of Ferrara for more than ten years: he did many surveys on the territory of Ferrara, especially along the course of the river Reno in the Po river plain stretching between the provinces of Bologna, Ferrara, and Ravenna. In 1759, Bonati and Bertaglia moved to Rome to attend conferences promoted by the Holy Congregation of Waters on the problem of the regulation of the river Reno, within a wider context of debates between Ferrara and Bologna. The Bolognese delegation was led by the famous mathematician Gabriele Manfredi (1681–1761). During his stay in Rome, Bonati,

Riccati (1709–1790), mathematician and son of Jacopo Riccati, has been recently published in Lugaresi (2020).

[6]"[...] dovrassi adunque cercare in favore della pratica, qual terzo principio, la legge, con cui si fanno le resistenze medesime; onde in ogni caso spariscano le differenze tra la teorica e la pratica. Ma il rinvenire codesta legge con i soli principj teorici è cosa troppo difficile, e forsi anche impossibile; dunque alle sperienze fare devesi ricorso. Già per questa via cominciarono a scoprirne alcuna cosa il Newton, Daniello Bernulli, ed. il Marchese Poleni; ma non poterono affatto svilupparla, né conoscerne l'estensione, appunto perché fare non poterono, che poche sperienze, e queste assai in picciolo, quantunque con somma accuratezza" (Michelotti 1767–1771, vol. 1, Preface, p. IV n.n.). The words of Francesco Domenico Michelotti (1710–1787), professor of mathematics and hydraulics at the University of Turin, highlights the difficulty but also the need for experiments to determine water measurements. According to this necessity, the laboratory of Parella was created in Turin in 1763. Several experiments performed throughout the years allowed Michelotti to demonstrate with facts "the combination of the three principles in the motion of waters". See Lugaresi (2021, pp. 155–156).

and Bertaglia were also consulted about the problem of the reclamation of the Pontine Marshes.[7]

In the dispute against the city of Bologna about the regulation of the river Reno, Bonati took part in the visit of the Cardinal Conti (the so-called *Visita Conti*).[8] The *Visita* officially began on 1761, May 12th, and ended on 1762, July 5th. The mathematicians representing the three provinces were, respectively, Jacopo Marescotti and Pietro Chiesa (Bologna), Romualdo Bertaglia and Teodoro Bonati (Ferrara), and Giuseppe Guarini (Ravenna). Another technician who took part in the visit was Ambrogio Baruffaldi, on behalf of the *Reverenda Camera Apostolica*.[9] During the visit, Cardinal Conti was assisted by the Tuscan Tommaso Perelli (1704–1783), an "impartial mathematician" (*matematico neutrale*), appointed directly by the Pope.

At the end of the visit, Perelli gave his opinion about the projects proposed by the delegates, "Report on the regulation of the waters of the three provinces" (*Relazione sopra il regolamento delle acque delle tre province*). The paper was written in 1763, but it was only published a year later. Soon after the publication of Perelli's report, a long and heated debate arose among the opponents. Teodoro Bonati (who was in Rome from October 1764 to May 1765) had the opportunity to explain his point of view during the debate in front of the Holy Congregation of Waters and in some printed works (1765).[10] Bonati was already known in the cultural context of Rome and had proved his skilfulness with hydraulic computations. In 1763, in fact, he had performed some hydraulic experiments in the garden of Villa Medici in the presence of other mathematicians like François Jacquier (1711–1788) and Thomas Le Seur (1703–1770).[11] These activities allowed Bonati to reject a thesis maintained by

---

[7] Bertaglia had been already involved in the affair of the Pontine Marshes under the Pontificates of Benedict XIII (1729) and Clement XII (1732) together with the Bolognese astronomer Eustachio Manfredi (1674–1739), but their surveys didn't produce any prompt effect. In the 1760s, Bertaglia was called back to Rome. On this occasion he worked with Gabriele Manfredi, brother of Eustachio (who died in 1739). During the inspections, the two mathematicians were accompanied by the technicians Andrea Chiesa, Angelo Sani, and Teodoro Bonati. Their report was printed in the "Report on the Pontine Marshes" (Manfredi and Bertaglia 1785).

[8] The so-called "visits to the waters" (*visite alle acque*) were scientific inspections carried out by hydraulic consultants in order to study a specific hydraulic problem in the field and to find, when possible, a valid and shared solution. The *visite* were often named after the Cardinal who chaired them. For instance, the Visita Conti owes its name to the Cardinal Pietro Paolo Conti (1689–1770), who was commissioned by Pope Clement XIII to superintend the visit to the waters between the provinces of Bologna, Ferrara, and Ravenna and to find a solution for their safety. Many *Visite* took place during the 17th and the 18th centuries: the long debate on this subject is addressed in Lugaresi (2014).

[9] In the reports of the *Visite* these scholars are labelled "hydrostatic mathematicians" (*matematico idrostatico*) or "mathematical practitioner" (*matematico perito*) or "hydrostatic practitioner" (*perito idrostatico*), confirming once again the relevant role played by the practical aspect in this matter.

[10] Bonati published four papers in support of the Ferrarese thesis: Bonati (1765a, 1765b, 1765c, 1765d).

[11] Villa Medici was built on Pincio hill in Rome. The 18-acre garden is divided into sixteen squares and six parterres. Served by a thorough watering system, the garden was adorned with several

Claude Leopold Genneté (1706–1782). The French physicist had performed several experiments on river flow that dated back to the second half of the 18th century (Genneté 1760). The thesis by Genneté was that

> [...] a large river can absorb all the waters of another river as considerable as itself, without this increase causing the waters of the first river to rise in any way, the width of the bed of which remains the same as before. This happens because the increase having doubled the quantity of water in the river, it has also doubled the speed of its flow.[12]

Some years later, the same experiments were repeated by Bonati. The final result, which refuted Genneté's interpretation, was described by Bonati in Bonati (1765–1774).[13]

Genneté's experiments are very fallacious and it is very true, as has always been maintained in the past, that a river merging into another significantly raises the surface of the container and that the branching of a large body of water significantly lowers the surface of the branched river.[14]

The experiments performed by Bonati on the speed of confluent rivers were aimed at refuting some assumptions on which the Bolognese deputies based their theses for the introduction of the river Reno into the river Po.

As member of the Ferrarese delegation, during the sixth decade of the 18th century Bonati travelled a lot between Ferrara and Rome, where together with Romualdo Bertaglia he represented the interests of the city of Ferrara in the debate on the river

pools and fountains. The experiments of Bonati in the garden of Villa Medici were first performed in Ferrara in September 1762. The French mathematicians François Jacquier and Thomas Le Seur, who belonged to the order of Friar Minor, were sent to Rome to complete their studies in the French convent of the order at Trinità dei Monti. They were both professors at the Studium of Sapienza University. Jacquier was professor of experimental physics, while Le Seur was professor of mathematics.

[12] "un grand Fleuve peut absorber toutes les eaux d'un autre Fleuve aussi considerable que lui, sans que cette Accrue fasse hausser en rien les eaux du premier Fleuve, dont la largeur du lit reste la même qu'auparavant. La chose a lieu, parce que l'Accrue ayant double la quantité de l'eau du Fleuve, elle lui a aussi doublé la vitesse de son écoulement". It means that if you double the quantity of water in a channel, the height of the water remains the same, being balanced by an increase of speed. See Genneté (1760, p. iv).

[13] The Italian translation of Genneté's paper, together with the work by Bonati, was published in two editions of the "Collections on the motion of waters" (*Raccolta d'autori che trattano del moto dell'acque*). The collections on the motion of waters were a *unicum* in the scientific European landscape of the 18th and the first half of the 19th century. They were published in Italy in five editions, two in Florence (1723 and 1765–1774), one in Parma (1766–1768), and two in Bologna (1821–1826 and 1823–1845), and consisted of thirty-six volumes in total. The collections on the motion of waters have been studied in Lugaresi (2015). For the contributions of Genneté and Bonati to the collections on the motion of waters see Lugaresi (2015, pp. 222–224; 213–232).

[14] "gli Sperimenti del Sig. Genneté sono fallacissimi ed. è verissimo, come si è sempre sostenuto in passato, che un fiume aggiunto ad un altro alza sensibilmente il pelo del recipiente e che la diramazione di un grosso corpo d'acqua fa abbassare sensibilmente la superficie del fiume diramato" (Bonati 1765–1774, pp. 501–502).

Reno.[15] After Bertaglia's death (1763), Bonati became the main representative of the Ferrarese delegation in the Reno affair.

During his long life, Bonati was in touch with important representatives of the Italian scientific community, as attested by his frequent journeys and his rich correspondences with notable scholars, who were active between the second half of the 18th and the beginning of the 19th century. Among them we find Paolo Frisi (1728–1784), Gregorio Fontana (1735–1803), Giambattista Nicolai (1726–1793), Anton Maria Lorgna (1735–1796), Francesco Domenico Michelotti (1710–1787), Giuseppe Calandrelli (1749–1827), Sebastiano Canterzani (1734–1819), Giambattista Guglielmini (1760–1817).[16] In particular, Bonati shared with Michelotti an interest in research and experiments on the motion of waters. The two mathematicians had a brief but rich scientific correspondence soon after the publication (1767) of the first volume of Michelotti's treatise "Hydraulic experiments" (*Sperimenti idraulici*).[17]

6.3 Governing the Waters: A Social Issue

The case study of Teodoro Bonati as hydraulic consultant for the city of Ferrara exemplifies a more general situation that interested many Italian states. Managing waters was not just a matter of problems in mathematical physics but also a political issue. Since the 16th century, hydraulic disputes were extremely frequent and dealt with the regulation of rivers and streams, the reclamation of wide marshlands, the sediment of debris at the mouth of harbours or in the lagoon of Venice. Many Italian states decided to appoint their own hydraulic consultants to investigate the specific hydraulic problems of their territories. Having been strongly influenced by the Galilean tradition, the science of waters combined the more theoretical aspects of mathematical (geometrical) studies with the more practical tradition of physics studies.[18] In this discipline, theory and practice had to coexist and two different

[15] See Lugaresi (2014, pp. 213–235). The manuscripts by Teodoro Bonati are collected in twenty-four volumes, preserved in the Ariostea Library of Ferrara. Most of these volumes deal with hydraulic themes. A list of these hydraulic manuscripts has been transcribed in Lugaresi (2014, pp. 325–331).

[16] The correspondence of Teodoro Bonati is preserved in the 24th volume. For a description of the manuscripts and their contents see Borgato and Fiocca (1980). See also Borgato et al. (1992).

[17] Michelotti (1767–1771). The correspondence between Bonati and Michelotti was recently transcribed in Lugaresi (2017). On the scientific relationship between the two mathematicians see also Lugaresi (2021).

[18] The problems of water management gave birth to a dynamic debate that involved the Italian scientific community between the 17th and the 18th centuries. Theoretical and practical issues about the science of waters were collected in five editions of the "Collection on the motion of waters" (*Raccolta d'autori che trattano del moto dell'acque*). The main case-studies were represented by the problem of the regulation of the river Reno between the provinces of Bologna and Ferrara and the problem of the silting up of the Venetian Lagoon. See Lugaresi (2015). The dispute

professional figures became relevant. A technical consultant, usually a mathematician, was required to find solutions to problems about practical hydraulics, while a practitioner—the so-called "expert" (*perito*)—had to put these solutions into practice. The first Italian hydraulic consultants were members of the Galilean school. Galileo was the pioneer in whose footsteps followed many of his direct or indirect pupils like Benedetto Castelli, Evangelista Torricelli, Giovanni Alfonso Borelli, Vincenzo Viviani.[19] The character of the hydraulic consultant was typical of the Italian scientific context between the 17th and the 18th century. These scholars, who had a deep knowledge of mathematics, came from different backgrounds and received different training. Some of them studied medicine, like Teodoro Bonati, Domenico Guglielmini, and Bernardino Zendrini; others completed juridical studies before devoting themselves to mathematics (Geminiano Montanari, Eustachio Manfredi). Still others were appointed hydraulic consultants after a military career (Domenico Corradi d'Austria, Anton Maria Lorgna). A group of scholars belonged to religious orders and aided the development of hydraulic science with their lectures and treatises (Famiano Michelini, Guido Grandi, Nicolò Cabeo, Ruggiero Giuseppe Boscovich, Giovanni Antonio Lecchi, Leonardo Ximenes, Paolo Frisi).[20]

The superintendent of the waters also received a social and administrative acknowledgement in many different Italian states. For example, in the Grand Duchy of Tuscany the institution devoted to the superintendence and maintenance of rivers and streets was the so-called "Magistracy of the Captains of the Guelph Party" (*Magistratura dei capitani di parte guelfa*) (1549). In this institution the persons who were directly involved with the government of the waters were the "Officials of the rivers" and the "Officials of the ditches" (*Ufficiali dei Fiumi* and *Ufficiali dei Fossi*). The office for the superintendence of the waters in the Papal State was the "Holy Congregation of Waters" (*Sacra Congregatio Aquarum*), created by Pope Sistus V in 1587. Among the tasks of this institution was the reclamation and maintenance of the riverbanks of the Papal States, but also the control of harbours, mills, ducts, gutters, and canals. In the city of Bologna, a hydraulic board—the so-called "Assunti to the waters" (*Assunti alle acque*)—was founded in 1518 with the aim of governing the maintenance of rivers and hydraulic devices of the Bolognese plain. In the duchy of Ferrara, the hydraulic legislation was provided in 1580 by the "Orders and commissions for Workers and Officers of the river Po" (*Ordini et provigioni sopra i Lavorieri del Po e Ufficiali a quelli deputati*) under the jurisdiction of the Duke of Ferrara, Alfonso II d'Este. These measures dealt with the restoration

between Benedetto Castelli and the Venetian authorities on the management of the lagoon was studied in Omodeo et al. (2020).

[19] See Lugaresi (2014). One of the most relevant problems was to measure the velocity of waters. Because of the lack of valid instruments, empirical and local measurement methods were developed. The practitioners had to apply certain empirical formulas by adapting them to suit concrete situations. On the Galilean school see Monti (2011).

[20] See Lugaresi (2021, p. 133).

of riverbanks and the excavation and maintenance of public drainage canals.[21] The administrative personnel collaborated with local architects and engineers.

As regards the training of technical personnel with suitable hydraulic skills, the case study of Ferrara offers an interesting perspective on how the "public" university teaching of mathematics intersected with "private" mathematical teaching in the Jesuit College of the city. The university program of mathematics covered three years and included the study of the first six books of the *Elements* by Euclid, the *Sphaera* by Sacrobosco, and the *Theorica planetarum*. Amid these topics, only the *Elements* could provide some rudiments of geometry that were useful to hydraulic technicians, but practical applications of the theory were still missing.[22] In 1675, a "private" mathematical lecture was held at the Jesuit College of Ferrara not only with a theoretical aim, but also with a practical and professionalising purpose. This private lecture aimed to train the so-called "notaries" and "judges of the riverbank" (*notai* and *giudici d'argine*), public technicians devoted to environmental management. A useful description of this professional role is featured in the book "The idea of the perfect judge of the riverbank" (*L'idea del perfetto giudice d'argine*):

> The judge of the riverbank was assigned to the management of riverbanks. He had also to pay attention to pipes, canals, ditches, streets, bridges, gutters, underground barrel vaultings, sea harbours, water direction and control. So he had to be very well instructed in mathematics, especially arithmetic and geometry. [...] As regards to arithmetic, he had to know how to add, to subtract, to multiply, to divide both integer and rational numbers, to extract a square or cubic root, to know the golden ratio and the rule of three. As regards to geometry, he had to know the first six books of the Elements by Euclid, because they enlightened on the knowledge and the practice of their profession. [...] He had to know how to measure surfaces and solids, how to create and read a topographic map. He should also know theory and practice of levelling. [...] In short, a perfect judge of the riverbank should be both a good practical and theoretical expert and he should be able to account for his conduct.[23]

---

[21] Before the devolution to the Papal States, in the Duchy of Ferrara the management of the waters was entrusted to a "ducal engineer". After the devolution (1598), the Pope Clemente VIII confirmed the statutory rules. See Lugaresi (2021, pp. 140–142).

[22] Lugaresi (2021, pp. 146–148).

[23] "L'Uffitio di Giudice d'Argine ha nella sua denominatione l'espressione delle sue incumbenze, essendo egli destinato a giudicare i bisogni de gli Argini, e successivamente a ripararvi [...] sotto la sua carica sta la cura de condotti, canali, fosse, strade, ponti, chiaviche, ponti canali, botti sotterranee, porti di mare, sbocchi di fiumi e di condotti pubblici, direzione dei fiumi, regolamento delle acque del paese, per lo che gli conviene essere molto bene instrutto e prattico delle Matematiche in universale, e particolarmente dell'Aritmetica, e Geometria. [...] Per quello riguarda l'Aritmetica, dovrebbero saper sommare, sottrare, moltiplicare, e partir di numeri sani, & intieri, e di numeri rotti, & intieri unitamente, e separatamente, schisare, cavare le radici quadrate, e cube, & essere molto prattici della regola Aurea, o del tre. Nella Geometria dovrebbero essere bene esercitati, & havere studiati, & intesi almeno li primi sei libri degl'Elementi d'Euclide, essendo incredibile, quanto lume apportino alla cognitione, e prattica delle cose di loro professione [...] dovrebbero ancora sapere misurare le superfitie, & i corpi solidi, pigliare in pianta i Paesi, e ridurne in carta i dissegni in misura, livellare i siti, e formarne i profili in carta [...] In sintesi, i perfetti Giudici d'argini dovrebbero essere non solo buoni pratici, ma anche teorici, per saper rendere ragione del loro operare" (Lambresagni 1692, pp. 1–8).

This description clearly highlights the complementarity of practical and theoretical qualities that a valid judge of the riverbank should have. This professional figure received further acknowledgement in the 18th century and this led to the emergence of a new role, the "hydrostatic mathematician" (*Matematico idrostatico*). Teodoro Bonati well exemplified these abilities.

6.4 Formalizing Practical Experiments into a Theory: Bonati's Theoretical Works About the Science of Waters

The expertise achieved by Bonati during his activity as hydraulic consultant was summarised in some important works published between 1785 and 1811 in the "Memoirs of Mathematics and Physics of the Italian Society of the Sciences" (*Memorie di Matematica e Fisica della Società Italiana delle Scienze*). In 1782, the mathematician and hydraulic engineer Anton Maria Lorgna founded in Verona the "Italian Society of the Sciences" (*Società Italiana delle Scienze*) and invited the forty foremost scientists from every part of Italy to join this academy.[24] The list included the mathematicians Gianfrancesco Malfatti and Teodoro Bonati from Ferrara. The Society was founded with the aim of promoting the work of its members at a national level. The first *Memorie* featured six contributions by Bonati, five of them dealt with hydraulic items and one concerned the nature of the solutions for literal equations of fifth and sixth degree.[25]

An overview of Bonati's hydraulic papers allows us to understand how strictly connected the theoretical problems are with their practical application in the science of waters. In this framework, what clearly emerged was the importance of practical experiments in the study of the motion of waters, but also the difficulty in realising them.

In the "Essay about a new Theory for the motion of waters and a new Method for finding by experiments the quantity of water flowing in a river" (*Saggio di una Nuova Teoria del movimento delle acque pei Fiumi, e Nuovo metodo per trovare colla sperienza la quantità dell'acqua corrente per un fiume*), Bonati recalled the thesis of Benedetto Castelli, Domenico Guglielmini, and Guido Grandi in order to explain their fallacy with the support of his experiments on the rivers.[26]

---

[24]The official name of the Society is now "National Academy of the Sciences or of the XL" (*Accademia Nazionale delle Scienze detta dei XL*), also known as "Academy of the XL" (*Accademia dei XL*) due to the number of its members. On the figure of the mathematician Anton Maria Lorgna see Piva (1992) and Lorgna (1998).

[25]Bonati (1799a). In the first part of the work Bonati talked about the nature of the solutions for an equation, establishing the number of real and imaginary roots. In the second part he dealt with the approximated calculus of the roots of a numerical equation by using Newton's method to approximate solutions. See Borgato and Fiocca (1980).

[26]Bonati (1784). One year later, Gregorio Fontana, professor of mathematics at the University of Pavia, decided to put this important paper as appendix in his Italian translation of the hydrody-

> There are two forces that tend to continually accelerate the water along the river, the intrinsic one derives from the weight of every particle, the extrinsic one derives from the pressure of the water pushing on the other water. If the action of such forces were not disturbed, we would see the streams going faster and faster as we approached the mouth of the river. But actually, this doesn't happen because the speed decreases in approaching the mouth. This effect derives from the resistance caused by the water, first by the tortuosity of the river, and then by the roughness of the bottom and of the banks.[27]

In his paper, Bonati also described the experiments performed by other mathematicians and reflected on the use of hydraulic instruments for measuring waters, like the Pitot tube or the "hydrometric jug" (*fiasca idrometrica*) adopted by Guido Grandi and Eustachio Manfredi. Bonati's considerations relate directly to a second article that he devoted to the description of a new hydrometric instrument that he himself invented and called the "ritrometric rod" (*asta ritrometrica*), a rod for flow measurement.[28]

> If we add a metal part CB to a wooden rod AB such as putting the entire rod in a stagnant water it floats with a part AD that stay perpendicularly and outside the water of one or more foot long, then we'll have a ritrometric rod. In my paper of 1784, I showed that if the same rod AB placed in a running water from M to N will be taken with the part AD leaning forward, the greater speed will be on the surface; otherwise, the opposite occurs if the part AD will lean backwards.[29]

Bonati's research on phoronomy[30] and the measurement of the speed of water coming out from a vessel was published in 1790.[31] In the paper, he recalled the results achieved by other important European mathematicians such as Edme Mariotte (1620–1684), Willem Jacob's Gravesande (1688–1742), Daniel Bernoulli

namic treatise by Charles Bossut, "Elementary treatise of hydrodynamics by Bossut" (Fontana 1785). For the paper of Bonati see pp. 499–570.

[27] "Queste due forze, una intrinseca, che deriva dal proprio peso di ogni particella, e l'altra estrinseca, che deriva dalla pressione dell'acqua insistente all'acqua, e nessun'altra sono le forze, che tendono ad accelerare l'acqua lungo il fiume continuamente, cosicché se la loro azione non venisse disturbata, noi vedremmo i fiumi sempre più veloci a mano a mano che ci avviciniamo allo sbocco. Ma questo realmente non accade, giacché allontanandoci dalla origine, ed arrivati per esempio alla pianura, dove il fondo è tuttora declive, benché meno, osserviamo la velocità scemata di molto, e talvolta la vediamo scemata anche vie più prima di arrivare allo sbocco. Quest'effetto deriva dalle resistenze, che fanno all'acqua prima le tortuosità del fiume, e poi le scabrezze del fondo, e delle sponde" (Bonati 1784, pp. 681–682).

[28] Bonati (1799b).

[29] "Se ad un'asta AC di legno più leggiero dell'acqua si aggiunga una tal porzione CB di metallo, che mettendo tutta l'asta in una acqua stagnante essa galleggi con una porzione AD di un piede o più, fuori dell'acqua ed. a piombo, si avrà una delle Aste ritrometriche da me proposte nel 1784 in questa Raccolta per iscoprire le velocità sotto la superficie dell'acqua nei fiumi. Mostrai, che se la stess'asta AB messa in un'acqua corrente da M verso N verrà portata colla porzione AD inclinata all'avanti, la velocità maggiore starà alla superficie; e che si avrà il contrario se la parte AD penderà all'indietro" (Bonati 1799b, p. 435). The description refers to Fig. 6.1 in this paper.

[30] Phoronomy is the theory that studies the efflux of water through small openings made on the walls of a vessel. The theory also gave the name to the famous treatise by Jacob Hermann (1716).

[31] Bonati (1790).

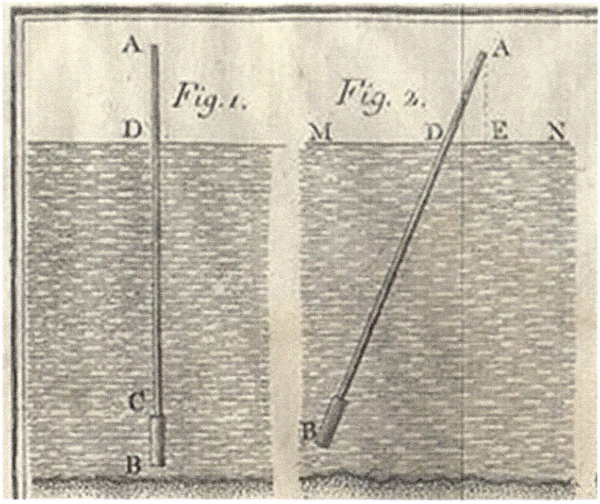

Fig. 6.1 The ritrometric rod (Bonati 1799b, p. 444)

(1700–1782), Charles Bossut (1730–1814); then he presented the experiments that he performed using different vessels and considering different kinds of apertures and fluids (not only water, but also air and mercury). As he remarked, the results had also a practical interest if applied to a furnace for melting iron ore.

> So, you know the artifice with which a violent and uniform breath is generated in some ironworks in order to keep the furnace turned on for melting iron ore. The water passes through a pipe AB to go into a recipient LNOM, from which it flows down through many pipes CE. When it reaches the plane F, it fills the space GK until the level IP and leaks out through the emissary KR. Air comes in through the holes u, made in the side of each canal (that's why it's called a flute), and it goes out from the mouth E because it is dragged by the descending water. When air reaches the plane F, it separates itself from the water. The chamber GQKH is completely closed except for two holes K, T. Air can't go out through the hole K that is always occupied by the water, so it goes out through the other hole T and through the pipe TV it violently and uniformly breathes inside the furnace.[32]

Another brief, but interesting essay was published by Bonati in 1804 and dealt with some experiments that he performed by using a small wooden channel whose banks

[32] "Si conosce quindi in che consista l'artificio, ond'eccitano con gran profitto in alcune Ferriere un soffio violento, ed equabile per mantenere accesa la fornace per fondere la vena del ferro. Per un condotto AB va l'acqua in una vasca LNOM, d'onde scende per più canali CE di tavole, divergenti, ed all'incontro dei sottoposti piani F spiccia tutt'all'intorno, empie la camera GK fino in IP, ed esce per un emissario KR. Pei fori u, fatti in un lato d'ogni canale (detto perciò flauto) entra l'aria per la ragione spiegata, e trascinata dall'acqua che discende, esce per le bocche E, ed all'urto coi piani F si separa dall'acqua: e trovandoli in una camera GQKH tutta ben chiusa a riserva di due fori K, T, né potendo uscire pel foro K sempre occupato dall'acqua, esce per l'altro T, e pel tubo TV va a soffiare con violenza, ed equabilmente nella fornace" (Bonati 1790, p. 524). The description refers to Fig. 6.2 in this paper.

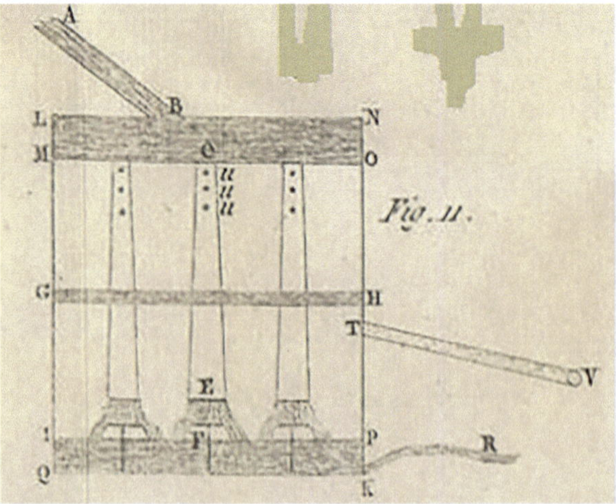

Fig. 6.2 The scheme of a conduct for the transport of a fluid into many pipes (Bonati 1790, p. 524)

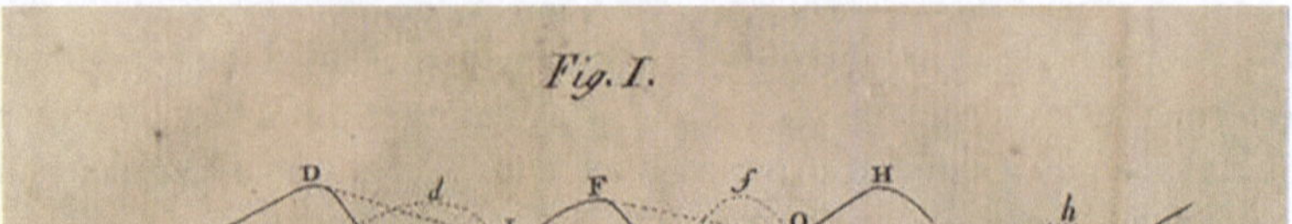

Fig. 6.3 The curve described by the sand in the channel (Bonati 1804)

were made of glass. Bonati placed a layer of sand at the bottom of the channel, then he let water flow and observed the effects. The sand settled in many hillocks (*CDE*, *EFG*, *GHI*) and after some time it described the curve *cdefghi* (Fig. 6.3).[33]

Bonati observed a similar effect on a large scale in the Castagnaro, a diversion channel of the river Adige, soon after a flood of the Adige occurred in September 1767:

> In September 1767, there was a great flood of the Adige. As a result, the channels Castagnaro and Canal Bianco flooded and two breaches occurred in the right bank of Castagnaro. This bank was made of sand, so the breaches became larger in a very short time. Thus, all the water of the river went towards the breaches leaving the inferior riverbed completely without water. The bottom was in the same condition as the hillocks CDE, EFG, previously described.[34]

[33] Bonati (1804). The label *e* should be where the diagram has the label *Q*.

[34] "Nel Settembre 1767 comparve l'Adige in grossa piena, che superò tutte le precedenti coll'altezza di mezzo piede. In conseguenza furono in grossa piena il Castagnaro, ed il Canal Bianco, e la mattina del dì 17 seguirono due rotte dell'argine destro del Castagnaro. Quell'argine era quasi tutto di arena, e perciò gli squarcj si fecero grandi in un tempo assai breve, cosicché tutta l'acqua del fiume

The flood of the Adige in 1767 was described by A. M. Lorgna in "Discourse on repairing the city of Verona from the floods of the river Adige" (*Discorso intorno al riparare dalle inondazioni dell'Adige la città di Verona*).[35] The control of the river Adige was included in the greatest water affairs of the Republic of Venice.[36] Many hydraulic scholars took part in this debate through the centuries: this item was also discussed by Bonati in his correspondence with Lorgna.[37]

Bonati was not only an able scholar but also a careful reader of the hydraulic works of other mathematicians. Some of his critical notes were published, others remained unpublished and are preserved together with his other manuscripts at the Ariostea Library in Ferrara. A detailed reading of the treatise "New principles of hydraulics" (*Nouveaux Principes d'Hydraulique*) encouraged Bonati to write and publish a paper to refute the theory of the French scholar Pons-Joseph Bernard (1748–1816) about phoronomy.[38] Bonati created a device aimed at introducing water into a vessel with a predetermined starting speed. Thanks to many other experiments, Bonati could prove that:

- a fluid can come out of an opening made in the bottom of a vessel that is kept constantly full of fluid with a significantly greater speed than that of a freely falling body from the same height;
- as the opening increases, the speed increases.[39]

---

prese corso per le rotte lasciando l'alveo inferiore affatto senz'acqua, onde si potè vedere il fondo nello stesso stato, in cui si era trovato nel colmo della piena prima delle rotte. Lo vidi pertanto con delle prominenze, o siano monticelli affatto consimili ai descritti CDE, EFG, ..., e con delle cavità simili alle E, G, ... profonde alcune due piedi, ed anche più" (Bonati 1804, pp. 680–681).

[35] See Lorgna (1768). Lorgna's paper was included in three editions of the Collection on the motion of waters. See: Lugaresi (2015, pp. 235–236; 254; 268).

[36] Water affairs always played a leading role for the Republic of Venice. The debate dealt with cuts and diversions of rivers from the lagoon, the maintenance of some harbours, the excavation of canals, the control of the rivers Adige, Brenta, and Piave, canalizations for internal navigation, and so on. Geminiano Montanari (1633–1687), who was professor of mathematics in Bologna and then professor of hydraulics in Padua, was the first mathematician who collaborated with the "Magistrate to the Waters" (*Magistrato alle acque*) of the Republic of Venice in the late 17th century. The Republic of Venice began to frequently ask for consultation by renowned scholars, engineers, and hydraulic experts in order to preserve the lagoon and the rivers that flowed into it (Domenico Guglielmini, Bernardino Zendrini, Giovanni Poleni). See Maffioli (1994, pp. 152–162; 273–296). On the history of the Po, Polesine, and Adige basin and its regulation, see Luzzini (2016).

[37] See Borgato et al. (1992, pp. 87–89). Lorgna was a valid interlocutor for Bonati, not only for his participation in the Society of the Sciences, but also for his experience in the field of hydraulics. The scientific relationship between Bonati and the mathematicians of the Republic of Venice has been studied in Borgato and Fiocca (1994).

[38] Bonati (1811). Pons-Joseph Bernard was the author of many papers in agronomy, botany, and hydraulic sciences and wrote a mineralogical description of Provence, which was partly published. Bonati's notes referred to the work entitled *Nouveaux Principes D'Hydraulique* (Bernard 1787).

[39] "due conseguenze spontanee ne discesero, che può da un foro praticato nel fondo di un vaso mantenuto costantemente pieno di fluido ad una data altezza sortir esso con una velocità sensibilmente maggiore di quella di un grave da pari altezza liberamente caduto. Che al crescer del foro cresce realmente la velocità" (Lombardi 1823–1825, p. CLXXXV).

The field experiments were performed by Bonati together with his pupil and assistant Luigi Gozzi, described by Bonati as a "skilled assistant" (*abile mio sostituto*).[40]

Bonati was involved in a scientific dispute with Francesco Maria Gaudio (1726–1793), professor of mathematics at the Studium of Sapienza in Rome. Gaudio wrote a polemical paper (Gaudio 1786a), to criticize Bonati's hydraulic theory. Gaudio's objections referred to Bonati's memoir on the speed of running waters in rivers, published in 1784. In reply to this work, Bonati published a paper (Bonati 1786a), where he proved his deep knowledge of the most important hydrodynamical works of his time—he quoted the treatises by Daniel Bernoulli, Euler, and Bossut.[41]

Among the manuscripts of Bonati in Ferrara there are some notes about the hydraulic works of Anton Maria Lorgna and Pierre Louis Georges Du Buat (1734–1809).[42]

6.5 Bonati's Academic Career at the University of Ferrara

Starting at the end of the 1760s, Bonati's scientific acknowledgement grew. He was often asked for hydraulic advice in various parts of Italy. Among the most important interventions was a project for the reclamation of some marshes (Burana) including those between the Secchia, Po, and Panaro rivers in an area parts of which were controlled by the Duchy of Modena and Reggio, Austrian Lombardy, and the Legation of Ferrara.[43] During the 1770s, Bonati was in Mantova (1771) to visit the lock (*sostegno*) of Governolo, a village at the confluence of the river Mincio with the river Po. He was also in Parma and Piacenza (1774) to settle a controversy regarding the streams Tidone and Luretta. Later on, he moved to Piombino (1779) to examine the boundaries between this Italian state and Tuscany.

At the apex of his professional career, Bonati also received academic acknowledgement with the appointment as professor of hydraulics and mechanics at the

[40] Luigi Gozzi was assistant of Bonati during his teaching at the University of Ferrara, from 1772 to 1804, then at the "Special School of Hydrostatics" (*Scuola Speciale di Idrostatica*) which was opened in Ferrara in 1804, soon after the dissolution of the Studium. The lessons of hydraulics began in November 1804. In the absence of Bonati, the first lesson was held by Luigi Gozzi. See Fiocca and Pepe (1986).

[41] The dispute between the two mathematicians continued with the publication of two other works: Gaudio (1786b) and Bonati (1786b).

[42] Bonati, Teodoro. *Osservazioni sulla Idrodinamica del Du Buat, accompagnate da esperienze e calcoli*. Autograph of c. 33. Idem, *Osservazioni sulle ricerche del Lorgna intorno alla distribuzione delle velocità dell'acqua nelle sezioni dei fiumi*. Autograph of c. 4. Ariostea Library of Ferrara, Manuscripts of Teodoro Bonati, vol. VII. Du Buat's major work was *Principes d'hydraulique* (Du Buat 1779); the treatise had a second edition (Du Buat 1786); a third edition was published posthumously (Du Buat 1816). As regards Bonati's notes to the work of Lorgna, he referred to the paper (Lorgna 1771).

[43] Between 1771 and 1774 many congresses took place in Bondeno to discuss the projects, but no executive decision was reached (Lugaresi 2014, pp. 323–324).

University of Ferrara. The chair was founded soon after the reform of the university in 1771. The study of mathematics in Ferrara developed in close connection with the study of hydraulics. Because of its geography, the city had paid particular attention to the advancements in hydraulics and technology throughout its history, in order to solve its many problems related to the management of rivers and land surveying. Developments in mathematics during the 17th and the 18th centuries led to the emergence of a new mathematical discipline, the hydrodynamics, and with it arose the need of a new professional figure, the "Hydrostatic Mathematician" (*Matematico Idrostatico*), to complement the simple technical expert. According to the plan for the reform of the university, the study of mathematics provided for a curriculum aimed to prepare high-qualified people for working on the regulation of waters. Gianfrancesco Malfatti presented an interesting proposal to renew the teaching of mathematics in Ferrara, taking into account both the more theoretical aspects of the discipline and the more applied ones related to the science of waters.

In the science of waters, Malfatti said, there are few general theorems. The main part of this subject was still uncertain and obscure so that every town had to conduct specific examinations of the surrounding rivers and streams in order to form the most likely conjectures from direct experiments and observations. For this reason, the teaching of mathematics should focus on the rivers in the province of Ferrara, so that the students could learn from direct experience in the field:

> And there were few general theorems about the science of waters, the rest remaining covered by a disgraceful uncertainty and obscurity so that each Province has to examine the nature of its rivers and streams in order to learn from experience and observations the more likely conjectures. Thus, after having dealt with flowing waters and rivers in general, we should examine in depth the rivers of the Ferrarese plain, and students should keep the map at hand in order to know the origin, the path, the fall, and the mouths of the rivers, the quality of the waters, the usual time of their overflows and everything else about the way to regulate them well, to provide them for good embankments &c. [...] It should be illustrated by means of physical and mathematical reasons, relying on the less doubtful theories.[44]

The training of the hydrostatic mathematician would last four years: the subjects were "algebra and mechanics" (plane and solid geometry, trigonometry, Cartesian algebra, conic sections, sublime calculus, and mechanics) in the former two years and "hydrostatics and hydrodynamics" in the latter two years. In the academic year 1771–72 there was a single lecture of mathematics (that included mathematics and hydrostatics), held by Gianfrancesco Malfatti. From the academic year 1773–74, there were three chairs, held by different professors. The chair of algebra and

[44] "E siccome pochi sono i Teoremi generali della Scienza delle Acque, rimanendo il restante involto in una deplorabile incertezza ed oscurità a segno, che ciascuna Provincia è obbligata, all'occasione, di fare degli esami particolari sulla natura de' suoi Fiumi e Torrenti, per ricavare dalla esperienza e dalla osservazione le più probabili congetture, così, dopo aver trattato delle Acque correnti, e de' Fiumi in generale, dovrebbe più particolarmente estendersi a parlare dei Fiumi del Ferrarese, instruendo i Giovani colle Mappe sotto gli occhi della loro origine, corso, caduta, sbocco, qualità di acque che portano, tempo ordinario delle loro piene, e di tutto ciò, che riguarda la maniera di ben regolarli, munirli di buone arginature &c: [...] illustrando ogni cosa con ragioni Fisiche e Matematiche, ed appoggiando tutto alle men dubbie Teorie" (Malfatti 1982, p. 10).

mechanics, held by Malfatti, covered Euclidean geometry, Archimedes' theorems, trigonometry, conic sections, Cartesian algebra, differential and integral calculus, mechanics. The chair of hydrostatics, held by Teodoro Bonati, covered mathematical formulas for hydrodynamics, general principles for running waters, the geography of rivers in the Ferrara province, rules for the drawing of maps and the building of embankments. The course on hydrostatics was also meant to include some practical exercitations on the field, and for this reason it was assigned to a professor who was very familiar with the territory. The teaching of the chair of practical geometry, held by Ambrogio Baruffaldi, focused on how to use geodetic instruments for measuring.[45]

The years spent in academia turned out to be the most scientifically productive for Bonati. In 1803, during the Napoleonic era, the University of Ferrara was closed. In April 1804, the so-called "Special School of Hydrostatics" (*Scuola Speciale di Idrostatica*) was established in Ferrara and Bonati was appointed professor in this institution. In 1809, Bonati retired. He spent the final years of his life in Ferrara, where he died in 1820.

6.6 Conclusions

In the science of waters, theory and practice were complementary. The problem of measuring the speed of waters required a great number of experiments that had to be performed with suitable technical instruments. Unfortunately, the motion of water could not be described with the equations of hydrodynamics, because they referred to the motion of ideal fluids and didn't take into account the whirling of waters in rivers and canals. In few cases the movement of water was similar to that of an ideal fluid, but a great number of problems couldn't be solved because of a lack of proper mathematical instruments. The hydraulic technicians, such as Bonati, had to apply certain empirical formulas that required adapting to concrete situations.[46]

In 1808, Jean-Baptiste Delambre presented to Napoleon a report on the progress of mathematics after 1789. In reference to hydraulics, he underlined the importance of this subject:

---

[45] See Lugaresi (2021, pp. 147–149). On the reform of the University of Ferrara see also Lugaresi (2022).

[46] During the 17th and the 18th century, many Italian scholars of hydraulics suggested and tested new instruments to measure flow velocity at different depths. Among these instruments we can quote the so-called "pendulum ball with quadrant" (*palla a pendolo con quadrante*) used by Domenico Guglielmini or "Michelotti's palmette wheel" (*ruota con palmette di Michelotti*), that was tested by the Piedmontese Michelotti. By counting the number of turns made by the wheel during a given time period, this device measured the velocity of the flow at the point where it hit the palmette of the wheel. One of the oldest floating hydraulic speedometers was the hydrometric rod. In this class of instruments there was also the so-called "ritrometric rod", invented and used by Bonati (Lugaresi 2021, p. 137).

> Hydraulics is one of the most important branches of mathematical and physical sciences applied to civil uses; it would require from those who devote themselves to it the deepest and most varied knowledge, the principles of calculus and rational mechanics, which are only found in the works of scholars, and the theory of the arts, which can only be acquired in the ateliers and through frequent communication with the workers who must be directed and monitored [...] thus the immense progress that rational mechanics had made thanks to Euler, d'Alembert, Lagrange and Laplace, remained without real application to construction art.[47]

Bonati exemplified this double education in the field of theoretical and practical hydraulics and his high reputation was very well known in Europe. For example, Gaspard de Prony praised him in the biography he wrote for the *Biographie Universelle*.[48] Furthermore, many Italian scholars of the 19th century, like Francesco Mengotti (1749–1830), Giuseppe Venturoli (1768–1846), Carlo Sereni (1786–1868) quoted Bonati's hydraulic work in their treatises.[49]

Bonati knew how to combine theoretical knowledge with the rules drawn from experience and in many cases the results of the experiments that he carried out proved to be effective in describing real phenomena.

References

Bernard, Pons-Joseph. 1787. *Nouveaux Principes D'Hydraulique*. Paris: chez Didot.

Bonati, Teodoro. 1765a. *Alla S. C. A. l'E.mo e R.mo Sig. Card. Conti ponente per l'Ill.ma Città e Ducato di Ferrara, Memoriale Idrometrico*. Roma: per il Bernabò.

Bonati, Teodoro. 1765b. *Alla S. C. A. l'E.mo e R.mo Sig. Card. Conti ponente Bononien., seu Ferrarien. Aquarum per l'Ill.ma Città e Ducato di Ferrara, Risposta idrometrica*. Roma: per il Bernabò.

Bonati, Teodoro. 1765c. *Alla S. C. A. l'E.mo e R.mo Sig. Card. Conti ponente Bononien., seu Ferrarien. Aquarum per l'Ill.ma Città e Ducato di Ferrara, Sommario della risposta idrometrica*. Roma: per il Bernabò.

Bonati, Teodoro. 1765d. *Alla S. C. A. per l'Illustrissima Città di Ferrara, Annotazione alla risposta del Sig. Marescotti*. Roma: per il Bernabò.

Bonati, Teodoro. 1765–1774. Sperienze del Signor Teodoro Bonati in confutazione delle sperienze del signor Genneté intorno al corso de' fiumi. In *Raccolta d'autori che trattano del moto dell'acque*, vol. 6, 501–552. Firenze: Cambiagi.

[47] "L'Hydraulique est une des branches les plus importantes des sciences mathématiques et physiques appliquées aux usage civils; elle exigeroit de ceux qui s'y dévouent les connoissances les plus profondes et les plus variées, les principes de calcul et de mécanique rationnelle, qui ne se trouvent que dans les ouvrages des savans, et la théorie des arts, qui ne s'acquiert que dans les ateliers et par un commerce fréquent avec les ouvriers qu'il faut diriger et surveiller [...] ainsi les progrès immenses que la mécanique rationnelle avoit faits entre les mains d'Euler, de d'Alembert, et de MM. Lagrange et Laplace, étoient restés sans application réelle aux arts de construction" (Delambre 1810, pp. 227–228).

[48] Prony (1835, pp. 535–537).

[49] See Borgato et al. (1992, pp. 32–34).

Bonati, Teodoro. 1784. Saggio di una Nuova Teoria del movimento delle acque pei Fiumi, e Nuovo metodo per trovare colla sperienza la quantità dell'acqua corrente per un fiume. *Memorie di matematica e fisica della società italiana* 2:676–719.

Bonati, Teodoro. 1786a. *Replica al discorso di F.M.G. pubblicato in Roma in quest'anno 1786 contro due memorie intorno ai fiumi.* Without typographical references.

Bonati, Teodoro. 1786b. *Seconda replica al P.F.M. Gaudio*. Without typographical references.

Bonati, Teodoro. 1790. Della velocità dell'acqua per un foro nel fondo di un vaso che abbia uno o più diaframmi: e della velocità pure dell'acqua per un tubo verticale cilindrico o divergente annesso a un foro nel fondo di un vaso semplice: e del soffio che si procura nelle fornaci di alcune ferriere col mezzo dell'acqua. *Memorie di matematica e fisica della società italiana* 5:501–524.

Bonati, Teodoro. 1799a. Natura delle radici delle equazioni litterali di quinto, e di sesto grado, e nuovo metodo per le radici prossime delle equazioni numeriche di qualunque grado. *Memorie di matematica e fisica della società italiana* 8 (1): 231–272.

Bonati, Teodoro. 1799b. Delle Aste Ritrometriche e di un nuovo Pendolo per trovare la Scala delle Velocità di un'Acqua corrente. *Memorie di matematica e fisica della società italiana* 8:435–444.

Bonati, Teodoro. 1804. Sperienze ed Osservazioni potamologiche. *Memorie di matematica e fisica della società italiana* 11:680–682.

Bonati, Teodoro. 1811. Alcune riflessioni critiche sui nuovi principi d'Idraulica di Mr. Bernard pubblicati in Parigi nel MDCCXXCVII. *Memorie di matematica e fisica della società italiana* 15:226–253.

Borgato, Maria Teresa, and Fiocca, Alessandra. 1980. Sugli scritti matematici di Teodoro Bonati. *Studi sulla Civiltà del secolo* XVIII:19–46.

Borgato, Maria Teresa, and Fiocca, Alessandra. 1994. Teodoro Bonati e i suoi corrispondenti veneti. In *Le scienze matematiche nel Veneto dell'Ottocento*, 101–150. Venezia: Istituto Veneto di Scienze, Lettere ed Arti.

Borgato, Maria Teresa, Fiocca, Alessandra, and Pepe, Luigi, eds. 1992. *Teodoro Bonati. Carteggio scientifico: Lorgna, Canterzani, Frisi, Saladini, Calandrelli, Venturi*. Firenze: Olschki.

Cormack, Lesley B., Walton, Steven A., and Schuster, John Andrew. 2017. *Mathematical practitioners and the transformation of natural knowledge in early modern Europe*. Cham: Springer.

Delambre, Jean-Baptiste. 1810. *Rapport historique sur les progrès des sciences mathématiques depuis 1789 et sur leur état actuel*. Paris: Imprimerie Impériale.

Du Buat, Pierre Louis Georges. 1779. *Principes d'hydraulique*. Paris: de l'imprimerie de Monsieur.

Du Buat, Pierre Louis Georges. 1786. *Principes d'hydraulique, vérifiés par un grand nombre d'expériences*. Paris: de l'imprimerie de Monsieur. 2 vols.

Du Buat, Pierre Louis Georges. 1816. *Principes d'hydraulique et de hydrodynamique, vérifiés par un grand nombre d'expériences*. Paris: chez Firmin Didot. 3 vols.

Fiocca, Alessandra, and Pepe, Luigi. 1986. L'Università e le scuole per gli Ingegneri a Ferrara. *Annali dell'Università di Ferrara* 32:125–166.

Fiocca, Alessandra, ed. 1998. *Giambattista Aleotti (1546–1636) e gli ingegneri del Rinascimento*. Firenze: Olschki.

Fiocca, Alessandra, Lamberini, Daniela, and Maffioli, Cesare, eds. 2003. *Arte e scienza delle acque nel Rinascimento*. Venezia: Marsilio.

Fontana, Gregorio, ed. 1785. *Trattato elementare d'idrodinamica del sig. abate Bossut tradotto dal francese*. Pavia: Stamperia del Monastero di S. Salvatore.

Gaudio, Francesco M. 1786a. *Discorso di F.M.G. professore nella Sapienza di Roma sull'azione delle potenze nelle macchine e su gli errori. Intorno alla medesima, ed altri contenuti nella Replica colla data dei 20 Marzo 1786 del Sig. Teodoro Bonati*. Roma: Stamperia Barbiellini.

Gaudio, Francesco M. 1786b. *Discorso di F.M.G. professore nella Sapienza di Roma sull'azione delle potenze nelle macchine e su gli errori. Intorno alla medesima, ed altri contenuti nella Replica colla data dei 20 Marzo 1786 del Sig. Teodoro Bonati*. Roma: Stamperia Zampel.

Genneté, Claude Leopold. 1760. *Expériences sur le cours des fleuves*. Paris: Lambert.

Hahn, Roger. 1965. L'Hydrodynamique au XVIIIe siècle. Aspects scientifiques et sociologiques. *Les Conférences du Palais de la Découverte*, Série D: Histoire des Sciences, n. 100.
Hermann, Jacob. 1716. *Phoronomia sive De viribus et motibus corporum solidorum et fluidorum libri duo*. Amstelaedami: apud Rod. & Gerh. Wetstenios.
Lambresagni, Petronio. 1692. *L'idea del perfetto giudice d'argine*. Ferrara: Camerale.
Lombardi, Antonio. 1823–1825. Elogio storico del Cavalier Teodoro Massimo Bonati. *Rendiconti dell'Accademia Nazionale delle Scienze detta dei XL* 19: CLXXII–CLXXXIX: CLXXIII.
Lorgna, Anton M. 1768. *Discorso intorno al riparare dalle inondazioni dell'Adige la città di Verona*. Verona: Stamperia Moroni.
Lorgna, Anton M. 1771. *Ricerche intorno alla distribuzione delle velocità nelle sezioni de' fiumi*. Verona: Stamperia Moroni.
Lorgna, Anton M. 1998. *Scienziato ed accademico del XVIII secolo tra conservazione e novità*. Roma: Accademia nazionale delle scienze detta dei XL.
Lugaresi, Maria Giulia. 2014. *Idrodinamica e idraulica. Le Raccolte sul moto delle acque. La questione del Reno*, Tesi di Dottorato, Università di Ferrara.
Lugaresi, Maria Giulia. 2015. Le raccolte italiane sul moto delle acque. *Bollettino di Storia delle Scienze Matematiche* 35 (2): 201–304. ISSN 0392-4432.
Lugaresi, Maria Giulia. 2017. *Vita scientifica di Giorgio Bidone. Torino dopo Lagrange*. Torino: Quaderni della Fondazione Filippo Burzio, Collana Studi e Ricerche.
Lugaresi, Maria Giulia. 2020. Il carteggio Gianfrancesco Malfatti – Giordano Riccati. *Bollettino di Storia delle Scienze Matematiche* 40 (1): 1–296. https://doi.org/10.19272/202009201001.
Lugaresi, Maria Giulia. 2021. Mathematics and hydraulics between Turin and Ferrara in the 18th century: the works by F. D. Michelotti and T. Bonati. *Revue d'histoire des mathématiques* 27 (2): 127–169. https://doi.org/10.24033/rhm.233.
Lugaresi, Maria Giulia. 2022. I professori ferraresi e la questione del Reno. In *L'Università di Ferrara nell'età del riformismo pontificio e in epoca napoleonica, Supplemento al Vol. 99 degli Atti dell'Accademia delle Scienze di Ferrara*, 141–162. Ferrara: ETS.
Luzzini, Francesco. 2016. "L'Itale Terre a vagheggiare inteso". La regolazione dell'Adige nel XVIII secolo: tra storia e scienza. In *Il fiume, le terre, l'immaginario. L'Adige come fenomeno storiografico complesso*, ed. Vito Rovigo, 287–312. Rovereto: Osiride.
Maffioli, Cesare. 1994. *Out of Galileo: the science of waters 1628–1718*. Rotterdam: Erasmus Publishing.
Maffioli, Cesare. 2010. *La via delle acque (1500–1700): appropriazione delle arti e trasformazione delle matematiche*. Firenze: Olschki.
Malfatti, Gianfrancesco. 1982. *Gianfrancesco Malfatti nella cultura del suo tempo, atti del convegno di studi*. Bologna: Monograf.
Manfredi, Gabriele, and Bertaglia, Romualdo. 1785. Relazione delle Paduli Pontine de' Sigg. Gabriele Manfredi, e Romualdo Bertaglia, da essi firmata in Roma l'anno 1761. In *Raccolta delle perizie ed opuscoli idraulici del sig. Abate Leonardo Ximenes*, ed. Leonardo Ximenes, 7–72. Firenze: stamperia di Pietro Allegrini.
Michelotti, Francesco Domenico. 1767–1771. *Sperimenti idraulici principalmente diretti a confermare la teorica, e facilitare la pratica del misurare le acque correnti*. Torino: Stamperia Reale.
Monti, Maria Teresa. 2011. *La tradizione galileiana e lo sperimentalismo naturalistico d'Età Moderna. Pratiche, teorie, linguaggi*. Firenze: Olschki.
Omodeo, Pietro Daniel, Trevisani, Sebastiano, and Babu, Senthil. 2020. Benedetto Castelli's Considerations on the Lagoon of Venice. *Earth Sciences History* 39 (2): 420–446.
Piva, Franco. 1992. *Anton Maria Lorgna: la biblioteca di uno scienziato settecentesco*. Firenze: Olschki.
Prony, Gaspard. 1835. Biographie universelle. *ancienne et moderne* 58:535–537.
Truesdell, Clifford. 1954. Rational fluid mechanics. 1657–1765. In: Euler, Leonhard, *Opera omnia*, ser. 2, vol. 12, pp. IX–CXXV.

Chapter 7
Sawing Measure: Thingking Wood Problems in Malayalam Kaṇakkatikāram and Other Indian Mathematical Treatises

Arun Ashokan

Abstract This paper shows how wood-related mathematical knowledge evolves through four interconnected networks of interaction—ecology, timber production, trade and use of wood, and pedagogy—which link humans (sawyers, carpenters, traders, teachers, etc.) and nonhumans (trees, timber, handsaws, etc.). To do so, this paper analyses measures, procedures, and rules in wood calculation problems in the medieval Malayalam mathematical treatises Kaṇakkatikāram and other Indian mathematical treatises. To understand the historical formation of mathematical knowledge, particularly in relation with wood calculations, this paper offers a theoretical framework that I call "Thingking". This framework shows the formation of mathematical knowledge through the interaction of bodily actions (motions and affects) of humans and the dynamics of materials (forces, properties, and affordances). This paper argues that mathematical knowledge of wood-related problems in these treatises is not a product of introspection, but a product of human practice in the material world. The nonhuman entities not only form or reform mathematical knowledge, but also decenter human agency in the formation of mathematical knowledge.

Keywords Network · Agency · Human · Nonhuman · History · Mathematics · India · Malayalam · Wood

Thingking, *Thingk*, and *Thingker* are three interconnected neologisms that emerged from the engagement with the materials of this paper. These terms are explained in the opening section.

A. Ashokan (✉)
Max Planck Institute for the History of Science, Berlin, Germany

Senthil Babu D., R. Wagner (eds.), *Normative Texts and Occupational Practices*, Boston Studies in the Philosophy and History of Science 350,
https://doi.org/10.1007/978-3-032-28419-8_7

7.1 Introduction

Let us start with a motivational anecdote that led me to the problem of this paper. The word *Īrccakkār* means 'sawyers' in Malayalam.[1] In Kerala, especially in some parts of Kozhikode, this word is colloquially used for connoting homosexuals,[2] in distinction from homosocials.[3] Whenever I heard this derogatory expression, it made me wonder about the emergence of this usage. My first guess was that it had emerged from the social context of double-handed sawing of wood. In Kerala, the act of hand sawing is an affair between two men, which can be conceived as partly reflecting the image of a homosexual relation. A big wood log cannot be sawn without the joint bodily effort of two male sawyers. Intimate coordination of their bodily acts and mutual understanding are necessary for performing this labor.

At that time, my thinking focused on social relations of labor (in this case, sawing) and its impacts on the emergence of the metaphor for homosexuality. But after engaging with wood problems in the Malayalam *Kaṇakkatikāram*—a genre of late medieval Malayalam elementary mathematical treaties which deal with calculations of land, rice, gold, wood, etc. (see Wagner and Ashokan 2024, 1–31)—I was induced to rework my labor-centric perspective. Since the names of the different units of wood and the designations of "handsaw" appear in these problems, I had to take into account their roles as well. They manifest themselves not merely as a material and a tool in sawing work, but also as conceptual categories, such as volumetric and surface area units of wood. These categories have an active role in wood calculation and thereby signify the relational agency of handsaw and wood in the forming of human thought.

Based on this reworked perspective, the handsaw and wood, too, play a significant role in fostering the mutual understanding between sawyers. They link the bodily forces of these two men and organize their motions and affects in sawing wood. In the absence of handsaw and wood, neither the connection of the intimate coordination of the sawyers' bodily actions, nor their shared understanding would exist. Moreover, since the handsaw, a big tool, demanded more than one sawyer's energy to handle it properly, the men had to synchronize their work trips, which made their lives almost inseparable.

Here we can perceive the role of the sawyers' action along with the dynamics of handsaw and wood in producing possible reasons for the emergence of the homosexual connotation of the word *Īrccakkār*. By drawing on clues from this reworked perspective, in the coming sections of this paper we pursue a more thorough discussion of the question of the relational agency of the handsaw and wood in the knowledge formation of wood problems. The wood problems in medieval Indian mathematical treaties generally deal with different kinds of calculations of volume, area, and wood pricing, and with the wages for sawing wood. For example, one of

[1] A language of southern India that is spoken by ca. 35 million people.

[2] This ambiguous signification is true for the word *īrccakkārepōle*, too.

[3] Homosociality is a predominant cultural character of Kerala society.

the wood problems in Malayalam *Kaṇakkatikāram* (see verse 2 in the Appendix) calculates the length of a piece of wood of a given volume and given depth and width.

In this paper, I argue that mathematical knowledge of wood-related problems is not a product of introspection, but a product of individual or collective[4] human practice in the material world that emerges out of correlations between human bodily actions (motions and affects) and material dynamics (forces, properties, and affordances). There is no action without interaction between human and nonhuman. In such a perspective, nonhuman entities (such as wood and handsaw) not only form or deform human agency and intentionality, but the very 'matter' of agency and intentionality are no longer solely human.

I refer to the motion and affect of humans as *thinking* and to forces, properties, and affordances of materials as *thinging*.[5] Therefore, knowledge of wood-related problems is an emergent product of the interactions between *thinking* and *thinging*. These two, thinking and thinging, do not exist separately. Thinging appears through thinking and thinking emerges through thinging. In short, this knowledge is what I would call a product of *Thingking*.[6] I developed this neologism through the engagement with material signs (length, surface and volume), measurement procedures, and rules of calculation in the wood problems of Malayalam mathematical treaties. But *Thingking*—the interactive knowledge formation through the synergy of bodily actions of humans and dynamics of materials—is applicable to any human-material interaction.

We will use the neologisms *thingk*, *thingker,* and *thingking* to express the mutual formation of mathematical knowledge that emerges from the networks of interaction between the human and nonhuman (things and material nature). In detail, we delineate how numbers, measures, procedures, and rules of wood calculation, present in the Malayalam *Kaṇakkatikāram* and other traditional Indian mathematical treatises, were formed by the complex web of interactions between human and nonhuman. This analysis connects water, forest, wood, rivers, human-labor, animal power, timber production, the handsaw, land, sea, building, shipbuilding, ships, trade, recreation, and pedagogy.

The synergy of networks, be they geographical (proximity and availability of wood resources), socio-technological (the use of timber in shipbuilding and in architecture), economical (production of timber and trade), or educational (textualization for pedagogy), shaped the wood problems as a form of mathematical knowledge dealing with the material wood. A non-discursive and discursive exchange of knowledge emerged around them. Thus, the wood problems in Malayalam

[4] Since I am focusing on the human-material relation, I do not develop in this paper the focus on the communal aspect of mathematical practices, but acknowledge that it provides an important perspective.

[5] This term derives from Martin Heidegger. He asserts this term as "The thing things. The thinging assembles" (1967, 8), but I use this term in a different sense, as explained above.

[6] I also benefited from Dignāga's notion of *hetu-cakra* (wheel of reason) to create this neologism. The notion of *hetu-cakra* delineates an indispensable interrelation between the mental world and the material world in the formation of meanings (Hattori 1968).

Kaṇakkatikāram offer ample material for case studies concerning the symbiotic relation between nonhuman (dynamics of things and material nature, such as the growth of the forest) and human (actions of building, shipbuilding, trade, and education) in the mutual constitution of new mathematical knowledge.

In the coming analysis, we are going to see, among other things, in which way wood problems emerged from the interaction of the sawyer/trader with wood/timber and handsaw/ship. Also, it will be shown how this knowledge was subsequently given a textual form as a matter of human knowledge (for example, in pedagogic and recreational contexts). Moreover, our arguments will illustrate the usefulness of the *thingking* framework set forth in this paper in highlighting not only the relational agency of things and material nature, but also in describing the correlation between the human and nonhuman in the formation of wood-related mathematical knowledge. Finally, the approach presented here uncovers the relational agency of the sawyer—the (slave) laborer—in the production of mathematical knowledge, whose name is completely absent[7] from the wood problems and who was traditionally considered an 'object in action' rather than a human subject.

This paper is divided into five sections. Following this introductory presentation of the argument, the subsequent sections develop the framework, content, and contexts. The second section discusses the theoretical framework of this paper by combining the perspectives of Jens Høyrup and Karenleigh A. Overmann on mathematical knowledge formation. The third section describes the content, language, and problems concerning wood in medieval Indian mathematical treatises. The fourth section is divided into four subsections that describe four interrelated networks of interaction from which wood problems emerge. The first subsection presents the possible grounds for the textualization of wood-related mathematical knowledge in educational and recreational activities. The second subsection illustrates how construction, shipbuilding, and wood-trade form a network for the distribution and use of wood, and how that network contributes to the formation of wood-related mathematical knowledge. The third subsection demonstrates how wood-related mathematical knowledge emerges from networks of timber production, and how wood, handsaw, and sawyer emerge, through their interaction, as *thingkers* of number, measure, and rules of wood calculation. The fourth subsection depicts the network of ecological, hydrological, and climatological forces of the natural environment of Kerala and its interrelations with wood-related mathematical knowledge. The fifth section reviews the arguments of this paper.

[7] The wood problems only represent the action/labor of the sawyer without explicitly mentioning him. Considering the historical circumstances of the medieval period, we can assume that sawyers were most probably coming from untouchable-castes (*Ircha Thachan* or *Tiyya* or *I̱ḻava*), slave-castes, slave-tribes, and other socially unprivileged groups. *Ircha Thachan (saw carpenter)* was a sub-caste among the traditional carpenter caste *(Marasaris)*. According to L. K. Anatha Krishna Iyer (1981, 342, 352), the superior sub-caste *Thachan* among *Marasaris* considered other subdivisions of *Marasaris* as untouchables.

7.2 Towards an Interactive Formation of Mathematical Knowledge

For theorizing the problem of this paper, we critically combine the perspectives of Jens Høyrup and Karenleigh A. Overmann on the historical formation of mathematical knowledge. The perspective of Jens Høyrup, which is largely inspired by the views of dialectical materialism and sociology of knowledge,[8] conceives mathematical knowledge as a product of the dialectics between human practice and material culture and as a construct of the interplay between the human and the social world. This perspective allows Høyrup to critically analyze the sources of pre-modern mathematics—thereby challenging the dominant paradigms that see mathematics as an independent mental construct of individual thinkers—by revealing the role of human material practice, and thus of human agency (collective or individual), in the historical formation of mathematical knowledge.

Through his in-depth analysis of the history of mathematics, Høyrup has not only delineated the role of "practical" traditions in constituting "pure" scientific or theoretical tradition of mathematics, but also mapped three types of partly autonomous pre-modern forms of mathematical knowledge ("sub-scientific", "scholasticized", and "scientific") and elucidated certain complex interconnections among them. He conceptualizes sub-scientific mathematical knowledge as being acquired and transmitted by specialist "practitioners taught in an apprenticeship network" (Høyrup 2017b, 207), such as accountants, merchants, artisans[9] and surveyors. Sub-scientific knowledge must be acknowledged as a partly autonomous body of knowledge that possesses its own rules and logics of practice, distinct from both "theoretical" knowledge as well as "applied science" (Høyrup 2007b, 7–8). The "*scholasticized*" is a type of mathematical knowledge that was "communicated in a school by masters whose own genuine practice is that of teaching, not the practical[10] use of the knowledge they teach" (Høyrup 2017a, 216). Høyrup argues that this type of mathematical knowledge emerged out of practical use and later turned into a "pure" form by ignoring real applicability (65f). The "scientific" is the type of knowledge "one to which historians of mathematics have dedicated most of their efforts" (Høyrup 2017b, 216).

---

[8] Høyrup (2017a, 216) claims that he is standing on the shoulders of two kinds of scholars. Firstly, the Soviet scholars at the London Congress of the History of Science in 1931, especially Nikolai Bukharin (2013 [1931]). Secondly, Edgar Zilsel, whose paper *The Sociological Roots of Science* (2000) was first published in 1942.

[9] For the role of artisans in the formation of early modern 'new science', see the studies of Pamela H. Smith (2012) and Pamela O. Long (2011). Smith theorizes how the artisans' approach towards nature and their bodily practices were formative for the 'new science'. Long describes how the 'new science' was shaped through the interaction of artisans' empirical attitudes and the scholastic-humanist concerns.

[10] The notion of "practical" is problematic, as seen from the different ways of approaching it in the introduction, the analysis in Sect. 3, and various papers in this volume.

Following this perspective, it is possible to identify the roles of the educational practice (of the teacher/scribe) and the mercantile practice (of the trader/accountant) in the formation of the wood problems. We can thus perceive how their practice-oriented intentionality produces their agencies in the formation of mathematical knowledge. However, if we depend exclusively on the theorization of Høyrup, we will fall into anthropocentricism that rejects the relational agency of things and material environments in the formation of mathematical knowledge. Because Høyrup's approach overemphasizes human practice and agency, it unknowingly reproduces the prominence of human agency and practice over nature and other things, and ignores the relational agency of nature and things in the formation of mathematical knowledge. To include the relational agency of things and material environment in the formation of mathematical knowledge, we will have to go beyond the theoretical tools offered by Høyrup. This endeavor may be supported by the perspective of Karenleigh A. Overmann (2019), which addresses the relational agency of matter in the context of mathematical knowledge formation.

Overmann studied counting devices of the Ancient Near East—fingers, tallies, tokens and numerical notations. Thereby, she delineated the material origin of numbers and problematized their conceptualization by introspectionists, who see numbers as a primarily mental phenomena. In her work, she shows "[h]ow materiality interacts with, and influences change in, behaviors and brains and how the cognitive system for numbers changes overtime, as attested by material change" (Overmann 2019, 9). However, before turning to her conceptual framework, it would be advantageous to recapitulate the genealogy of Overmann's theory.

The development of the theoretical stock of Overmann's work substantially depends on Lambros Malafouris's Material Engagement Theory (MET). Malafouris's primary concern for 'material agency' emerged mainly from Bruno Latour's Actor-Network-Theory (ANT). In his ANT, Latour (1999) theorized agency as a product or effect of a network of relations between human and nonhuman. Agency is not seem only as the product of individual or collective human free will, as put forth by dialectical materialists and social determinists. This perceptive not only denies the primacy of the human actor over the nonhuman realm, but also places agency "in the blind spot in which society and matter exchange properties" (Latour 1999, 190). Their agencies do not lead a separate existence and actions of these actors always emerges from an association of human and nonhuman elements (Latour 1999, 180–3).[11]

[11] Tim Ingold's work establishes a critique of the hylomorphic model, according to which humans impose their mentally thought-out forms on the material world to create things. As Ingold (2013, 6) states: "… the conduct of thought goes along with, and continually answers to, the fluxes and flows of the materials with which we work. These materials think in us, as we think through them." He critically engages with analytical categories such as "agency", "network", and "interaction", which are used by Bruno Latour and Lambros Malafouris in their work, and replaces these categories with his concepts such as "animacy" (94–107), "meshwork" (132–136) and "correspondence" (105–108) in his theory of "thinking through making".

In his works, especially in his "At the Potter's Wheel: An Argument for Material Agency", Malafouris (2008) carried Latour's framework forward. By discussing the making of pottery as a case study while "focusing on the brain-artifact interface" (22), he explored the causality of human action and the question of agency. In order to address this question, he introduced a new concept named "material agency" (ibid.). "Agency", in his argument, "is a property or possession neither of humans nor of nonhumans. Agency is the relational and emergent product of material engagement. It is not something given but something to become realised" (34). The emergence of intentionality and agency, he further elucidates, result from the "network of material engagement" (ibid) in which mind, body, things and material structures interact with each other. Hence, intentionality and agency are neither properties of things nor the properties of humans; they are the properties of "material engagement" (34).

Malafouris further delineates the nonhierarchical relation of intention and action and its processual relation with agency. He argues that agency emerges from "intention in action", not from any "prior intention", taking up a claim put forward by John Searle. "[I]ntention no longer comes before action but it is *in the action.* The activity and the intentional state are now inseparable. [...] That means that "intention in action" is not an internal property but a component of extended cognition. Consequently it constitutes and is constituted both by persons and things" (31). In this perspective, things and material structures become parts of the mind, and intentionality is seen as "distributed, emergent and interactive phenomenon rather than as a subjective mental state" (33). Therefore, as Malafouris makes clear, "in certain cases, human intentionality identifies with the physical affordance. The mediational potential of a certain artefact in a quite significant way shapes (both in the positive and negative sense of enabling and constraining) the nature of human intentions" (33).

Overmann clarifies her research framework by describing "three commitments" of MET:

> First, *cognition is extended and enactive*. That is, cognition is a system that includes materiality as a constitutive component, and cognition is the interaction of brain, body and materiality. [...] Second, *materiality has agency*: Its properties influence our behaviors and psychological responses. [...] [T]he material forms used to represent and manipulate numbers influence how we use them and how we conceive numerical content, structure, and organization. [...] Third and finally, materiality has a semiotic function (or a meaning potential), different from that of language, which makes it a *material sign*. Simply, things are meaningful in virtue of what they are and what we do with them. Such enactive signification is important in numerical cognition, where material meaning precedes language for numbers, and change in material form informs change in the way numbers are conceptualized (Overmann 2019, 9, emphasis in the original).

This capacity of matter to form intentionality by working together with the human mind and body is referred to as the relational agency of matter. Our analysis in this paper selectively combines these two perspectives and uses the theoretical insights of these scholars for substantiating the interactive formation of mathematical knowledge of wood problems by the *thingking* of human and material.

7.3 Content of Wood Problems in Medieval Indian Mathematical Treatises

We will start our analysis by first unraveling the wood problems in the Malayalam Kaṇakkatikāram,[12] Kanakkusāram, and Kanakkusāram *Bālaprabōdham.*[13] These are elementary mathematical treatises. The Malayalam version of the *Kaṇakkatikāram,* our focal register of this paper, is a genre of mathematical texts that experienced considerable circulation in medieval and colonial Kerala. It is difficult to know anything accurate about the author(s) or the period of composition of these treatises, which—as far as we could tell—bear no colophons. Moreover, based on its form and content, it seems that the *Kaṇakkatikāram* was a popular and authorless genre in Tamilized Malayalam, in the *pāṭṭu* style, which consists of Dravidian poems or songs. The linguistic, stylistic, and internal evidence of these treatises suggest that the different verses were formed in different times and places (Wagner and Ashokan 2024).

Unlike Sanskrit mathematical treatises, which are generally considered to be lofty Brahmin compositions, it is almost sure that Malayalam *Kaṇakkatikāram* consists of contributions by people from a non-elite and lower caste background with diverging poetic skills and distinct professional involvements. The verses of the *Kaṇakkatikāram* were recombined, re-edited, and renewed by local teachers and practitioners, thus creating new *Kaṇakkatikāram*. From internal evidence, we can presume that the preliminary composition of the verses of these treatises was probably happening in a time span between the 14th and 16th centuries, although it was sometimes continued until the nineteenth century. For instance, in 1863, Māvanān Māppiḷa Seyitu Muhammadu prepared a new *Kaṇakkatikāram* collection.[14]

Texts of the *Kaṇakkatikāram* genre display an extensive diversity in content and internal order. Most verses can only be found in some versions and certain sections are completely absent in some treatises. Wood problems in Malayalam *Kaṇakkatikāram* collections are no exception. For example, in the palm-leaf manuscript 817C available at the Calicut University Thunchan Manuscripts Repository (TMR), we do not have any wood related problems. The wood problems we are going to discuss in this article are based on the palm-leaf manuscript 1307B (TMR), on the manuscript 14633A, available at the Kerala University Oriental Research Institute and Manuscripts Library (ORIML) in Trivandrum, and on the paper

---

[12] The *Kaṇakkatikāram* treatises exist in both Tamil and Malayalam. Senthil Babu (2022) has already analyzed the Tamil versions. Arun Ashokan, Naresh Keerthi, P. M. Vrinda and Roy Wagner have prepared Draft of an edition of the Malayalam *Kaṇakkatikāram* based on four palm-leaf manuscripts and a paper transcript (see Ashokan et al. 2024). For a detailed discussion about the content and context, see Wagner and Ashokan (2024).

[13] The analysis is based on the editions by C. Achutha Menon (1950a, 1950b) from palm-leaf manuscripts, authored in Malayalam around the seventeenth century.

[14] This book mentions the Malayalam year 1038 (equivalent to the year 1863 CE) on page 61 as the date of preparation. The year of printing is not clear, however, and the publisher was CMS Kottayam (Govi 1998, 164).

transcript of another manuscript, T365 (ORIML).[15] We also depend on wood problems present in the printed version of *Kaṇakkatikāram* that was prepared by Māvanān (1863).

The subjects of *Kaṇakkatikāram, Kanakkusāram, and Kanakkusāram Bālaprabōdham* are classified, in the introductory verses, according to topics such as land, gold, paddy, and rice. One manuscript of the first treatise (TMR 1307B),[16] and the two latter treatises mention wood as one of the subjects (Menon 1950a, 1; 1950b, 1). In three of the Malayalam *Kaṇakkatikāram* manuscripts that we study (TMR 1307B, ORIML14633A, and ORIML T365), wood problems come up as a theme in the word-problem sections. This demonstrates the significance that the *Kaṇakkatikāram, Kanakkusāram* and *Kanakkusāram Bālaprabōdham* attributed to wood problems. Even in the *Kaṇakkatikāram* of Māvanān, wood problems appear as a subject, together with their associated units of measure (Māvanān 1863, 59, 94–119).

The inclusion of wood in introductory verses highlights the important role wood-related problems played in medieval Malayali society. This fact reflects Senthil Babu's observation on the vernacular Tamil mathematics of medieval South India. In the context of Tamil *Kanakkathikaram* treatises, he describes how mathematical knowledge emerged from practical calculations concerning such objects as grain, gold, and land (Babu 2022, 82–6).

Contrary to this situation, Sanskrit mathematical treatises of medieval India, such as *Brahmaguptagaṇitam,*[17] *Pāṭīgaṇitasāra* (also known as *Triśatika*),[18] the *Līlāvatī,*[19] and the Apabhramsa[20] treatise *Gaṇitasārakaumudī,*[21] each present only a very small number of problems related to wood. In these canonical Indian mathematical texts, wood problems are included in the sections *Krākacavyavahāra* (Heroor 2014, 146–9), *Krākaca* (Heroor 2015, XXX–XXXI) and *Karkaca-vyavahāra* (Sakhya 2009, 71) that generally refer to "saw" or "mathematical-saw-procedures". These sections cover various calculations concerning the wage for

[15] This paper transcript is entitled *Kanakkupāṭṭu*, which means song of calculation. It was copied by Rāmanuṇṇi śāstri from a copy owned by Nārāyaṇan Nīlakaṇṭhanuṇṇi Nārāyaṇamaṅgalaṁ Kunnattukara Āṟanmuḷa at the Malayalam date of 07.08.105- (the last digit is missing, placing it between 1875 and 1884 CE). The copier noted Nārāyaṇan as the owner.

[16] See Verse 1 and its commentary in the Appendix.

[17] This treatise was written by Brahmagupta in the year 598 A.D. (Heroor 2014, XXI).

[18] The period of this treatise and of its author, Sridhara, is under dispute. B. Datta and A.N. Singh dated the period of Sridhara as 750 CE (Heroor 2015, XXIII–XXVI, 168–71).

[19] Bhāskara II wrote *Siddhānta Śiromaṇi* in the year of 1150 CE. *Līlāvatī* is part of this treatise (Rao 2014, XII–XIII).

[20] Apabhramsa "is the name of the standardized literary language that was in use onwards from about the sixth century by the side of Sanskrit and Prakrit". Even "after the rise of new Indo-Aryan languages (onwards from about 1000 CE) it was still used by some specialized groups (such as Jains) for lyrical and narrative usage compositions". For detailed discussion see H.C. Bhayani (2012, 1).

[21] *Gaṇitasārakaumudī* is a mathematical treatise composed by *Thakkura Pheru* during the first quarter of fourteenth century (Sakhya 2009, XI, 71f, 157f).

sawing wood, the surface area of a sawn section of a cylindrical or rectangular wooden log, and the number of wood planks that can be produced from a wood log. Not only is the number of problems centered around wood much smaller in these treatises in comparison to the Malayalam *Kaṇakkatikāram,*[22] but these problems also share certain commonalities with other calculation procedures related to area, depth, and volume of land excavations, mounds of grain, and plane figures.

Our analysis of wood problems in Malayalam *Kaṇakkatikāram*, *Kanakkusāram,* and *Kanakkusāram Bālaprabōdham* reveals that seven out of nine types of problems are about different kinds of volume calculation. They neither show any derivative connection to the above-mentioned problems in Sanskrit and Apabhramsa treatises, nor do they stem from the Tamil *Kaṇakkatikāram*, as it does not include any problems related to wood. The volume calculations in Malayalam *Kaṇakkatikāram* follow various distinct procedures, depending on the given dimensions.[23]

The other two types of wood problems in the above treatises concern the calculation of the sawn area when cutting a log of a given dimensions and the wages for the sawing given the ratio of sawn area to wages. In the former, the area of a given number of straight sections of a log is calculated from the dimensions of a log. This kind of problem appears in the *Pāṭīgaṇitasāra*, the *Līlāvatī*, and the *Gaṇitasārakaumudī* as well. Moreover, one problem of *Gaṇitasārakaumudī* calculates the number of wood planks that can be produced form a given volume of a log (Sakhya 2009, 71). The same problem type appears in some *Kaṇakkatikāram* manuscripts as well. However, it is not clear whether the Malayalam problems depend in any way on the Sanskrit ones.

The wood problems in the *Kanakkusāram* and *Kanakkusāram Bālaprabōdham* are identical in content and form. They are given in a Maṇipravāḷam style, that is, a Sanskritized version of Malayalam. These treatises state that they are based on the *Kaṇakkatikāram and* the *Līlāvatī*, although the relation to the *Līlāvatī* is weak.[24] Most probably, the majority of their wood-related problems are derived from the same source as that of the *Kaṇakkatikāram* (especially the volume calculations), as they show a close similarity. One of the calculations in the *Kanakkusāram* and *Kanakkusāram Bālaprabōdham* determines the volume of a cylindrical log from a given circumference and length via the ratio 22/7 between circumference and diameter.[25] These two treatises also include a few exceptional and unique entries, such as

---

[22] *Ganitadhyaya of Brahmasphutasiddhanta* has 6 problems (including rules and examples), *Sridhara's Patiganitasara* has 5 problems (including rules and examples), Bhāskarācārya's Līlāvatī has 4 problems (including rules and examples), and Thakkura Pheru's *Ganitasarakaumudi (The Moonlight of the Essence of Mathematics)* has 9 problems (including rules and examples).

[23] For example, calculate the volume of wood from a given circumference or width and a given length, or calculate the volume of a log after removing its bark from a given circumference or width and a given length (see verses 3 and 4 and their commentaries in the Appendix).

[24] For a detailed discussion about the relation between *Kaṇakkatikāram* and *Kanakkusāram,* see Wagner and Ashokan (2024).

[25] This same ratio is also used to calculate related volumes.

calculations of the volume of wood pieces in the form of a frustum, a cone, or a pyramid, and a problem that calculates the ratio of exchange between two pieces of wood with different circumferences (Menon 1950a, 66–73; 1950b, 43–50, 127–141).

7.4 Wood Problems and the Four Networks of Interaction

7.4.1 Network of Circulation and Erudition: Education and Recreation

The verses on wood problems and the accompanying commentaries in Malayalam *Kaṇakkatikāram*, *Kanakkusāram,* and *Kanakkusāram Bālaprabōdham* have many direct and indirect layers of meaning. Among those, we can identify layers related to pedagogy, recreation, trade, building, shipbuilding, labor, the handsaw, ecology, the production of timber, the status of the tree in society as well as linear and volume measure. These issues are, however, not merely contextualizing practical circumstances. Instead, we propose to conceive them also as a reflection of networks of interactions and practices that produce and circulate the knowledge contained in the wood problems.

Even though the wood problems are presented in a practical context, some of them do not actually fit the circumstances faced in practical reality. For instance, a handful of them are asking to calculate the length of a piece of wood, while its width, depth and volume is given.[26] However, in a real practical situation, such as in the context of timber production, one cannot know the volume of a piece of wood without already knowing its length, width, and depth. There are also problems that ask to calculate, given a log's circumference and length, its volume after removing the bark, the volume of a quarter of the removed bark-shell, and the price of the log given the price of a quarter of the bark-shell.[27] Though merchants might wish to calculate or predict the volume of a log, there are three reasons why such calculations deviate from practical needs. Firstly, from the carpenter's point of view, carving work proceeded through action-oriented-cognition rather than calculation. Secondly, in medieval Kerala, wood was widely available, making the value of the shell negligible. Thirdly, teak wood was the major product in the medieval Kerala timber trade, but its bark-shell is very thin, and had practically no value in trade.

Hence, we can infer that these kinds of problems were created by local schoolteachers or scribes for pedagogic and perhaps recreational purposes.[28] A considerable number of wood problem verses pose a challenge to the listener by asking to

[26] For instance, see verse 3 in the Appendix. The actual practicality of the mathematization of various seemingly practical contexts could be questioned in many problems in the *Kaṇakkatikāram* this collection.

[27] For instance, see verses 3 and 4 in the Appendix.

[28] For a discussion of reactional problems, see (Høyrup 1990, 66).

reveal "…the good calculation, if you can…" and to "state properly … by means of correct calculation…".[29] The appearance of the student or expert (*ceṭṭar*, see, for example, Māvanān 1863, 106) as an addressee and the adoration of the Guru (see, for example, Māvanān 1863, 109) further corroborate the close connection to pedagogic action/practice.

Moreover, the presentation, procedure, and solution of these problems seem to be deliberately convoluted. This might have been done to prevent easy communication of the knowledge without the intervention of a teacher or a scholar. From this perspective, the wood problems are not only a means to express the virtuosity of the teacher and other authorities by posing challenges to listeners, but also a way to articulate their power.[30] As Jens Høyrup (1990) has observed in another context, one of the main functions of these kinds of problems was "the formation of professional identity and pride" (9) of teachers and other authorities. Even in the context of more clearly practical problems, the commentary of the verse is not only adding an example of calculation and solution, but is also producing and asserting the authority of the author, teacher, or scholar through wood-related mathematical knowledge. If we follow Høyrup's classification, these aspects of the wood problems reflect a "*scholasticized*" type of mathematical knowledge. Let us therefore turn to another network of interaction that shaped the production of wood-related mathematical knowledge, in order to examine their embodiment of practical use.

7.4.2 Network of Distribution and Utilization: Building, Shipbuilding and Trade

Another network of interaction that is evident in the wood problems concerns building, shipbuilding, and trade. General references in the problems, particularly to the felling, measuring, debarking, and cutting methods of wood, as well as to the calculation of the area and volume of wood in different shapes and sizes, reflect their possible connection to the built environment of medieval Kerala.

Wood was one of the main materials used for construction in medieval Kerala.[31] Temples and palaces were "built of wood, brick and stone", according to Stella Karamrisch (1961), while houses were "built originally of mud, lightwoods and thatch" (21). At least until the middle of the sixteenth century, temples and humble dwelling houses were mostly made of clay. In this period, according to traditional law, only the houses of kings and the temples of deities could be constructed from wood and stone (ibid., 24). The structure of Kerala architecture typically had

[29] See verses 3 and 2 in the Appendix.

[30] For a detailed discussion of the relation between the authority of the teacher and the arcane nature of knowledge of *Kaṇakkatikāram*, see Wagner and Ashokan (2024).

[31] William Logan (1951) was quoting Ibn Batuta, who visited Kerala 1342–47 CE, when he asserted that at that period "… at the distance of every half mile there is a house of wood" (288).

centered on the wooden construction and woodcarvings (ibid., 14; Schildt 2012, 93–7). Furthermore, Ibn-al-Majaawir, a medieval Arab writer of the thirteenth century, discusses the 12,000 houses of Soher, the capital of Oman. Those were made of brick, mortar, and teakwood, the latter of which, in the opinion of Dionisious A. Agius (2008, 85f), was brought to Oman from the Malabar Coast of South India.

There is another feature of the wood problems in Malayalam *Kaṇakkatikāram* that corroborates the existence of a relation between building and wood problems. The problems display names of the units of volume measures such as *vaḷa* (1 *viral* × 1 *viral* × 1 *kōl*), *paleka* (equal to 24 *vaḷa*), *tūmaram* (equal to 144 *vaḷa*), and *kaṇṭi* (equal to 576 *vaḷa or* 1 *kōl* × 1 *kōl* × 1 *kōl*).[32] The term v*aḷa* denotes the crossbars that support the rafters of a roof or a wooden needle driven into the rafters, respectively, while *paleka* means plank. *Tūmaram* can be read as "pure-wood" or "pillar-wood". Moreover, problems in *Kanakkusāram* and *Kanakkusāram Bālaprabōdham* refer to Malayalam names of wooden parts of the structure of traditional Kerala buildings such as *paṭṭika* (lath of wood), *paṭi* (plank of wood), *uttaraṁ* (beam of wood) in to the context of the calculation of the volume of these pieces of wood (Menon 1950a, 71f; 1950b, 48f).

The presence of these names reveals a connection between the way wood was used in medieval Kerala society and the units of volume measures that show up in wood problems. All of them, except for *kaṇṭi,* are structural parts of traditional buildings[33] and ships.[34] These allusions already indicate a possible linkage between the material demands of people (to construct a building or a ship) and the emergence of the wood problems. This connection, however, is put into question by the absolute silence concerning traditional buildings (palaces, temples, and mansions), woodwork, and wood calculations for building construction and shipbuilding.

We can only speculate about the reasons for the omission of these kinds of references in the wood problems. Firstly, by taking the historical circumstances into consideration, we can assume that this is probably the consequence of deliberate exclusionary measures employed by the composers or copiers to convert it into a "*scholasticized*" type of mathematical knowledge. Secondly, during the medieval period, the traditional practical mathematical knowledge of Malayali carpenters might have circulated mainly through apprenticeship channels based on kinship and

---

[32] See *Kaṇakkatikāram* palm leaf manuscripts 1307B and 14633A and paper transcript T365. We can also see *tūmaram* and *kaṇṭi* in Māvanān's *Kanakkatikaram* and in *Kaṇakkusāram.*

[33] Strangely, the books I have examined on the medieval architecture of Kerala do not use these wood measure terms for parts of traditional buildings. Instead, Sanskrit names for the wood parts of buildings are used in the traditional Kerala texts of *Vasthu* and *Thachushasthram*. For instance, Henri Schildt's (2012) book does not have any Malayalam names of wood parts; even *vaḷa* (crossbar) appears in this book as *valaya*, that is, in its Sanskrit form. However, in Kerala, even today carpenters use these words for wood parts of the superstructure of traditional buildings (96).

[34] The medieval Malayalam ballad *Payyannūrppāṭṭu* (Antony 1994, 12f) and K. S. Mathew's (2018, 35) book mention *palaka*/*ottupalaka* as a plank that is fixed by a carpenter on the *paṭṭika-thattu* (keel) of the ship.

caste. Hence, they wanted to keep the knowledge to themselves.[35] Thirdly, trade-oriented computational knowledge, which is very evident in the wood problems, was probably not a matter of concern to the medieval elite-castes (the Malayali Brahmin *Namboothiris* and the ruling *Rajas*). At this time, we can assume that most of the *svarūpaṁ* (principality), *dēvasvaṁ* (property of temple), *brahmasvaṁ* (property of Brahmins), and property of the *nāṭuvāḻi* (territorial ruler) included private forests or the right to cut down trees from public forests. Even in eighteenth century Malabar (Ganesh 2016, 174) and nineteenth century Travancore (183), the greater part of the land was covered by forests. In the case of these elite castes, the context of the free availability of wood might have prevented the commodification of wood.

What we can find in the wood problems, however, apart from the area and volume calculations, are mentions of the rate of payment for cutting wood[36] and the price of wood planks.[37] They already signal a possible connection to the timber trade and shipbuilding. Even though logistics, transportation, and categorization of wood are an inevitable part of the timber trade and shipbuilding, there is again an absolute silence in the verses of the wood problems about these processes. Nevertheless, to get a more detailed picture about its role in the emergence of the wood problems, let us examine more closely in what way that the timber trade operated in Kerala during this period. To do that, we rely on medieval native records and colonial accounts, which provide us with proofs and insights about the use of wood in shipbuilding and the timber trade.

Al-Muqaddasi, an Arab geographer of the tenth century, remarked on the export of teak wood, along with sandalwood, ebony, and other commodities, from India to Soher, the capital of Oman (Agius 2008, 85). Many of these products originated from Kerala (Gurukkal and Varier 1999, 181–3). The medieval Cairo Geniza provides further evidence about the timber trade. For instance, one of the items Abraham Bin Yiju, a Tunisian Jewish merchant of the twelfth century, shipped from the Malabar Coast to Aden was teakwood boards (Goitein and Friedman 2008, 337). Timber of different kinds was available in the Malabar coastal hinterlands at that time. The trade and export of these items to Arabic regions goes back to antiquity, with Calicut having been the active node of centers of timber trade (Agius 2008, 87).

Furthermore, before the arrival of the Portuguese in the fifteenth century, there was a famous indigenous shipbuilding yard in Malabar (Mathew 2018, 31). Concerning this medieval center of trade and technology Agius (2008, 142) states:

> "On the southwest coast of India, south of Calicut, the Beypore shipyards were well-known as they were located at a crossroads between the Western Indian Ocean and the Far East; their importance grew in terms of the varied types of timber the region could supply and in the number of shipbuilders with workshops dotted all along the coast. In addition, the Southwest Indian coast was the right spot for ships to stop over because of the monsoonal

[35] For a discussion of the knowledge of traditional Malayali carpenters, see Nedumpally Kizhakke (2012).

[36] See verse number 4 in the Appendix, page 42 of transcript T365, and Māvanān (1863, 106: 21–31).

[37] See the manuscript ORIML 14633A, 25r:1–8.

winds and/or to repair their ships. Raw materials such as teak planks, coir and *nūra* (lime and fat substance) were exported from here (to medieval Arabian countries)".

Apart from the teak wood, which was ideal for shipbuilding, Agius presents a list of the kinds of timber that were exported, including *aini*, jackfruit, mango wood, and ven teak, that were part of the biota of the Western Ghats of Kerala (Agius 2008, 148). The medieval Malayalam ballad *Payyannūrppāṭṭu*[38] (Antony 1994, 12f) discusses the process of indigenous shipbuilding by native carpenters using timber available in Kerala forests for native Chetty traders who were engaged in sea trade.

In the sixteenth century, timber from Kerala—especially teak and *aini*—was used by colonial powers for the building of ships in the Cochin dockyard (Pilla 2017, 22f; Subrahmanyam 1990, 228), while they also imported timber from south Malabar to Jaffna (105). This usage finds an expression in the works of contemporary Portuguese writers, who stated that "nature created these woods (teak and *aini*)" for naval architecture (Mathews 2018, 26). Hence, they understood Kerala mainly as a land of fine timber suitable for shipbuilding.

Moreover, in the sixteenth century, Kerala merchants exported timber to Sind by ship.[39] In the seventeenth century after the conquest of Cochin by the Dutch East India Company (VOC), officials of the VOC realized the abundance of fine teak timber along the Malabar Coast (Odegard 2019, 485). They constructed a shipbuilding yard at Cochin, where they employed day laborers such as native sawyers, carpenters, and woodcarvers, and exported timber to Coromandal and Ceylon (490).

Finally, *Kulatturayyan Kappal Pattu,* an old Tamil song, describes the use of wood in shipbuilding. It refers to the search of suitable trees in the Western Ghats of Kerala, the selection of trees for wood,[40] and the use of different types of timber[41] for shipbuilding by craftsmen (Mathews 2018, 28). Moreover, in the late medieval Malayalam ritual song *Kannakiyum Cheermakkavum,*[42] we can find a narration of the selection of trees made by a carpenter (*Tachan*), who went into the hills in order to build a vessel for a goddess (Mathews 2018, 30).

The content of this section substantiates the assertion that wood problems emerged from the practical contexts of building, shipbuilding, and trade, among others. The coastal and overseas trade together with its key apparatus, the ship, connected medieval Kerala to the Coromandal coast, Orissa, Gujarat, Ceylon, Java, Indonesia, China, as well as to the Arab, African, and European regions. The needs of the people living in these regions were met through the flow of goods, people, and

[38] This ballad narrates the revenge of native *Chetty* trader's wife on her husband.

[39] Irfan Habib (2010) wrote, "Ships from Malabar or Kerala called at the main port in Sind ('Diul') with rice, sugar, spices (pepper presumably), timber and canes, to return with cotton, cloth and horses" (128).

[40] They selected wood such as *vembu, ilupai, punnai, krimaruthu, sirunangu, aini,* and *kongu.*

[41] They used timber such as *venteak* and *teak* for side planks and supporting beams.

[42] This is a song related to the worship of Kannaki in the Sreekurumba shrines of Malabar.

skills.[43] The object of the mathematical wood problems—timber—was not only a commodity, but also the crucial material for the construction of ships, and thus was a fundamental floating force within this network. In this context, both natives and foreigners relied on wood-related mathematical knowledge to enable their engagement in the wood trade.

Timber in the form of crossbars, rafters, planks, beams, pillars, and blocks was used to construct buildings, bridges, boats, ships, or dams, and to make furniture in medieval Kerala. The fact that one had to deal with different sizes of timber is reflected in the volume and surface area calculations of the wood problems and further corroborated by the names of the measures included, such as *vaḷa*, *paleka/tūvaṭa, tūmaram, kaṇṭi/oṭṭam, īrcca*, *vāḷ* and *catiram*. The fact, however, that the problems contain no explicit reference to the specific kinds of wood,[44] their quality, or the type of article to be made still demands an explanation. The omission of details, dispositions, and aims might be a reflection of the process of adapting the knowledge used in practical context to an educational and recreational context.

However, many of the calculational techniques presented in the wood problems were useful in practice. The direct references (to wood trade, numbers, measurement units, and procedures) and indirect allusions (to building and shipbuilding) in the verses reflect the practical use of wood problems and suggest similarities between these problems and Høyrup's concept of sub-scientific mathematical knowledge. In light of Høyrup's perspective, it is visible here how human actions, such as trade, housebuilding, and shipbuilding, together with the agency of traders and carpenters, participate in the formation of wood-related mathematical knowledge. However, as we discussed above, Høyrup's overemphasis on human practice and agency not only prioritizes it over things and material nature, but also overlooks their relational agency in the formation of mathematical knowledge. Therefore, in order to demonstrate the relational agency of things and material environment in the formation of knowledge, let us examine the references to things and material nature in the wood problems, which invoke another network of interaction concerning the production of timber.

7.4.3 Network of Production: Tree, Wood, Timber, Carpenter, Handsaw, and Sawyer

Tree, wood, carpenter, sawyer, timber, and handsaw are parts of another network of interaction that is exhibited by wood problems in Malayalam mathematical texts. Such a network is in turn connected to measurements, human labor, and price. These connections become tangible in palm-leaf manuscript verses such as: "A

[43] For additional details on such exchanges between Kerala and other regions through coastal and overseas trade, see Subrahmanyam (1990).

[44] Except for two vague instances of names of woods such as *irūḷ* and *piccamaraṁ*.

quarter part of its circumference is sawed off precisely",[45] "Hero! If the circumference of a piece of wood is given, how much [volume remains] after sawing precisely one quarter of the shell outside the internal-square?",[46] and "Note the *paṇam* (i.e. money) for the precise shell obtained by sawing".[47] Michael Mann (2001) provides us with a general picture about medieval market-oriented timber production in Kerala. He summarizes four main processes by drawing information from native[48] and colonial sources:

> "Firstly, in certain area called *kup* by the locals, trees fit for cutting were marked by a master carpenter whenever an amount of trees or *kanties* have been ordered. [...] Secondly, the trees were cut on a special day with a ceremonial or, rather, consecrated saw. The master carpenter attended the whole procedure because only he knew about the quantity of the trees, the quality of the timber and the techniques of felling. [...] Thirdly, the transport of the logs was organised by the owners of draught animals such as elephants, buffaloes and bullocks. [...] Finally, when the logs had reached a nearby riverside, they were tied together and floated to market places" (407-8).[49]

The relations that enable the production of timber appear with many nuances in two verses of the above-mentioned paper transcript T365 (ORIML) and in Māvanān's *Kaṇakkatikāram.*[50] We can see two interconnected kinds of content in these verses: First, the special ritual status of the tree, and second, the interplay between measurements of wood and measurements of human labor. We will discuss them in two steps, beginning with the opening lines of the verses. These lines illustrate the first kind of content, that is, the ritual import of trees:

> "To cut and fell a good tree with circumference four *kōl*, at the gathering of the carpenter and slave/worker, look around the present bottom circumference!
>
> He considers the manner of looking properly for defects and rot in the beam, bowing at the root, you, by breaking a coconut, ask for permission!".[51]

We can immediately connect these lines to the already quoted passage by Michael Mann, specifically to the second step of his description. We recognize that the carpenter's knowledge is at play since the tree appropriate for this particular use has to be four *kōl* in circumference. The lines from the verse above all show that, in Medieval Malayali society, the tree is deemed to be a 'sacred subject', having animistic status and possessing a special influence over the life of humans (Mateer 1981, 206–207; Swami 2011, 72). This becomes visible in the act of asking the tree

---

[45] See verse 3 in the Appendix.

[46] See verse 4 in the Appendix.

[47] See verse 4 in the Appendix.

[48] *Navoi Sattiram*, a medieval Tamil text, discusses the techniques for felling timber, the quality of timber, and the color of a newly cut section of the log. The text is extant in a fifteenth-century Tamil palm-leaf manuscript in the McKenzie collections of the Madras Archives, and has been edited by Soundarapandian. For further discussion see (Mathew 2018, 28).

[49] See K. S. Mathew (2018, 32–61) for further detailed discussions on medieval Kerala's (pre-Portuguese) production of timber, shipbuilding methods, types of ships, and navigation techniques.

[50] See verse 5 in the Appendix.

[51] See lines 1–4 of verse 5 in the Appendix.

for permission at the end of the quoted lines. Cracking a coconut and bowing down are two rituals to obtain the grace of a divinity in India. These ritualistic acts fit well with Mann's account, in which a special day for felling is chosen and a consecrated saw is being used.

The second part of the verses illustrates connections between logging, removing the bark, marking the log with a colored rope, and sawing it into the desired shape, depth, and width. These relations have to do with measurements both of labor and of wood.

> "After asking for permission, cut and hack, it takes a workday to fell [the tree].
> Arduously, for each *kōl* of circumference, the slave/worker removing the bark needs a workday.
> Cut five *kōl* in length! Take the cut and scraped [wood]!
> Scrape from top to bottom! To lift [the wood] is work for two, mark [it] with a rope!
> For this marking [use] a blackened rope. Using a saw, it is turned over around the middle for getting the desired shape, depth and width. The rope set along a line, if you tighten it while measuring, you may mark [the measurement] with ease.
> By multiplying [the measured width] by half the *aṅṅuḷa*[52] that occur in the depth, obtain the *īrcca*[53], [54]

The four stages of timber production as described in the lines are as follows. First, one day of labor is needed for the removal of the bark in one *kōl* circumference of the log. Second, the log is cut at a length of five *kōl* after marking it with a rope. This is the prerequisite for being able to lift the log for sawing. Third, the handsaw and the rope markings are used to saw the log into the desired shape with the right depth and width. Finally, to get the *īrcca* (surface area), the width of the log is multiplied by half the *aṅṅuḷa* of its depth. The interdependence between the measure of wood and the measure of human labor becomes apparent here. The interaction of the laborers' body with the properties of the material wood determines the extent of work, and both together constitute the estimate of the quality and quantity of work required. The material wood is present both as a medium and a gauge of human labor. This knowledge is a product of *thingking*, because, in this knowledge formation, the bodily action of laborers correlates with the dynamics of wood (and axe).

Wood problems in *Brahmaguptagaṇitam*, *Pāṭīgaṇitasāra, Līlāvatī,* and *Gaṇitasārakaumudī* also express the mutual influence between wood, human labor, and handsaw. For instance, in *Brahmaguptagaṇitam,* the wage for sawing a piece of wood is calculated by dividing the total measure of the surface sawn by the given unit of hardness of the wood (this ratio is known as *hasta-karma*, meaning "handwork") (Heroor 2014, 146f). *Gaṇitasārakaumudī* (SaKHya 2009, 71, 157) calculates the wage for sawing by multiplying the length of the log (in *gajas*, meaning "hand") by the number of cuts made by the saw. Then, for each width-range (e.g., 1 to 1½ visuvas, where a *visuva* is 1/20 of a *gaja*), a list provides the length in *gaja* for which the wage-rate is one *koḍī*. Another list then states a factor that adjusts the wages according to the hardness of the wood. Here the interaction between the

[52] This is the Sanskrit measurement unit that means "finger".

[53] A surface area unit, equal to 1/12 of *kōl* by *kōl*.

[54] See lines 5–12 of verse 5 in the Appendix.

properties of the wood and the actions of the worker informs the wage-rate, which might in turn be considered as a measure of labor and wood. Thus, these Indian mathematical treatises provide sufficient examples for the *thingking* nature of mathematical knowledge formation.

In the Malayalam *Kaṇakkatikāram,* the actions involving wood, human worker, labor, handsaw, and other tools (rope and charcoal) are rendered visible via the forming of the measure of wood and the measuring of labor in terms of the wood's surface or volume. What does it mean to say that these *actions* are not a property of human intentionality alone? In what way do the relational agencies of thing (handsaw) and material (wood) collaborate in the emergence of perception, intentionality, and action? How do the relational agencies of the handsaw and wood function as actors in the very production of quantities of wood pieces, units of wood measure, and different sizes of timber? To address these questions, we proceed with a thorough analysis of the process of wood sawing as it used to take place in Kerala.[55]

The first step in sawing a large-sized log consists in the construction of a platform on which the log is going to be placed for sawing. The form and functioning of these platforms are adapted to the specific topography of Kerala, a large part of which is made up of hilly landscapes providing many trees suitable for use. These hills were terraced by humans into vertically descending plots of land, one above the other, for agricultural purpose. So, every plot is on one side bordered by a plot lower than itself. The construction of the sawing platform starts by vertically ramming two wooden poles into the lower of the two plots, maintaining a certain distance between them as well as from the side wall leading up to the adjacent higher-up plot (see Fig. 7.1). The upper ends of these poles should end up situated higher than the height of the side wall. Next, a strong wooden bar is fixed between the upper ends of the two poles, and it is fastened tightly using coir ropes. It should run parallel to the upper end of the side wall leading to the higher-up plot. Then, two strong wooden bars are fixed between the tops of the poles and the side wall. The resulting platform, demarcated by the three bars and the upper end of the side wall of the higher-up plot, has the same height as the upper plot. It is known as *īrccathatam.*[56]

To saw a log, which has been previously debarked and bucked, in a particular length, the sawyers roll it from the higher-up plot onto the platform with the help of rollers, a crowbar, and rope. They place it across the platform in between the two wooden bars leading onto the higher-up plot of land and fix it firmly. Now, the task consists in sawing the log to a length suitable for a particular use. The sawyers mark a line on the log with a measuring rope that has been dipped into a paste made of turmeric or charcoal. While sawing at the demarcated place, one of the sawyers stands on the log and the other one underneath it, on the lower of the two plots of land. They use a two-handed saw having the right length for cutting the diameter of

[55] The narrative of the platform and the process of wood sawing is based on the memories of myself and my parents, and on a video (Eyyanchery 2021) concerning the old style of wood sawing.

[56] Alternatively, in some parts of Kerala, a sawpit might be used as workplace for sawing, especially in plain land such as Palakkad. A reference to a sawpit and to the process of double-handed sawing appears in a Malayalam short story (Radhakrishnan 2013, 176.)

Fig. 7.1 Wood, Handsaw, and Sawyer: Three Thingkers of Number, Measure, and Rule, drawn for this publication by Kannan Prakrithi, whom I thank for his support

the log in question. Each of them holds fast one of the handles at the end of the saw. Then they alternatingly push and pull the saw through the mark on the log. If the kerf begins to close, they insert wooden wedges into the kerf to keep it open until the log is fully sawed through. Subsequently, they again mark the desired width lengthwise and repeat the process of sawing.

If we view the action of sawing solely as a result of human intentionality (or mental causation, or free will), it is possible to describe the process as a linear causal chain like the following: "wood log—eyes—brain—muscles—saw—sawing—wood log".[57] Such a view understands human action as the result of commands issued by the brain and therefore places the human brain in the position of the sole agent of cognition. However, I am going to argue that deeper engagement with the

[57] In other words, this description follows the order: perception—signaling—brain—command of brain—bodily motor functions.

actual material stages of the process reveals this brain-centric perspective to be insufficient. A close look at the processes involved actually leads to an overcoming of the traditional binary between mind (or the human) and materiality (or the nonhuman), which supports the perspective just described. To begin with, every lengthwise mark on the log arguably results in a distribution of cognition by extra-neural materials. This is because the alleged command of the bodily motor functions by the brain (especially concerning eyes and hands) is in truth insufficient to saw the log straightly and cleanly to the desired measure. To do so, the eyes and hands of the sawyer need to follow the mark made by the rope and the charcoal- or turmeric-paste. The properties of wood, charcoal or turmeric, measuring-rope, wooden wedge, as well as the relational abilities of the sawyers, cooperate in letting emerge an 'outsourcing' of cognition. The specific color of charcoal sticking on and standing out from the surface of the log is one example of the properties of charcoal that provide the extra-neural materials for the distribution of cognition. Another example would consist in the properties of the measuring-rope; for example, the flexibility of the measuring-rope is necessary to distribute the color across the bumpy wood surface.

More importantly, the actual process of sawing is not well captured by a description focusing only on the interplay between the brain and the eyes of the individual sawyer. This is because neither the sawyer standing above the log nor the one positioned below it are able to completely keep in view the log or the sawyer on the opposite side. Their positioning and the large size of the log prevent any holistic vision of the object they are working on and of the actions of their partner. In the absence of clear visual perceptions, the sawyers cannot pass on any comprehensive visual signal to the brain in order to coordinate their bodily actions. This implies that, with regards to sawing on the *īrccathatam*, the source of perception, cognition, and action must be sought in the interaction between the log, handsaw, body, and brain. This entails that the sawyers do not have a 'prior perception' nor a 'prior intention' that would neatly lead them to their actions. We rather observe the emergence of perception and intention in action. The sawyers perceive (in action) the relative hardness of the wood, as well as the difficult knots and knot-holes, via the grip and the movement of the handsaw and through the sawing sound. These properties of the wood exist beyond vision; they appear only through the actions of the workers, as does the sound. The sense of touch is extended through the saw, or, to put it into the reverse order, the sense of touch of the handsaw is extended to the sawyer. Analogously, the sawing sound is a product of the contact between the wood, the handsaw, and the sawyer's body.

As we hinted at the outset of this paper, interactions among sawyers, handsaw, and wood not only mediate, connect, and coordinate the bodily actions of the sawyers, but also produce the very motion and affect of sawyers in sawing. The properties of the saw and the affordances of the wood are indispensable in enabling the sawyers to bring their actions into effect. The sawyer standing on top of the log does so only by the help of the saw in his hands and his feeling both for the wood and the moves of the other sawyer beneath him. The upper sawyer has to muster skills similar to the ones evinced by a tightrope walker following a dangerous rhythm between a rope, a balance pole and his feet. The flexibility, strength, and sharpness of the iron blade together with the divisibility and slight springiness of the wood help the sawyer balance his body, just as they make it possible to overcome the resistance he faces while sawing. More significantly, the vertical positioning of the bodies of the sawyers amplifies the force of their actions with the help of the gravitational downward pull on the saw. Here the process of sawing is not an implementation of a prior idea of sawyers' mind into action, but a formation of knowledge through the interaction of bodily actions of sawyers and the dynamics of wood, handsaw, measuring-rope, and charcoal. In short, sawing is a process of *thingking*. Wood, handsaw, and sawyers are collaborative *thingkers* of this process.

So far, we have described wood sawing as a distributed process bringing together many different entities, all with their respective relational agencies. We now proceed to show how sawing is connected both to counting and to measurements. The Malayalam words formed from the root *īr* point to a close connection between the actions of dividing wood and counting. From this root, different words besides the already-mentioned *īrccathatam* are derived, such as *īrcca* ("saw" or "cut") as well as *īrāyi* and *īrccakkāran* meaning "sawyer", or, more literally, "man who saws" or "man who splits into two" (Padmanabha Pillai 1998: 333–4). As becomes clear from the last two examples, the root *īr* does not only signify "cutting", "sawing" or "splitting" but also "two". We see here that sawing lends itself naturally to the formation of a certain mathematical knowledge. Sawing a wood log splits this log first into two sections and then possibly into arbitrarily many more. Seen from a material perspective, the resulting numerically countable segments of wood are the result of the complicated *thingking* process described in the preceding paragraphs. The sawing is also a procedure for generating material numbers.

In this procedure, counting and measuring are interlinked. This runs counter to received conceptions which tend to keep counting and measuring apart, and which are described by Emanuele Lugli (2019) as follows:

> "[A]s Aristotle noticed, counting and measuring operate according to different regimes. Counting identifies a collection of discrete entities, such as the number of marbles in a bag. Their number cannot be questioned, and arithmetic is beyond doubt. But measuring deals with continuous entities that must be broken to be quantified. […] Measuring relies on external agents who, so to speak, make the cut. While the objects to be counted are independent from each other before the counting takes place, measuring constructs the very units it pretends to simply record" (96).

Contrary to this description, the process of sawing, measuring and marking the log with color takes precedence over the production of countable pieces of wood (and, by extension, over the production of material numbers.) Then, counting the stacked-up sawn pieces of wood leads to calculating the measure of the width of the wood. Both the measuring and counting are subjected to the interactions among the sawyer, the wood, and the handsaw—neither of them is given beforehand.

We can perceive the interaction of wood, handsaw, and sawyer in the formation of units of linear, area, and volume measurements of wood. *Viral*, *cāṇ*[58] and *kōl* are the linear measures that appear in the wood problems. *Viral* means "finger", *cāṇ* means "span", and *kōl* means "stick". *Viral* and *cāṇ* are parts of the human body, and *kōl* is four times bigger than *cāṇ*. Semantically, these linear measures are situated between being extensions of the human body and being extensions of material nature. The basic unit of square measure for wood was the *īrcca* ("saw"/"cut") being made up of 12 *viral* × 24 *viral*. So, this area measure is conceived in analogy to a handsaw and it may originally have referred to a small piece of wood that a handsaw can cut.

The next bigger square measure, *vāḷ* is directly named after the Malayalam word for "handsaw"; it comprises 12 *īrcca*. Cubic measures for wood have already been mentioned in this paper. These measures include the *vaḷa* (1 *viral* × 1 *viral* × 1 *kōl*), the *paleka* (equal to 24 *vaḷa*), the *tūmaram* (equal to 144 *vaḷa*), and the *kaṇṭi* (equal to 576 *vaḷa or* 1 *kōl* × 1 *kōl* × 1 *kōl*). As their names indicate, these terms also designate crucial components in carpentry and construction work in Kerala. Linear, square, and cubic measures provide ample evidence for the distribution of cognition beyond the brain. So-called cognitive terms are embedded in the affordances of the human body, the handsaw, and wooden structures. The interaction of the wood log with the handsaw and with the sawyer is a *thingking* that produces mathematical concepts. These concepts can be regarded as resulting from a semiotic function of materiality instead of as purely conceptual terms.

Certain volume problems throw a spotlight on the interaction between the actions of sawyers and the dynamics of the handsaw and the wood. These verses and their commentaries center on calculations of the volume of a piece of wood. Without the actual shape of the piece being mentioned, a formula is used which applies to calculations of a geometric body with a square base. It is conspicuous that applications of the proportions between the circumference and the diameter are absent from these problems, whereas they are used in other wood-related problems in the *Kaṇakkatikāram*[59] and in *Kanakkusāram* as a step to calculate the size of circular pieces of wood (e.g. "diameter multiplied by 3 + 4/20 is circumference").[60] The mathematical assumption that a piece of wood has a square base must have a connection to the handsaw and to the use of wood in medieval Kerala society.

A probable answer to the question of how this connection worked can be found in the affordances of the handsaw and in the use of wood.[61] A two-handed saw shapes wood only into flat, rectangular, and square shapes; it cannot shape wood

[58] This unit only appears in the wood problems of Māvanān's *Kanakkatikaram*.

[59] See verses 3 and 4 in the Appendix.

[60] We can see the use of this proportion in land-related problems in Māvanān's *Kaṇakkatikāram* as a step to calculate the size of a circular piece of land. Māvanān employed this proportion to calculate the dimension of a circular object in his volume section.

[61] One may also argue that the bark having almost no value leads to approximating a circular piece of wood as a square-shaped piece of wood. However, most of the fine Kerala timbers such as teak have very little bark.

into a circular form. On the one hand, this is due to the shape, sharpness, strength, straightness, flexibility, and temper of the saw's iron blade and teeth, as well as to the positioning of its handles. On the other hand, it is due to the capacity of wood to be sawn only along flat planes (when using a handsaw), that is, due to its "sawability". To draw a contrast: We cannot easily saw granite into flat shapes using a handsaw. This is due to the lack of sawability of granite. The interaction among motions of the handsaw, motions of sawyers, and forces of the wood produce flat surfaces and cubic forms from the cylindrical shape of the tree trunk. These dynamics of wood, handsaw, and sawyers shape the wood into boxes and planks, and generally into pieces that have square or rectangular sides. We can therefore argue that the typical linear, square, and volumetric units of wood are formed by its sawability, and, moreover, that this sawability expresses itself in specific problems contained in the medieval Malayalam mathematical corpus.

As we have already noticed above, in Malayali society, wood was predominantly being employed in flat and cubic forms, e.g. as the crossbeam, the plank, the block, and as square wood pillars (the latter supported the roofs of traditional buildings). Measures for regular flat and cubic shapes such as *vaḷa, paleka, tūmaram* and *kaṇṭi*[62]—which are related to the construction of buildings, bridges and dams, boats and ships, as well as of furniture—testify to this history. Circular pieces of wood, such as round pillars, were used to support buildings, but less commonly than pieces with square bases. Circular pieces were possibly produced using the turning tools of a carpenter. We might legitimately surmise that the lack of an advanced, widespread technology for such tasks points to an uneasy relationship between the material wood, heavy turning tools, and the use of wood in circular shapes in the society of medieval Kerala. Furthermore, as we have shown in this subsection, the specific ways of working with wood, the handsaw, and the many wood-related units of measurement form a tightly knit network. The mathematical problems dealing with wood abound with specific terminology related to felling, cutting, measuring, marking, debarking, sawing, exchanging, and buying that refer to processes of timber production and trade, and above all to the process of sawing. Just as cognition is distributed in the process of sawing; the numerals, measures, procedures, and rules in the mathematical problems emerge through *thingking*—through the interaction of bodily actions of human and the dynamics of material.

In the next subsection, we will turn to the questions of why and how wood became such a predominant material in medieval Kerala. For this, we are going to examine the geographical and ecological contexts in which wood was widely accessible.

[62] This was a very common cubic measure of wood not only in medieval Kerala but also in many other medieval south Indian regions, see Mann (2001, 407) and Pilla (2017, 147).

7.4.4 *Network of Nature: Ecological, Hydrological, and Climatological*

Ecology, hydrology, and climatology are parts of another network of interaction indirectly present in the wood problems. One of the main reasons behind the predominance of wood in the construction of buildings must be sought in the environmental peculiarities of Kerala. Kerala was not only characterized by a shortage of granite and rock-stone, but its ecological, hydrological, and climatological conditions also resulted in an abundance of wood, long coastal lines, the profusion of rivers and rivulets, the concomitant difficulty of the transportation of granite and rock-stone, and the obverse easiness of the transportation of wood.[63] The transport of the available granite and rock-stone from the Western Ghats into the midland and to the coastal lands of Kerala was very difficult, even by elephant, due to the fragmentation of the land-mass of Kerala by forty-one rivers and their tributaries. Transporting such items on the rivers with the help of large rafts was equally challenging due to both the high costs of this method and the precipitous character of the landscape. Contrary to this, the abundance of wood in the forests of Kerala and its easy transportability on the rivers turned wood into a widely available material. As mentioned above, even in eighteenth-century Malabar and nineteenth-century Travancore, the greater part of the land was covered by forests. These forests contained various types of fine timber trees such as teak, ironwood, and jack tree. Shipbuilding and the construction of the traditional buildings of Kerala relied on these kinds of timber.

Both M. G. S. Narayanan and H. Sarkar write on the utilization of certain materials for construction at the detriment of others. The historian of medieval Kerala Narayanan (2013, 50) lucidly deals with the ecological specificities of Kerala when he writes:

> "The supply of solid workable quality of granite is extremely limited on the West coast except in the extreme south of former Travancore. At the same time perishable materials like laterite, which is incapable of being a good medium for sculpture, and timber, which is excellent for a variety of woodcraft, are available in plenty here. This appears to have prompted men to employ granite sparingly, restricting its use to the foundations of the temples alone. The upper portions of all buildings including these temples were mostly built in brick, laterite and wood. [...] The landscape is full of ups and downs and the fact that it is often broken by rivers and rivulets must have made the transport of heavy building material, except through water, extremely tedious and costly in early times".

Sarkar (1978, 132) has explained the extensive use of timber in the construction of the structures (and even in the construction of the walls) of medieval Kerala temples as being due to "high precipitation on the west coast, resulting into the growth of lush vegetation and timber-yielding forest. It was again the heavy rainfall that conditioned the external and internal forms of Kerala's temple architecture".[64]

---

[63] For a detailed discussion about the environment of Kerala see (Gurukkal and Varier 1999, 31–52).

[64] Additional details can be found in Sarkar (1978, 2, 13).

The observations made by the two scholars make clear to us how the forces and flows of nature silently contribute to the formation of the wood problems. Significantly, as we saw in the earlier discussion, in the Sanskrit mathematical treatises from other parts of India we do not find the same kind and quantity of wood problems as in Malayalam collections. The exceptions are the very few wood problems we find in some Sanskrit and Prakrith mathematical treaties.

Senthil Babu's theorization points to the reason for this phenomenon. In considering South Indian mathematics as practice oriented, rather than as a derivative of Sanskrit knowledge, Babu (2022) notes the presence of problems related to rock and stone in the Tamil *Kaṇakkatikāram*, whereas my examination confirmed the absence of wood-related problems there. The Malayalam *Kaṇakkatikāram* has considerable genealogical connections and similarities to the Tamil *Kaṇakkatikāram.* The fact that in the Tamil version problems related to calculation of rock occupy the place of the problems related to sawing, and calculation of wood in the Malayalam version can be plausibly explained by the affordances of the natural landscape of Tamil Nadu, which clearly differs from that of Kerala. We can conclude that the affordances and properties of the natural environment of Kerala play a decisive role in shaping the formation of wood-related mathematical knowledge. This is nothing but the relational agency of the natural environment of Kerala.

The relational agency of nature highlighted in this section shows that knowledge production is not spatially bounded and fixed. On the contrary, we see an array of sometimes subtle forces stemming from different relational actors that are connected trans-regionally in the constitution of mathematical knowledge. If we carefully follow the symbiotic relation between nonhumans and humans, we can understand how this complex trans-regional interaction of relational actors moves the context of formation of the mathematical wood problems beyond the physically fixed boundaries of Kerala. The networks of this trans-regional interaction complicates the very notions of geographical boundedness and fixity that we tend to ascribe to "regionally" made vernacular knowledge. Therefore, the historical formation of wood problems stemmed from knowledge shaped by the shared interests and interactions of natives, foreigners, forests, rivers, woods, handsaws, timber, seas, ships, and so on.[65]

[65] A related argument concerning the natural landscape of Kerala might go as follows: The landmass of Kerala is composed of a conglomerate of multiple ecosystems that have been formed long periods of marine progression and regression. Therefore, the ecology of Kerala, which produces such an abundance of wood, is supported by minerals that have largely been deposited by the sea and which originate in different geographical areas of the world. For a detailed discussion of the formation of the Kerala landmass, see Gurukkal and Varier (1999, 31–56).

7.5 Conclusion

To sum up, we have shown how four interactive networks—(a) the growth of the trees in the forest, (b) the production of timber from trees, (c) the trade of timber for building and shipbuilding that processes timber into wood, and (d) the educational use of this lived knowledge by textualization—transform the relations between trees and humans. This interaction is mediated through wood, measured and numbered timber, and mathematical signs, among others, and gives rise to wood-related mathematical knowledge. We have demonstrated that the wood problems in Indian mathematical treatises can be described as having been shaped by *thingking* that emerges from networks of interaction between the bodily actions (motions and affects) of humans, such as sawyers, carpenters, traders, and teachers, and the dynamics (forces, properties and affordances) of nonhumans, such as forest, river, tree, wood, chisel, handsaw, palm leaf, stylus, land, ship, and sea.

Vastly different networks of interaction are interwoven together through these problems. First, ecological, hydrological, and climatological networks. Second, networks of logging and sawing, in turn interlinked with specific processes of counting and measuring. Thirdly, the networks of trade together with house-building and shipbuilding. And lastly, the use of the problems for educational and recreational purposes. These four networks have four different kinds human and nonhuman interactions. Wood problems reveal themselves as knots between all these heterogeneous beings and becomings.

Acknowledgements I would like to acknowledge the regular interactions with my postdoctoral supervisor Professor Roy Wagner concerning various aspects of this paper. I also want to thank Professor Senthil Babu who co-directed the project "History of Vernacular Mathematics in Medieval South India (Malayalam and Tamil, ninth to sixteenth centuries)" along with Professor Wagner. This project was funded by the Swiss National Science Foundation, and employed me as a postdoctoral researcher. I thank the two anonymous reviewers of this paper for their valuable comments. Finally, I thank Damian Moosbrugger and Niklaus Mueller who proofread this paper.

Appendix[66]

The following verses are copied from the manuscripts, maintaining the original spelling conventions, but introducing word division (where no *sandhi* is broken) and line division. Numeral symbols are underlined. The translations are a compromise between maintaining the literal content and producing a text that would be readable in English.

[66] The five verses in this Appendix are translated and interpreted by Roy Wagner and Arun Ashokan.

Verse 1: TMR manuscript no. 1307B, 1r:9-1v:1.

ആതി നിലം പൊന്നെല്ലരിശി അകലിടത്തിൽ
നീതി തരും കാൽകല്ലും നെരിഴയായി
ഒതയറുപ്പതുമാകച്ചമച്ചേൻ ഒന്നൊഴിയാതവണ്ണം
മെതിലറുപതു കാതൽക്കെ അടെത്തു പൊന്നെനെന്നവാറെ ।

ഇതിൻപൊരുൾ
നിലക്കണെക്കിൻവഴി പൊൻകണക്കിൻവഴി നെൽവഴി അരിശിവഴി
കാൽവഴി കൽവഴി പൊതുവായുള്ള മരക്കണെക്കിൻവഴി
എന്നിങ്ങിനെ ൬ വക കണക്കിനുടെ ആധാരത്തെ ച്ചൊന്നവാറ ।

I have created the complete chant, without omission,
Beginning with land, gold, paddy, rice, in the entire world,
Justly measuring capacity and weight, O bejeweled girl, [or: in the right sequence.]
Enclosing the world's core in but sixty verses, I proceed—so it is said.

This means:
Procedures of land calculation, procedures of gold calculations, paddy procedures, rice procedures,
capacity procedures, weight procedures, general timber calculation procedures—
altogether 6 kinds of procedures. Thus the foundations of calculation are said.

Verse 2: TMR manuscript no. 1307B, 9r:10-9v:2.

This verse and commentary calculate the length of a piece of wood of a given volume and a given depth (which is assumed to equal the width).

എഴു വിരൽ കനമാന മരത്തിനു തൂമരമാനതു ൧൨൳
വീഴുകിലമ്മരമെത്തിന നീളം വടിവൊടു ചൊല്ലുക നല്ല കണക്കാൽ
കൊഴ്ഴയറുംപടി തൂമരവിരലതു വർഗ്ഗിപ്പൂ മരമതു തമ്മിൽ
ശെഷം ചൊല്ലിന കനവിരൽ വർഗ്ഗം കൊണ്ടു കിഴിച്ചാൽ
വന്തതു നീളം

ഇതിൻപൊരുൾ
എഴു വിരൽ കനമുള്ള മരത്തിന ൧൳ തൂമരം കണ്ടാൽ
നീളം എത്തിനാ
എന്നാൽ ചൊല്ലുംപടി
തൂമരക്കനെ വിരൽ വർഗ്ഗിക്ക ൱൪൰൪
അത തൂമരം പത്തെകാലിൽ പെരുക്ക ൲൱൹൰൪൰൪ അത നിർത്തി
൭ വിരലിനെ വർഗ്ഗിക്ക ൪൰൯
൲൱൹൰൪൰൪നെ ൪൰൯ൽ കിഴിക്ക ൩൰൬
അതലാൽ നീളം ൩൰൬ എന്നറിക

From a tree of seven *viral*[67] depth, if 12¼ *tūmaram*[68] are cut down,
State properly the length of this tree by means of correct calculation.
The steps are: Square the *viral* of one *tūmaram,*[69] times [the volume of] this wood;
If the result is divided by the depth-*viral* squared, the length arises.

---

[67] Finger, a length unit.

[68] A volume unit.

[69] A standard *tūmaram* block has depth and width 12 *viral*. The square is 144, as in the calculation below.

This means:
If one has a tree of 10¼[70] *tūmaram* with seven *viral* depth, what is the length?
The steps are:
Square the *viral* of a *tūmaram*: 144.
Multiply by the ten and quarter *tūmaram*: 1764.[71] This is kept.
Square the 7 *viral*: 49.
Divide 1764 by 49: 36.
Therefore, the length is 36. Know that!

Verse 3: TMR manuscript no. 1307B, 9v:6-10r:3.

The verse presents the following problem. A piece of wood has 24 *kōl* length and 3 *kōl* circumference. If the piece is carved off so that the circumference is reduced by a quarter, one is asked to calculate the volume of the remaining square beam. The commentary further instructs to cut one quarter of the outer shell and calculate its volume. It also verifies the results by adding the pieces back together in a different order, using sub-units instead of fractions.

The calculation of the side of the carved square beam involves two steps. First, the circumference is multiplied by 1/4+1/20+1/80 = 5/16, and then multiplied again by 1/2+2/20 = 3/5. This can be interpreted in two ways.

According to one interpretation, the original piece of wood is cylindrical, and the carved piece is a square beam, whose square face is the square inscribed in the circular face of the cylinder. The first step then corresponds to finding the diameter of the original cylinder, which is obtained by dividing the circumference by an approximation of pi. Then the side of the circumscribed square beam is obtained by dividing the diameter by an approximation of square root of 2. The two-step procedure, as well as a statement in the commentary of the next verse, support this interpretation.

However, the total product of the two factors is 3/16, which equals 3/4 × 1/4. This would also fit the scenario where the original piece of wood has an arbitrary shape, and it is carved to yield a square beam whose square face has a circumference lower by one quarter than the original circumference, without necessarily touching the original circumference. Multiplying by the combined factor 3/16 then yields the side of this internal beam. This interpretation fits the volume calculation of the original piece of wood in the commentary of the next verse.

This interpretation is less likely, as it does not explain why 3/16 is broken into 5/16 × 3/5. However, over-complication may have its own pedagogical and social motivations. Either way, we see how the same calculation might end up fitting two different practical situations.

ചുറ്റും മൂന്നിരിപത്തു നാലിതിനീളമിമ്മരം തന്നുടെ
ചുറ്റു നാലിനിലൊന്നു കൂറിനു നെർ പെളിച്ചതഴിഞ്ഞുതെ
മറ്റു ശെഷമഴിതിന്നു തൂമരമുറ്റു നീയുരചെയ്കെടൊ
കാറ്റും നല്ല കണക്കു വല്ലകിലാവൊളം വീരെഞ്ഞത്തന്നൊടെ ।

The piece of wood has circumference three and length twenty-four.
A quarter part of its circumference is sawed off precisely.
Hey, you, say the *tūmaram* of the other part, which remains,
Revealing the good calculation, if you can, having risen so far with me!

---

[70] Note the discrepancy between the commentary and the problem.

[71] This is the product of 12¼ *tūmaram* (as in the verse) by 144.

ഇതിൻപൊരുൾ
ഒരു മരത്തിന നീളം വ്യ്ധർ ചുറ്റ ന്ന അതിൻറെ നാലൊന്ന ചുറ്റിന
നെരെ വാൾ വെച്ച പെളിച്ച പൊയീശെഷത്തിന തൂമരം
എത്തിനായെന്നാൽ
ചൊല്ലുംപടി
ചുറ്റിനെ നാലിൽ കിഴിക്ക വിരൽ ധ്വ്യ
അതിനെ വർഗ്ഗിക്ക ന്നന്വ്യധർ
ഇതിനെ നീളം വ്യധർൽ പെരുക്ക ൌന്ൗൌന്ദൗധന്ദ
ഇതിനെ ഫന്ദർധർൽ കിഴിക്ക ദ്ര്യധർ
ഇതിൻറെ ഉൾച്ചത്തിരത്തിന തൂമരം അറിയുംപടി
ചുറ്റ വിരൽ ൗധവ്വനെ ൯ഫ്മയിൽ പെരുക്ക വ്യധവ്വൾ
ഇതിനെ ൾ[72]ഡ്യയിൽ പെരുക്കുക ധന്നൾ
ഇതിനെ വർഗ്ഗിക്ക ഫന്ദവ്യധവ്വ൯
ഇതിനെ നീളം വ്യധർൽ പെരുക്ക ർന്ൗന്നന്ദൗധർ
ഇതിനെ ഫന്ദർധർൽ കിഴിക്ക ന്ന്യധ൯ഹ്നു
ഉൾച്ചതിരത്തിനൊ തൂമരം ന്ന്യധ൯ഹ്നു എന്നറിഞ്ഞ
ഉൾത്തൂമരം ദ്ര്യധർൽ ന്ന്യധ൯ഹ്നുഠം പൊക വ്യധന്നൾഹ്നു
ഇതിൽ നാലൊന്നിന ദ്രൾഡഡധ പൊക ധൗൾര്ൗമധ
ഉൾത്തൂമരം ന്ന്യധ൯ഹ്നുൽ ധൗൾര്ൗമധയും കൂട്ടത്തൂമരം ർധവ്വ
വിരൽ വ്വ തൊര വ്വ നിന്നകെ
പൊയ തൂമരം ദ്ര വിരൽ വ്യധഫ തൊര ന്ന ആകെ ദ്ര്യധർ
ഒക്കുമെന്നറിക ।

This means:
From a piece of wood with length 24 and circumference 3, a quarter of the circumference is cut off precisely with a saw, what is the *tūmaram* of the remainder?
The steps are:
Divide the circumference by four: 18 *viral.*[73]
Square that: 324.
Multiply this by the length, 24: 7776.
Divide this by 144: 54 [*tūmaram*].
The known steps for its internal square are:
Multiply the circumference in viral, 72, by 1/4+1/16[74]: 22½.
Multiply this by 1/2+1/10[75]: 13½.
Square this: 182¼.
Multiply this by the length, 24: 4374.
Divide this by 144: 30+1/4+1/8.
The *tūmaram* for the internal square is 30+1/4+1/8—this is known.
Subtract from 54, the internal *tūmaram*, 30+1/4+1/8: 23+1/2+1/8.
Subtract its quarter, 5+3/4+3/20+1/160 from 23+1/2+1/8: 17+1/2+1/5+1/80+1/160.
Adding the internal *tūmaram* 30+1/4+1/8 and 17+1/2+1/5+1/80+1/160, the total is 48 *tūmaram*, 2 *viral*, 2 *tōra.*[76]
With the removed 5 *tūmaram*, 21 *viral*, 6 *tōra* the total makes 54—know that!

[72] In the manuscript: erroneously ൭.

[73] 3 *kōl* divided by 4 is 3 × 24/4 = 18 *viral*.

[74] This number is an approximation of 1/pi, which serves to derive the diameter from the circumference.

[75] This number is an approximation of the inverse of the square root of 2, which serves to derive the side of the internal square from tis diagonal (which is also the diameter).

[76] A *viral* here is 1/24 *tūmaram* and *tōra* 1/8 *viral*.

Verse 4: TMR manuscript no. 1307B, 10r:3-10v:3.

The verse expands on the problem from the previous verse with different data. The piece of wood has 20 *kōl length* and 4 *kōl* circumference. It is carved so that its circumference is reduced by a quarter, and one quarter is cut from the remaining external shell. The problem is to calculate the volumes of the carved beam and of the quarter of the shell. Moreover, if the price of the quarter of the shell is given as 13¾ *paṇam*, one is to calculate the price of the entire piece of wood.

The calculation of the internal square follows the same procedure as in the previous verse. Then, the volume of the entire piece of wood is calculated, so the difference is the volume of the external shell. After calculating the volume of a quarter of the external shell, the rule of three is used to derive the cost of the entire piece of wood.

Referring to the interpretations of the procedure noted in the previous verse, we find ourselves in an odd situation. The calculation of the volume of the side of the internal square beam is performed in two steps as above, noting explicitly that the first multiplication derives the side of the circumscribing square (which equals the diameter). This fits the interpretation of the original piece of wood as cylindrical.

However, when calculating the volume of the original piece of wood, one simply divides the circumference by 4, squares the result, and multiplies by the length. This fits the situation where the original piece of wood is actually a square beam, and a smaller square beam is extracted from it. The text does not reflect on this tension, and never explicitly notes the shape of the original piece of wood. Given the approximations involved in the calculation, both methods might be considered practically identical or exchangeable.

വീരാ മരമതു ചുറ്റെയുരത്താൽ ഉൾച്ചതിരത്തിനു പുറമെയുള്ള
നെരെ നാലൊരു പുറമീർന്നതു പൊയി പിന്നെതെത്തിര എന്നാൽ
പാരാ ഉൾച്ചതിരത്തിന തൂമരമറിവൂ ഒക്കെക്കുള്ളതുമറിവൂ
വീരാ ഉൾച്ചതിരത്തിനെ നീക്കിയ ശെഷത്തിൽക്കുറെയും
നാലൊന്നും വഴി കൂറു കെൾ മുന്നെ വഴി ।
നീളമൊരിരിപതു ചുറ്റതു നാലായിക്കണ്ട മരത്തിന്നുൾച്ചതിരത്തിന
കെളിയതെറുംപടി
നെരെ ഒരു പുറമീർന്നു കൊടുത്തെനതിനു പണം കെൾ
ആയിനതെട്ടൊടഞ്ചാമരെക്കാൽ അതിനുടെ തൂമരവും
ശെഷത്തിനു നീളം കുറയാതെ തൂമരവും അതിനു പിടിക്കും
പണവും ചൊല്ലെ ।

Hero! If the circumference of a piece of wood is given, how much [volume remains] after sawing precisely one quarter of the shell outside the internal-square?
Observe! Know the *tūmaram* of the internal square! Know the entire joint [volume] content [of the piece of wood]!
Hero! Say the method for the remaining portion after removing the internal square and for its quarter, hear the previous method!
Given the length twenty and circumference four, listen to the pleasantly clear steps:
Note the *paṇam* for the precise shell obtained by sawing.
If it is eight with five, a half, and quarter [i.e. 13¾], say its *tūmaram* and the *tūmaram* of the remainder without neglecting the length, and the *paṇam* obtained for that.

ഇതിൻ പൊരുൾ
നാൽകൊൽ ചുറ്റിൽ ൨൰ കൊൽ നീളമുള്ള മരത്തിന
ഉൾച്ചതിരം കഴിച്ച
പുറവട്ടിന ഒരു പുറത്തെ മരത്തിന ൧൩൵ൽ പണം ആനാൽ
മരത്തിന തൂമരം എത്തിനാ എന്നും പണം
എത്തിനായെന്നു ചൊൽക

എന്നാൽ ചൊല്ലുംപടി
ചുറ്റിനെ വിരലാക്ക ന്ദ്ധന്നു അതിനെ കാലെമാകാണിയിൽ പെരുക്ക
ന്ന്ധ അത പുറച്ചതിരം
അതിനെ അരെയിരുമയിൽ പെരുക്ക ധവ്വ
അതിനെ വർഗ്ഗിക്ക ന്നൃവ്ധർ
അതിനെ നീളം വ്ധൽ പെരുക്ക ന്നന്ദ്ര്ന്നവ്വധ
അതിനെ ഫന്ദ്ധർൽ കിഴിക്ക ദ്ധദ്ര അത ഉൾത്തൂമരം എന്നറിഞ്ഞ
ന്ദ്ധന്നു വിരൽ ചുറ്റിനെ ദ്ൽ കിഴിക്ക വിരൽ വ്ധർ
അതിനെ വർഗ്ഗിക്ക ദ്രന്നൗധന്നു
അതിനനെ നീളം വ്ധൽ പരെുക്ക്ക ധഫന്ദ്രന്നവ്ധ
അതിനെ ഫന്ദ്ധർൽ കിഴിക്ക വ്വധ
അതിൽ ഉത്തൂമരം ദ്ധദ്ര◌ം പൊക ന്ധദ്ര
ഇതിൽ നാലൊന്നിന വ്വഞ്ൽ തൂമരത്തിന ധന്നഹ്നൽ
പണമെന്നറിഞ്ഞ
ഇനി തള്ള പിള്ള പെറ്റുവാൾ വഴി
വ്വധം ധന്നഹ്ന പരെുകക ഫന്ദ്രധ
വ്ധദ്ര[77]നെ വ്വഞ്ൽ കിഴിക്ക ഫന്നവ്ധ
അതലാൽ ഫന്നവ്ധ പണമെന്നറിക ।

Its meaning:
From a piece of wood with four *kōl* circumference and 20 *kōl* length, the internal square is removed.
If for the wood of one side [i.e. quarter] of the outer shell one gets 13+1/8 *paṇam,*[78] please say how much is the *tūmaram* [volume] and the *paṇam* for this wood.
The steps are:
Multiply the total *viral* of the circumference, 96, by a quarter, a twentieth and an eightieth: 30. This is the [side of the] outer square [i.e. the diameter].
Multiply this by half and two twentieths: 18.
Square this: 324.
Multiply this by the length, 20: 6480.
Divide this by 144: 45. This is internal *tūmaram*—this is known.
Divide the circumference, 96 *viral*, by 4: 24 *viral.*
Square this: 576.
Multiply this by the length, 20: 11520.
Divide this by 144: 80.
Subtract from that 45 *tūmaram*: 35.
For its quarter, 8 ¾ *tūmaram*, 13+1/8 *paṇam*, this is known.
Then, according to the rule of three [literally: mother, child, and expecting mother]
Multiply 80 by 13+1/8: 1050.
Divide 25 [should be: 1050] by 8¾: 120.
Therefore, this is 120 *paṇam*—know that!

Verse 5

This verse is based on Māvanān's verse (pages 107:26–108:7, Māvanān prepared this in 1863) and supported by a verse in T365 (pages 41:3–42:3, this was copied in a year in between 1875–1884).[79] These verses, which deal with the measure of wood and labor in

---

[77] Should be ഫന്ദ്രധ, as in the line above.

[78] This is a monetary unit.

[79] T365 has three other lines. They discuss the process of cutting and sawing, the calculation of sawing labor, and the relation of the hands and the sawn wood.

timber production, do not have commentaries. These two verses are very similar, but some of their phrases are hard to interpret.

M: നാലു കോലതു ചുറ്റു തങ്കിന നന്മരം മുറി-വീഴ്ത്തുവാൻ,
ആളുമങ്ങ്ങൊരു തച്ചൻ, ചന്നെതിൻ-മൂടു-ചുറ്റു ചുഴന്നു-കാൺ!
T: നാലു കോലതു ചുറ്റും തങ്കിന നന്മരം മുറി-മുറി-വീഴ്ത്തുവാ-
നാളുമുണ്ടുങ്ങ തച്ചു ചെന്നതിൽ മൂടു-ചുറ്റുമുഴന്നു-കാൺ
To cut and fell a good tree with circumference four *kōl*,
a carpenter with a slave/worker, look around the present bottom circumference! [or: at the gathering of the carpenter and slave/worker, look around the bottom circumference!]

M: ഊന-മങ്ങുഴലയ്ക്ക നൽ-വരി കണ്ടതിൻ-പരിചൂന്നിനാൻ
വേണി നിന്നു വണങ്ങി, നീയടി നാളികേരമനുജ്ഞ കെൾ!
T: ഊതമങ്ങുഴലക്കു നൽ-വരി കണ്ടതിൽ പരിചിന്നിനി
വേണി നിന്നു വണങ്ങി നീയറി നാളികേരം മനിജ്ഞകൾ
He considers the proper manner of looking carefully for defects and rot in the beam,
Bowing at the root, you, [ceremoniously] break a coconut, ask for permission!

M: അനുജ്ഞ കേട്ടു തുടങ്ങി, വെട്ടി-മുറിക്ക, വീഴ്ത്തുക തച്ചിനാൽ
അരിപ്പം, ചുറ്റിനു കോലോരോന്നിനു തച്ചു വേണ്ടും എറിക്കുന്നാൾ
T: അനിഞ്ഞു കേട്ടു തുടങ്ങി വെടി മുറിച്ചു വീഴ്ത്തുക തച്ചിനാൽ
അരിപ്പം ചുറ്റിന കോലൊരോന്നിനു തച്ചു വേണ്ടുമെറിക്കുന്നാൾ
After asking for permission, cut and hack, it takes a workday to fell [the tree].
Arduously, for each *kōl* of circumference, the slave/worker removing the bark needs a workday.

M: മുറിക്ക അഞ്ചതു കോലു നീളം മുറിച്ചെറിച്ചതു കൊണ്ടിടാ
ഉയർച്ച കാലിഴ, വച്ചതേറ്റുക തച്ച രണ്ടിനു, നൂൽക്കുറി!
T: മുറിക്കയഞ്ചതു കോലു നീളം എറിച്ചെറിച്ചതു കൊന്നിട
ഉയർച്ച കാലിഴ-വർച്ചതേറ്റുക തച്ചിരണ്ടിനു നൂൽക്കുറി
Cut five *kōl* in length! Take the cut and scraped [wood]!
Scrape from top to bottom! To lift [the wood] is work for two, mark [it] with a rope!

M: [കുറിച്ചതിന്നു കറുപ്പു-നൂൽ], വാൾ കൊളുത്തി, നടുപ്പുളാ
മറിച്ചു, വേണ്ടുമുരുവിനും കനം വീതിയും കൊടുവിന്നു. നൂൽ
അടുക്കി-വൈത്തതു അളന്നു-തീക്കിൽ, എളുപ്പമിച്ചു-കുറി
ക്കവെ!
പെരിക്കിലക്കനം പാടുമങ്ങുള-പാതിയങ്ങതിലീർച്ച കൊൾ!
T: കുറിച്ചതിന്നു കറുപ്പു-നൂൽ ക്കുറി വാൾ കൊളുത്തി നടുപൂള
മറിച്ചുരുവിനു വേണ്ടുമകനം വീതിയും കൊടുവിന്നു നൂൽ
ഇരിക്കിലക്കനം പാടും മഞ്ഞളു പാതിയങ്ങതിനീർച്ച ചൊൾ
അടുക്കി വച്ചത അളന്നു തിക്കിൽ എളുപ്പമിച്ച കുറിക്ക
For this marking [use] a blackened rope. Using a saw, it is turned over
around the middle for getting the desired shape, depth and width. The rope
set along a line, if you tighten it while measuring, you may mark [the measurement]
with ease.
By multiplying [the measured width] by half the *aṅṅuḷa*[80] that occur in the depth, obtain the *īrcca*[81]!

[80] This is the Sanskrit synonym for *viral*.

[81] A surface area unit, equal to 1/12 of *kōl* by *kōl*.

Bibliography

Agius, Dionisius A. 2008. *Classic Ships of Islam. From Mesopotamia to the Indian Ocean*. Boston: Brill.

Antony, P., ed. 1994. *Payyannūrppāṭṭu pāṭhavum pāṭhanaṅṅaḷum*. Kottayam: DC Books.

Ashokan, Arun, Naresh Keerthi, P. M. Vrinda, and Roy Wagner. 2024. *Draft of an edition of Kaṇakkatikāram - Research Collection* (ethz.ch). ETH Zurich, Switzerland.

Babu, Senthil D. 2022. *Mathematics and Society. Numbers and Measures in Early Modern South India*. Oxford: Oxford University Press.

Bhayani, H. C. 2012. *Apabhramsa Language and Literature*. New Delhi: Rashtriya Sanskrit Sansthan.

Bukharin, Nikolai I. 2013 [1931]. Theory and Practice from the Standpoint of Dialectical Materialism. In Nikolai I. Bukharin et al. (eds.). *Science at the Cross Roads. Papers from the Second International Congress of the History of Science and Technology 1931*, pp. 11–33. Abingdon: Routledge.

Eyyanchery, Anvar. പഴയകാല മരം ഈർച്ച (video). Upload date: 24.01.2021. https://www.youtube.com/watch?v=FhvtbAKU5C4. Accessed: Feb 1st 2023.

Ganesh, K. N. 2016. *Reflections on Modern Kerala*. Thrissur: Cosmo.

Goitein, Shlomo Dov, and Mordechai Akiva Friedman. 2008. *India Traders of the Middle Ages. Documents from the Cairo Geniza ("Indian Book")*. Boston: Brill.

Govi, K. M. 1998. *Ādimudraṇam Bhāratattilum Malayālattiḷum*. Thrissur: Kerala Sahitya Academy.

Gurukkal, Rajan, and M. R. Raghava Varier. 1999. *Cultural History of Kerala*. Vol. 1. Thiruvananthapuram: Department of Cultural Publications.

Habib, Irfan. 2010. *Economic History of Medieval India, 1200–1500*. Delhi: Pearson.

Hattori, Masaaki. 1968. *Dignäga, On Perception, Being the Pratyaksapariccheda of Dignaga's Pramänasamuccaya from the Sanskrit Fragments and the Tibetan Versions*. Massachusetts: Harvard University Press.

Heidegger, Martin. 1967. *What Is a Thing?* Indiana: Gateway Editions.

Heroor, Venugopal D. translated and annotated. 2014. *Brahmaguptagaṇitam: Ganitadhyaya of Brahmasphutasiddhanta*. Ernakulum: Chinmaya International Foundation.

Heroor, Venugopal D. translated and annotated, 2015. *Sridhara's Triśatika/Pāṭīgaṇitasāra*. Ernakulam: Chinmaya International Foundation.

Høyrup, Jens. 1990. Sub-scientific Mathematics. Observations on A Pre-modern Phenomenon. *History of Science* 28 (1): 63–87.

Høyrup, Jens. 2007. Sub-scientific Mathematics. Undercurrents and Missing Links in the Mathematical Technology of the Hellenistic and Roman World, pp. 1–49 [unpublished manuscript written in 1990 for *Aufstieg und Niedergang der römischen Welt* II, vol. 37(4) which remains unpublished].

Høyrup, J. 2017a. Practitioners – School 'Mathematicians': The Divisions of Pre-Modern Mathematics and Its Actors. In J. M. Steele, & M. Ossendrijver (Eds.), Studies on the Ancient Exact Sciences in Honour of Lis Brack Bernsen (pp. 207–224). Edition Topoi. Topoi - Berlin Studies of the Ancient World Vol. 44 https://doi.org/10.17171/3-44-11

Høyrup, Jens. 2017b. Practitioners – School Teachers – 'Mathematicians'. The Divisions of Pre-Modern Mathematics and Its Actors. In *Studies on the Ancient Exact Sciences in Honour of Lis Brack-Bernsen*, ed. Mathieu Ossendrijver and John M. Steele. Berlin: Topoi.

Ingold, Tim. 2013. *Making. Anthropology, Archaeology, Art and Architecture*. New York: Routledge.

Iyer, L. K. Anatha Krishna. 1981. *The Tribes and Caste of Cochin*. Vol. I. New Delhi: Cosmo Publications.

Kizhakke Nedumpally, Sunandan. 2012. *Ways of Knowing. Asaris, Nampoothiris and Colonialists in Twentieth Century Malabar, India*, PhD thesis. Emory University.

Kramrisch, Stella. 1961. *Dravida and Kerala in the Art of Travancore*. Ascona: Artibus Asiae.

Latour, Bruno. 1999. *Pandora's Hope: Essays on the Reality of Science Studies*. London: Harvard University Press.

Logan, William. 1951. *Malabar Manual*. Vol. 1. Madras: Government Press.

Long, Pamela O. 2011. *Artisans/Practitioners and the Rise of the New Science. 1400–1600*. Corvallis: Oregon State University Press.

Lugli, Emanuele. 2019. *The Making of Measure and the Promise of Sameness*. London: University of Chicago Press.

Malafouris, Lambros. 2008. "At the Potter's Wheel: An Argument for Material Agency", in: Carl Knappett/Lambros Malafouris (eds.). Material Agency. Towards a Non-Anthropocentric Approach, New York: Springer, pp. 19–36.

Mann, Michael. 2001. Timber Trade on the Malabar Coast, c. 1780–1840. *Environment and History* 7 (4): 403–425.

Mateer, Samuel. 1981. *The Land of Charity. A Descriptive Account of Travancore and its People*. London: John Snow and Co.

Mathew, K. S. 2018. *Shipbuilding, Navigation and the Portuguese in Pre-modern India*. New York: Routledge.

Menon, C. Achutha. critically ed. 1950a. *Kanakkusāram*. Madras: Government of Madras.

Menon, C. Achutha. critically ed. 1950b. *Kanakkusaram Bālaprabōdham*. Madras: Government of Madras.

Narayanan, M. G. S. 2013. *Perumals of Kerala*. Thrissur: Cosmo Books.

Odegard, Erik. 2019. Construction at Cochin: Building ships at the VOC-yard in Cochin. *The International Journal of Maritime History* 31 (3): 481–494.

Overmann, Karenleigh A. 2019. *The Material Origin of Numbers. Insights From the Archaeology of the Ancient Near East*. Piscataway: Gorgias.

Pilla, Lakha. 2017. *Making the Colonial Timber Empire in Travancore. A critique of Colonial Forest Policy*, Ph.D. thesis, Mahatma Gandhi University.

Pillai, Sreekanteswaram G. Padmanabha. 1998. *Sabdataravali*. Kottayam: National Book Stall.

Radhakrishnan, Methil. 2013. *Mētilkathakaḷ*. Kozhikode: Mathrubhumi.

Rao, A. B. Padmanabha. Translated and edited, 2014. *Bhāskarācārya's Līlāvatī*, Ernakulam: Chinmaya International Foundation.

SaKHya. Edited with Introduction, Translation, and Mathematical Commentary, 2009. Gaṇitasārakaumudī. The Moonlight of the Essence of Mathematics by *Ṭhakkura Pherū*, New Delhi: Manohar Publishers.

Sarkar, H. 1978. *An Architectural Survey of Temples of Kerala*. New Delhi: Archaeological Survey of India.

Schildt, Henri. 2012. *The Traditional Kerala Manor. Architecture of a South Indian Catuḥśāla House*. Pondicherry: French Institute of Pondicherry.

Seyitu Muhammadu, Māvanān Māppiḷa. Year of preparation 1863. *Kaṇakkatikāram*. Archived at: https://archive.org/details/kanakkadhikaram1863mavananmappilaV1. Another partially preserved copy is archived here: https://archive.org/details/kanakkadhikaram1863mavananmappilaV2

Smith, Pamela H. 2012. *The Body of the Artisan. Art and Experience in the Scientific Revolution*. Chicago: The University of Chicago Press.

Subrahmanyam, Sanjay. 1990. *The Political Economy of Commerce. Southern India 1500–1650*. Cambridge: Cambridge University Press.

Swami, V. V., ed. 2011. *Ceṅṅannūrādi*. Kottayam: D.C. Books.

Wagner, Roy and Ashokan, Arun. 2024. "The content and context of Kaṇakkatikāram Manuscripts: Pre-Modern Malayalam Elementary Mathematics", History of Science in South Asia 12: 1–31 https://journals.library.ualberta.ca/hssa/index.php/hssa/article/view/103/146

Zilsel, Edgar. 2000 [1941]. The Sociological Roots of Science. *Social Studies of Science* 30 (6): 935–949.

Chapter 8
Evaluating Prestige Commodities in Early Modern South India: Calculating with Pearls

Muthu V. Prakash and Tathagata Sengupta

Abstract In this essay we attempt to reconstruct some of the measuring practices associated with the evaluation of pearls in late medieval/early modern south India, from a set of available Tamil texts called *Muttukaṇakku*—'pearl computations' or 'pearl arithmetic'. Such practices of computation, involving memorization, calculation, comprehension, the use of physical instruments, and the teaching and learning of such practices, were key activities that enabled the highly competitive and speculative economy of prestige commodities like pearls. Paying close attention to such texts, including their silences, such as about the pearl divers themselves, allow us to specify the role of such texts in the economies of competition over prestige and of labor obligations.

Keywords Pearl trade · Economic value · Accounting · Labor · Indebtedness · Speculative capital · Tamil Nadu · Indian Ocean · Accounting texts · Mathematics · Education

8.1 Introduction

This paper is an attempt to reconstruct some of the practices associated with the valuation of pearls in south India through a set of available Tamil texts called *Muttukaṇakku*—'pearl computations' or 'pearl arithmetic'. Evidence related to the actual practices of computing the value of pearls is sparse, limited to the few colonial reports that obliquely refer to these calculations. Hence, a text like the *Muttukaṇakku* becomes important in a reconstruction of the calculation practices. However, like in any such attempt at reconstructing actual practices based on

M. V. Prakash (✉)
International School for Jain Studies, Pune, India
e-mail: prakash.v@ifpindia.org

T. Sengupta
Independent Researcher, Kolkata, India

Senthil Babu D., R. Wagner (eds.), *Normative Texts and Occupational Practices*, Boston Studies in the Philosophy and History of Science 350,
https://doi.org/10.1007/978-3-032-28419-8_8

normative texts available to historians, we can only be tentative given that such texts need not be faithful records of the actual practices. Moreover, the texts focus on measuring pearls for the markets without mentioning the realities of the labor and social relations involved in producing or procuring the objects of measurement. Still, we believe certain observations can be made from the logic of the text, and in relation to existing secondary literature on the history of pearl fisheries and trade in these coastal landscapes.

We begin with a brief introduction to these texts, their possible provenance and relation to the pearl trade in southern India. Then we introduce the various stages in the measuring of pearls as reconstructed from the text, along with the physical apparatus allowing such computations—from sieves and weight systems, to the counter's fingers and knuckles. For each stage of measurement, such as grading, sorting, weighing and measuring, followed by algebraic calculations of the final value (*'cevvu* value'), we try to provide appropriate explanations as interpreted from the text. Such practices of computation, involving memorization, calculation, comprehension, the use of physical instruments, and the teaching and learning of such practices, are a form of labor in and of themselves—a 'computational or accounting labor', geared towards the market systems in place. We also compare the *Muttukaṇakku* computations with those in other traditional pedagogic texts such as the *Kaṇakkatikāram*. Finally, we briefly comment on how such deterministic arithmetic accounting algorithms and practices relate to questions of labor, dignity, and uncertainty in the lives of the pearl divers themselves that such texts do not talk about.

8.2 Geopolitical and Economic Contexts of Pearl Fisheries and Trade

The Gulf of Mannar has been historically one of the key pearl fishing regions in the world, tightly controlled by the rulers and big traders. The long history of pearl fisheries and trade in the Gulf has been recorded and studied across the centuries. We have information available from as early as medieval and early modern sources, such as travelers' accounts (Hornell 1905, 2), inscriptions and copper plates by local rulers,[1] as well as colonial reports. Officials of the colonial governments of Ceylon and India, such as James Steuart (1843), G. Vane (paper presented at a conference in 1886 and published in 1888), Edgar Thurston (1894), W.A. Herdman (1903), James Hornell (1905), and S. Sivalingam (1961), also published extensive reports on pearl fisheries in the Gulf of Mannar. Apart from such foreign notices and reports, S. Arunachalam's *The History of the Pearl Fishery of the Tamil coast* (1952) is a pioneering work on pearl fisheries. In more recent times, Deckla has written an

[1] For the Cholas see Hultzsch (1991, 21; 1892, 178), for the Pandiyas see Sastri (1972, 31), for the Nayakas see Steuart (1843, 102), Hornell (1905, 3–11), for the Sethupatis see Kamāl (2017, 348–349).

exceptional thesis on Portuguese pearl fishing in Thoothukkudi (2004). However, very few of these accounts discuss how the value of pearls was computed and the role of the accountants in such a trade. It is in this context that the hitherto unpublished manuscript, the *Muttukaṇakku* (literally meaning, "pearl computations", hereafter MK), becomes significant as a normative ready reckoner of such computations.

South Indian pearl diving has been traditionally associated with the Parawa fishing community of Thoothukkudi, a coastal region in southern Tamil Nadu (Cāminātaiyar 1935, 24; Hornell 1905, 2; Sastri 1972, 31). Arab Muslims are also known to have been involved in pearl trade from early on, as well as in the grading and valuing of pearls (Vane 1887). When international maritime trade in the Indian Ocean was disrupted in the later period (fourteenth century onward), particularly by the Portuguese and other European raiders and companies, leading to new business conflicts, a commonly agreed and textually 'transparent' protocol for computation seems to have become necessary. This seems to be the time when the MK was composed, between the sixteenth and eighteenth centuries, by compiling and standardizing already existing practices.

Speculative Prestige and Indebted Labor in the Indian Ocean Our primary sources, as well as existing literature on the history of pearl trade, point to the markets of speculation and gambling within which such trade and its labor practices were located: from the trade in uncertain economic futures and profits above the seas, through the deep dives under the uncertain waters, to the uncertainties and gambles in the economic lives of the people involved, particularly the pearl divers (Ostroff 2016; Fernando 2022; Deckla 2004). Writing about nineteenth century commerce in the "maritime Indian Ocean *bazaar*", Bishara and Wint note how at the core of such an economy "were the bonds of debt and credit—of obligation, more generally—that bound economic actors together, from Basra all the way down to Zanzibar" (Bishara and Wint 2021, 52). These bonds of debt were 'most visible' on board the pearl *dhows* that marked the seascapes, carrying the pearl divers around the Persian Gulf, who were indebted and bonded to the captains (*nakhoda*).

Based on the labor of such debt-bonded divers, a high-risk high-profit economics of speculation by rich financiers worked as the main driver of the trade. The key investors for pearling dhows were merchants invested in the economies of speculation and prestige commodities. For example, East African ivory traders. In the seventeenth century travelog *Ship of Sulaimān (Safīna'i sulaimānī)*, the Persian scribe Muhammad Rabi'bin Muhammad Ibrāhīm described pearl fisheries off Ceylon as the "world market of fate" (O'Kane 1972, 168–69).[2] Early British administrators also characterized the pearl fisheries as a site of "gambling", by this time acquiring new significance by being mapped onto wider colonial discourses about the

[2]A discussion of the text is found in Ostroff, *The Beds of Empire*. Echoes of this can also be found in the language used by present historians such as Sanjay Subrahmanyam, who describes the scene as "The Great Lottery", and the fishing episode itself as "Part fair and festival, part industry, and part pageant".

economic behavior of Indian merchants as "reckless", "rapacious" and "superstitious" (Ostroff 2016, 217). The presence of pearl gamblers and other "loose characters" at the fishing parties have been noted by even later colonial ethnographers and officers, such as in the correspondences of Ceylon Governor Henry Arthur Blake (Subrahmanyam 1996), or in reports by Leonard Woolf of the Bloomsbury group of London intellectuals, who visited and described the Mannar fisheries in 1906 (Woolf 1961, 87–98). The highly speculative and risk-laden nature of the industry is known to have often brought "even the most prudent and well-heeled renters to the brink of ruin" (Ostroff 2016, 208). Fishing rights were decided through bids. Successful bidders usually brought together groups of investors for the purpose of renting a pearl fishery, while also engaging in other acts of social and cultural entrepreneurialism that augmented their status and reputation – such as through charity contributions to religious institutions, involvement in community and caste affairs, and in general the economies of "honor, trust, and friendship" that mediate interpersonal and business relationships. In other words, "the rent of the pearl fishery demanded the expenditure of capital in all its forms", as noted by Samuel Ostroff in *The Beds of Empire* (Ostroff 2016, 209). Tamara Fernando calls for tracing pearls to their animal/mineral origins as a way to "read capitalism in context", a shift from the language of commodities to "ecologies", as a means of reading beyond the "simplifying logic of capitalism"—that of singling out and simplifying in order to scale up (Fernando 2022). Our reading of pearl computations suggests how paying proper attention to the political economy of risks, uncertainties and violence could provide a way to understand the interplays of capital, labor and ecologies, going beyond such simplifying logic of sorting and scaling. For instance, the dynamic 'slot machine' nature of the otherwise static-looking arithmetic sorting apparatus of the colander system, as we shall describe, becomes clearly visible when we consider its role in 'stochastic wage' payment systems for the divers. This is important to note given the impression we get from such texts about the measured value of the pearls being 'intrinsic' or mathematically 'self-evident', based on their sizes and weights. Our point is, to reiterate, that the 'value of pearls' as measured in such texts should be read *in relation* to the value of social lives and labor, such as that of the divers, that the text itself does not speak of. But for now, we turn to the texts in question.

8.3 The Texts and the Sorting Apparatus

Introducing the Texts For this essay, we studied two available *Muttukaṇakku* manuscripts, one in the French Institute of Pondicherry (IFP) (Mss.No. RE 33705) and another in Oriental Manuscript Institute Library (ORI), GOML, Thiruvananthapuram (Mss.No. 8086). The author of the *Muttukaṇakku* is someone called Pachaimuthu Suthan Subramanian. No other information is available to us at this point about him, other than his name. Another printed text, more prosaic in nature, called the *Muttu Cevvu Kaṇakku* (MCK) by Kudavatti Mohammad Leppai, published in 1881, which is also a compilation of pearl arithmetic practices that prevailed in the nineteenth century, allows us to further understand the *Muttukaṇakku* manuscript (Leppai 1881).[3] The MK manuscript copies are in ornamental verses whereas Leppai's MCK is a more business-like list of the calculations. The details furnished in this essay however are based directly on the MK, rather than the MCK. The MCK helped in a standalone sense to understand the MK text better.

The provenance and period of the MK are not clearly known. However, the editors of these texts contend that the available copies could be from the nineteenth century, even if the texts themselves could be older, as far back as the sixteenth century (Prakash and Babu 2024). These texts were oriented towards the needs of the accountant, equipped with a set of instruments and tables, possibly working for the pearl trader. Though from the information directly available in these texts, it is difficult to ascertain to what extent such practices and guidelines were strictly followed, they nonetheless do provide a window to the ways of calculating, involving bodily actions using fingers and digits as counters, as well as the heavy use of orality and performativity in the act of counting, anchored in training of the mnemonic arts of computation in the early modern Tamil region. Typical of such texts, they also reveal how the technical computational operations were embedded, justified and coded using aesthetic forms and values through ornate literary compositions (Babu 2022). The performative and literary elements in these texts together denote a pedagogic culture of teaching computation through cultivation of recollected memory, enabled through such literary forms. Before the numerical calculations are applied, qualitative graded hierarchies between different classes of pearls were established through literary descriptions. Such descriptions attributed to the bigger pearls unique anthropomorphic characters, aura and charisma, honorific epithets and adjectives, modeling the identity and prestige of the bigger pearls after existing hierarchies in society. On the other hand, smaller pearls were reduced into generic, lackluster objects, devoid of special individual identities, anonymized into groups and bunches. The lowest level pearls were mostly sold in bulk, by weight, instead of individually passing through the intricate measuring apparatus intended for the 'high quality' pearls.

The MK text primarily contains tables of pearl values, called the *cevvu*, measured in terms of the traditional units of weight measures in the Tamil system, such as the *kaḻañcu* and the *mañcāti*. The text follows steps starting with sorting the pearls, classifying them into grades, followed by computation of weights, and eventually algebraically calculating the *cevvu* value—a multiple of the *square of the weight* of a pearl. In all this, physical instruments such as the pearl colander, or sieve, played a central role. The colander is a graded, hierarchical arrangement of sieves whose purpose was to classify the pearls in terms of their sizes. The first part of the text explains the geometric construction of the sieves constituting the colander. After this, the text moves on to the arithmetic around weight measures, backed

[3] Apart from the sole information from the text he published, no other information is available on this author. However, the orthography of the *Muttukaṇakku* manuscript suggests that it must be older than this 1881 text. So, it is unlikely that Mohammad Leppai is the same person mentioned in the manuscript.

up by tools such as the 'rule of three', for cross verification. The text itself operated as a kind of ready reckoner in which the arithmetic procedures for different weight units are repeated separately, even if the underlying arithmetic logic stayed the same. It is almost as if nothing could be left for interpretation, extension or even generalization. In this essay, we follow the above-mentioned structure of the text to take the reader through the consecutive stages of calculation, which forms a crucial stage in the long value chain of the pearl, from the ocean beds to the marketplace of the empires. Once the pearls were brought ashore, dried, cleaned and prepared for measurement and valuation, they passed onto the supervision of the accountant, who began by first sorting the pearls using the sieves.

The Colander System Colanders, as a graded tower of sieves, performed the function of classifying pearls according to their sizes. Size-based segregation of the pearls would be eventually tied to the weight-based computations, leading to their final valuation. The number of holes in a sieve seems to have varied across different markets, though the MK text itself tries to standardize such numbers. The colanders were central to the valuing of pearls, and seemed to have provided the valuers with a quick and consistent way of estimating the total value of groups of pearls, based on their places in the sieve hierarchy. Qualitative descriptions of the colander system, as in the textual verses, seem remarkably self-conscious about the graded hierarchy that they embody. The first four sieves were called "*cevvu* sieves" (*cevvu peṭṭi*). They contain the best pearls, in terms of their size. The next three sieves of sizes 100, 200 and 400, are called *vaṭivu peṭṭi*, or the "beautiful sieves". These contain the middle classes of pearls. The last three (sizes 600, 800 and 1000) are called *tūḷ peṭṭi*—"very small" in size. These contain pearls of the least computation-worthy value (Fig. 8.1).

The Geometry of Holes in a Sieve While there must have been local variations, the MK nonetheless attempts at prescribing standardized distribution of holes across various sizes of sieves. The number of holes on a colander would be the same as the number of pearls that would get 'stuck', or 'stand' in it when the catch of pearls is passed through it. One of the initial verses of the text prescribes a way to compute the number of holes given the size of the colander. For this purpose, the colander is assumed to be a regular hexagon in the sense of a 6-*cornered* figure (*aṟukōṇam*).

The verse asks one to calculate the number of holes along one such "corner" or angle as we would call them today. Next, one needs to multiply this number with itself ("square it"). The result thus obtained needs to be scaled ("reduced") by a factor of $\frac{3}{4}$, and the result then needs to be increased by adding a $\frac{1}{4}$. This is supposed to give us the total number of holes in a colander.[4]

Like generally seen in the Indian canonical tradition, the *Muttukaṇakku* also doesn't explain its algorithms. It is not supposed to be a justificatory text, but a

[4]The original verse reads: "Aṟukōṇat torukōṇat tatilē nērē, Yaṉpākat tamartaṉai eṇṇik koṇṭu, Caṟukāmal kaṇṭataṟkup patilē vaittut, Tayavāka iruvakaiyum perukkic cērttuc, Ciṟukavē mukkālil kaḻittu, niṉṟa Cēkarattil kāl kūṭṭit tukaiyē".

Fig. 8.1 Pearl colanders in the Sharjah Maritime Museum. Photo: Haupt & Binder, Universes in Universe

compilation of computation protocols. Before we proceed with the text itself, it is helpful to try to understand the physics of the computation here. The easiest way to see it is to consider the corner created as a central angle rather than as the one created by two neighboring edges. It is clear that this does not change the arithmetic of the computation.[5] To take an example, the number of holes in one such 'corner' (or central angle) in Fig. 8.2 is 9. Squaring 9 gives 81. Reducing this by $\frac{3}{4}$ gives $60\frac{3}{4}$. Adding $\frac{1}{4}$, we get 61. Thus, according to this rule, the total number of holes in a colander with 9 holes along one corner ought to be 61. Now, the number of holes along one corner of such a hexagon is equal to the number of holes along a diameter, as shown in Fig. 8.3.[6] Hence, we can rephrase the protocol described in the verse as $\frac{3}{4}d^2 + \frac{1}{4}$, which does provide a circular area approximation for the hexagonal sieve.

Such a rule of computational tool—a rule that simply says "$\frac{3}{4}d^2 + \frac{1}{4}$", where d is the number of holes in one corner—would have helped the artisan to drill the correct number of holes into a circular flat disc of a standard size, appropriately spaced and therefore each hole appropriately sized. The number of holes in a particular

[5] We are not claiming this was how the verse was thought of or compiled. In all likelihood the algorithm in the verses emerged through recognizing repeating patterns while working regularly with the sieves.

[6] We only need to think of the magenta arms in Fig. 8.2 below as two hands of a clock. The number of holes along the arms would stay the same if we could rotate one down by two notches to bring them in diametrically opposite directions to each other.

Fig. 8.2 One 'corner' of a sieve

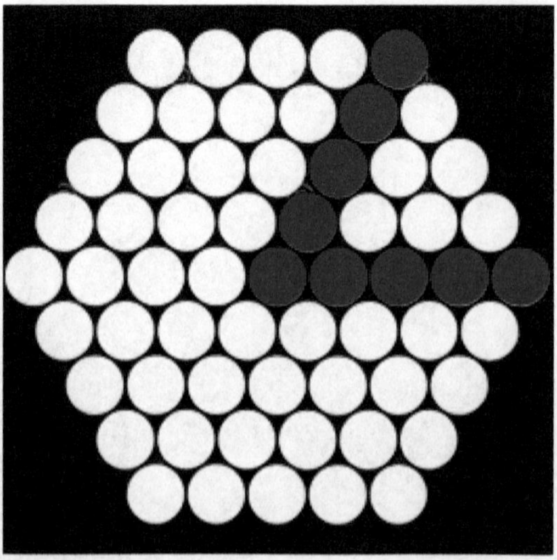

Fig. 8.3 'Diameter' of a sieve

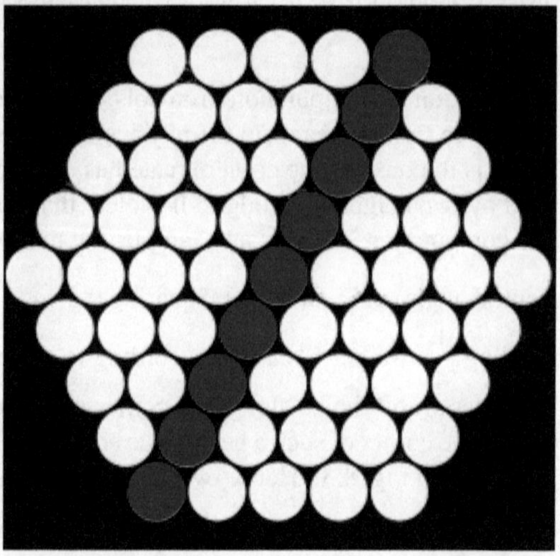

sieve would be the same as the number of pearls that could settle on the sieve at a time. The colanders are named with numbers 20–1500 (variously, also numbered 1 to 10)—but these seem to be just some numerical tags without any obvious connection with the actual number of holes.[7] The next few verses recite the prescribed number of holes in one corner of a standard sieve of a given size, and the corresponding total number of holes, according to the above algorithm. We have listed

[7] Not unlike present-day football or shoe sizes.

Table 8.1 Number of holes in the sieves

Colander 'size'	Number of holes in one corner	Total number of holes
20	11	91
30	13	127
50	15	169
80	17	217
100	21	331
200	25	469
400	29	631
600	33	817
800	39	1141
1000	41	1261

these values in Table 8.1 above.[8] We note that as the colander size and thus the number of holes grows, the size of each hole shrinks. Thus, the biggest sized sieves are stacked at the lowest rung, for the smallest and therefore cheapest pearls.

The physicality of the counting practices around pearls go beyond the colanders and other instruments, and includes the measurer's body itself. The texts themselves recognize the centrality of the calculator's body, and try to codify this. A couple of curious verses in the MK deal with finger-counting methods, while computing with the pearls. Verse 19 says,

> With the fingers, to say the cost of [a thing],
> hey girl listen to this:
> to gain the profit, take the index finger away after touching, it is 9.
> also 100 *vīcam*, 10 and also 1,
> middle finger is 2; take the middle finger away after touching it is 8;
> the [adjacent] ring finger is 3, take the next finger away after touching it is 7;
> the little finger is 4, take the little finger away after touching it is 6;
> if you shake the fingers inside that is 5 [?].

Verse 20 follows up with,

> Rubbing the upper part of the finger is ¾.
> Touching the middle part of the finger is ½.
> Upper part of the finger is ¼.
> [Holding] two fingers together is ⅛ and 20 and 200 [?].
> [Holding] three fingers together is 3 vīcam [i.e. 3/16] and 30 and 300.
> [Holding] four fingers together is 4 vīcam [i/e/ 4/16] and 40 and 400.

We haven't been able to reconstruct satisfactorily the counting practices described in the above verses. However, these do highlight a key feature of the text: the physicality of counting practices that often gets otherwise drowned in our textual imagination of counting as a primarily 'mental' process. Such physicality is palpable

[8] Other than these ten sizes, two more sieves—sizes 1500 and 2000—are mentioned in the text *Muttu cevu kaṇakku* published in 1881. Later, in the *Muttukaṇakku* text, while discussing the *cevvu* per *mañcāṭi* for each sieve the author mentions the 1500 sieve, but it wasn't mentioned while listing the sieves earlier in the text.

throughout this text. To take one example, as one goes through the numerical values listed in the verses, one gets the distinct feeling of physically pushing 'real' tokens to go from one row of the table to the next (without really worrying about the 'general algebraic principle' which typically precedes such numerical lists in the verses). For instance, if we have 4 *muntiri* (4/320's) in some long string of addition, they are immediately clubbed together or 'bunched', to simplify to 1 *kāṇi* (1/80), thus moving up one denomination, and thereby shrinking the length of the additive expression at regular intervals, making it easier to remember and compute with. Such base-2 denominators, allowing for regular periodic clubbing and 'pushing up' the denominator levels in powers of 2, makes it feel as if one is simply pushing actual currency tokens (such as 'quarters'), grouping them (into say 'dollars'), thus replacing groups of smaller denomination by a single token of a higher denomination, and so on. One can only imagine how much of the logic of the computations and memory shortcuts around such arithmetic logic would be encoded in the physicality of the tunes and rhythms and modulations of such recitations, as poems to be sung out loud as a mnemonic aid to the calculations—all that physicality of the oral which gets lost the moment such songs are transcribed into written text.

8.4 From the Colander to the Balance and the Balance Sheets

The units of weight used to measure pearls are *kaḻañcu* and *mañcāṭi* (20 *mañcāṭi* is equal to 1 *kaḻañcu*, according to the traditional Tamil measurements) ("Eṇ Cuvaṭi," 1926; Kāri 2007). *Kaḻañcu* is usually equated to 72 grains (approx. 4.5 g) in the metric system. One *mañcāṭi* is equated to 3.6 grains (approx. 0.24 g). The exact measuring values of both of course, in practice, seem to have varied. For example, in the colonial records, James Steuart equates $7\frac{3}{4}$ grains to 1 *mañcāṭi*, while Vane records $3\frac{1}{3}$ or more exactly $3\frac{35}{100}$.

From Size to Weight After describing the size and layout of the sieves, the *MK* verses go on to convert the number of pearls into weight measures. Table 8.2 is compiled from the next few verses that relate the colander sizes with the average (or aggregate) weight of the corresponding pearls.

We do not find any explicit explanation in the verses about how these numbers were computed. We can only guess that they are speculated from averaging the numbers obtained from long traditions of actually dealing with the different sized pearls and measuring their weights on a day-to-day basis. The actual weights must have varied in each individual instance, and across time and place. One Chola inscription from eleventh century CE, for instance, much earlier in time that the MK edition we are considering, measured the weight of 37 pearls as 8 *mañcāṭi* (Hultzsch 1892, 43). This means around four and a half such pearls should add up to 1 *mañcāṭi*

Table 8.2 Number of pearls per mañcāṭi weight

Colander size	# of pearls needed to sum up to a weight of 1 *mañcāṭi*	# of pearls needed to sum up to a weight of 1 *kaḻañcu*
20	1	20
30	1½	30
50	2½	50
80	5¼	105
100	8½	170
200	15	300
400	24	480
600	40	800
800	60	1200
1000	120	2400
1500	240	4800

of weight, which does not correspond exactly to any of the listed numbers above (the rank it comes closest to seems to be that of the size 80 colander).

From Weight to the Final Cevvu Value The final market value of the top and middle class pearls, called the '*cevvu*' value of the pearl, was computed on the basis of *the square of the weight* of the pearl (up to a scalar index). Vane describes the *cevvu* 'weight' as "a conventional artifice for assigning the proper value by weight to a very valuable article of very small weight. Its effect is much the same as if valuers agreed to assign to these classes of pearls increased weights above their real weights, and then valued the pearls at these fictitious weights" (Vane 1887, 36). The *cevvu* value, as a multiple of the 'weight squared', makes the valuation of pearls a nonlinear (quadratic) computation. This is most likely because the total value of two smaller sized pearls had to be kept lower than that of one big pearl of the combined weight—a distinction that a linear computation model would fail to make.[9]

Verse 16
of the MK says:

Mañcāṭi taṉṉil patilvait tuṭaṉmāṟi
Miñcā mukkālil mikakkaḻittu—añcāmal
Kaṇṭatukai yaikkaruti muttataṟ kīyak
Koṇṭatō cevvē kuṟi (MK 16)

"square the weight in *mañcāṭi*, reduce by ¾ and divide the balance with the number of pearls, which will give the *cevvu* [of each pearl]".

In other words, the 'formula' for computing the *cevvu* of a pearl with weight w is:

$$c \cdot w^2$$

[9] For instance, the total value of 2 pearls weighing ½ *mañcāṭi* each cannot be equal to that of 1 pearl weighing 1 *mañcāṭi* which is in a whole different category of value than the former ones.

where c is the '*cevvu* index' or the scaling factor. Assuming there is only one pearl, this simply means squaring the weight of the pearl and multiplying it with a scaling index of ¾. The total *cevvu* value of n such pearls is $(c \cdot w^2) \cdot n$. It is worth pointing out that the *cevvu* calculation is quadratic in weight and linear in the number of pearls. That is, if one pearl's *cevvu* is x, that of n such pearls is $(n \cdot x)$.

In the MK text that we have looked at, c is typically taken to be ¾ (when the computations are in terms of *mañcāṭi* units of weight). It is not clear why ¾ is adopted as a scaling factor, but such indices mostly indicate the qualitatively different computational logic that was used for different grades of pearls. On the one hand, these high quality pearls that made their way into the market systems (through the colanders, etc) were possibly seen as slightly below the unique 'topmost pearls' that defied the simple logic of measurement based on weight or size alone. For instance, a pearl with a strikingly rare color or pattern or shape was kept aside for non-market tributes to the emperors or to the gods (Hornell 1905, 12–14). If these top pearls were assumed to have a *cevvu* value index of 1, then it is plausible that the 'rest of the pearls' meant for the more secular market and its arithmetic measurement apparatus, would have to be valued at slightly less than 1. Hence an index of $\frac{3}{4}$ or '*mukkāl*', which is the next highest standard denomination after one, '*oṉru*', in the Tamil number system. On the other hand, the lowest pearls did not seem to deserve a *cevvu* index at all, and did not make it to the *cevvu* computations. They seem to have been simply valued in terms of their bulk weight. In this sense, the index of *mukkāl* could have played a role in qualitative classification, marking out the pearls meant for the market – sitting *in between* the two extreme types, both of which were beyond the *cevvu* metric in opposite ways. However, all this is speculative and we do not have any confirmation of this from the texts we reviewed. We note that this question of 'Why is ¾ used as a scaling factor?' remains open for us, and needs further inquiry. Whatever the rationale behind such a choice for an index, the best way to think of the index itself is to think of it as the *cevvu value per mañcāṭi-squared.* The next few verses in the text recite the *cevvu* values for unit weight of pearls (in *mañcāṭi*, and then in *kaḻañcu*) across the various sieves, illustrating the above rule, as summarized in Table 8.3.

To take an example, for the 80-sized sieve we know that on an average $5\frac{1}{4}$ pearls weigh a total of 1 *mañcāṭi* (Table 8.2). This in turn means the weight of each such pearl is $\frac{4}{21}$ *mañcāṭi*. The *cevvu* value for 1 such pearl is then $\frac{3}{4} \times \left(\frac{4}{21}\right)^2$. But then we need $\frac{21}{4}$ such pearls to make up 1 *mañcāṭi*. Hence the *cevvu* value for 1 *mañcāṭi* of such pearls is:

$$\frac{3}{4} \times \left(\frac{4}{21}\right)^2 \times \frac{21}{4} = \frac{1}{7}$$

The *Verse 16* itself of course simply instructs directly, without such explanations: 'square the weight in *mañcāṭi*, reduce by ¾ and divide the balance with the number

Table 8.3 Cevvu value of unit weight of pearls

Colander size	*Cevvu* value of 1 *mañcāṭi* weight of pearls (here, 1 *mā* = $\frac{1}{20}$)	*Cevvu* value of 1 *kaḻañcu* weight of pearls (here, 1 *vīcam* = $\frac{1}{16}$)
20	$\frac{3}{4}$	15
30	$\frac{1}{2}$	10
50	$\frac{1}{4}+\frac{1}{20}$	6
80	$\left(2+\frac{3}{4}+\frac{1}{8}\right)$ *mā*	$2+\frac{3}{4}+\frac{1}{8}$
100	$\left(1+\frac{3}{4}\right)$ *mā*	$1+\frac{3}{4}$
200	1 *mā*	1
400	$\left(\frac{1}{2}+\frac{1}{8}\right)$ *mā*	10 *vīcam* = $\frac{10}{16}$
600	$\left(\frac{1}{4}+\frac{1}{8}\right)$ *mā*	6 *vīcam* = $\frac{6}{16}$
800	$\frac{1}{4}$ *mā*	4 *vīcam* = $\frac{4}{16}$
1000	$\frac{1}{8}$ *mā*	2 *vīcam* = $\frac{2}{16}$

of pearls, which will give the *cevvu*'. Here the weight is 1 *mañcāṭi* of pearls. Thus we have:

$$\frac{\frac{3}{4}\times(1)^2}{\frac{21}{4}}=\frac{3}{4}\times\frac{4}{21}=\frac{1}{7}$$

Of course, we do not find the figure '$\frac{1}{7}$' in the table. But that is because there is no way to exactly express a fraction with denominator 7 in terms of fractions whose denominators are multiple of 2's and 5's (which are the kind of fractions being used in the Tamil number system). The table entry corresponding to the sieve 80 reads $\left(2+\frac{3}{4}+\frac{1}{8}\right)$ *mā*, that is, $\left(2+\frac{3}{4}+\frac{1}{8}\right)\times\frac{1}{20}=\frac{23}{160}=0.14375$, which is a close approximation to $\frac{1}{7}$. For sieve 800, likewise, we know that 60 pearls (on average) make up 1 *mañcāṭi* weight in total. Thus, each such pearl weighs $\frac{1}{60}$ *mañcāṭi*. The total *cevvu* value for 1 *mañcāṭi* of such pearls is:

$$\frac{3}{4} \times \left(\frac{1}{60}\right)^2 \times 60 = \frac{1}{80} = \frac{1}{4} m\bar{a}$$

Or, more directly,

$$\frac{\frac{3}{4} \times (1)^2}{60} = \frac{3}{240} = \frac{1}{80} = \frac{1}{4} m\bar{a}$$

And so on.

Before moving on, we note that when computing in *kaḻañcu* units (rightmost column in Table 8.3), the scaling factor itself needs to be appropriately scaled. The cevvu index for 1 *mañcāṭi*-squared of pearls in the top sieves is $\frac{3}{4}$. For 1 *kaḻañcu* (= 20 *mañcāṭi*) of such pearls, the *cevvu* index (which is a value per 'weight squared') becomes $\frac{3}{4} \times (20)^2$, that is, 300 per kaḻañcu-squared of pearls.[10] The *cevvu* index in *kaḻañcu* units is suggestively called the *peruñcevvu* (*perum* means bigger), and its value is 300. For instance, for the 20-size sieve, we need 20 pearls to total a weight of 1 *kaḻañcu* (Table 8.2). Therefore, the *cevvu* value of 1 *kaḻañcu* of such pearls is

$$300 \times \left(\frac{1}{20}\right)^2 \times 20 = 15$$

And so on.

The final step is to then multiply this *cevvu* weight computed directly from the colander, with the market rate of *currency units per cevvu weight*. From ethnographic accounts of the eighteenth and nineteenth century pearl fishing in the Gulf of Mannar region, such market rates seem to have been decided based on competing bids by gangs of wealthy merchants who flocked to the fishing sites during the season. This gives us the final market price of the pearl. The computations often involve rounding off to the nearest significant value by ignoring the smallest units—by "giving or taking a few *muntiri* ($\frac{1}{320}$) here and there", as the text puts it. At times the weights are expressed in denominations of the *muntiri*—traditionally the smallest

[10] We can think of this as something like land valuations. Land is measured in 'square-length' units. The 'chain' is a unit of length used in England since the 16th CE. It is still in use in particular contexts across various countries in the world, such as in forestland measurement in the United States, or in the making of cricket pitches (1 chain = 22 yards.) Now, if the value of a particular kind of land was ¾ units per square-yard, then the value of the same kind of land per square-chain would have to be ¾ times 22 times 22. This means if we are to find the value of 5 square chains of land, we can either convert it to square yards first and then multiply by ¾. This gives (5 × 22 × 22) × ¾. Or we could stay with the square chains unit itself, in which case the value would be 5 × (22 × 22 × ¾). The final value is the same regardless. The same thing is really going on with the *cevvu* and *peruñcevvu* calculations.

denomination used in Tamil counting systems. This helps with small weights that are tiny fractions of a *mañcāṭi* or a *kaḻañcu*. It allows one to simply work with the numerators (when it comes to squaring, etc), and then finally divide the results by squares of 320. This is aided by ready-to-use multiplication tables in 320, which we know to have been part of regular computational practices in such traditions (Babu 2022; Prakash and Babu, 2025).

A Brief Note on the Use of 'mā' We note that in the pearl-weight measurement system, 8 *mā* equals 1 *mañcāṭi*. That is, 1 *mā* is $\frac{1}{8}$ th of a *mañcāṭi*. This is different from the standard usage of a *mā* in the Tamil counting system which typically stands for $\frac{1}{20}$. We find this dual usage in the *cevvu* tables too. While 1 *mā* equals $\frac{1}{20}$ on the right hand columns that list the numerical *cevvu* values, it equals $\frac{1}{8}$ *mañcāṭi* on the left hand columns that list the weight values. The convention of taking 8 *mā* as equivalent to 1 *mañcāṭi* is explicitly mentioned in the MK in a specific instance.

Mañcāṭik koṉṟuk keṭṭu māvataip pāti ceytu.
Miñcavē patilmā vaittu mēviyē perukkic cērttuc.
Ceñcol muntiriyai vaittuc ciṟappuaṭaṉ kaḻitta cēṭaṅ.
Koñciyē muttuk kīyac cevveṉak kuṟikka lāmē (MK 17).

This verse gives the procedure to calculate the value of *cevvu* for one *mañcāṭi* pearl weight. It says, "For one *mañcāṭi*, 8 *mā*; halve it, which is 4 *mā*, square this to get 16 *mā* and reduce it by *muntiri* (1/320) to get a sum which has to be divided by the number of pearls, which will yield the *cevvu* for one *mañcāṭi*, as per this procedure".

In fact, in another context, ⅛ of a *mañcāṭi* is called a *vīcam* (the weight of a paddy grain) and also a *mā*.[11] The literal meaning of *mā* is 'grain'. Thus the two meanings of *mā* seem to correspond to a literal and an arithmetic meaning, a grain and 1/20 respectively.

This concludes the arithmetic computations as prescribed by the *Muttukaṇakku* text. In the next section, we consider some other texts contemporary to MK that also deal with similar computations. These texts in fact take us closer to the practice of computations associated with pearl valuations, since they also carry a distinct pedagogical voice in them, indicating that these were products of apprentices training for the occupation of accounting. We see this in their focus on various kinds of word problems, which anticipate different situations in the trade. It is here that we find the play of arithmetic techniques such as the rule of three, making the text more prosaic and pedantic than poetic.

[11] This is from the *Kaṇakkatikāram* text in Malayalam, as pointed out in Wagner and Ashokan (forthcoming).

8.5 Practical Computational Situations and the *Kaṇakkatikāram*

Kaṇakkatikāram The *Muttukaṇakku* computations resonate with problems in the *Kaṇakkatikāram* texts that deal with pearls.[12] These problems have actually been appended to the main body of the *Muttukaṇakku* text that we have considered. This part of the text is from the *Kaṇakkatikāram* manuscript (TR2877), GOML, Chennai. Usually, in generic *Kaṇakkatikāram* texts, there is no explicit section that deals with pearl arithmetic. Problems related to pearls appear in the general problems, but those are meant for training in problem solving in arithmetic progressions, rather than pertaining to the pearl trade per se. However, in this specific *Kaṇakkatikāram* text, we find a unique occurrence of pearl arithmetic as a standalone section, appearing as a fragment that stands apart from the structure of the generic *Kaṇakkatikāram*. The accommodation of pearl computations in such more general pedagogic texts possibly also indicates the growth in the industry and the need to train more technicians in these measurements. The 'question and answer' word-problem nature of the verses further points to the text's pedagogic value. As does the fact that the verses here are followed by commentaries, unlike in the MK text that only sings the verses without any attempt at explanations.

The fragment contains ten verses and a list of information relating units of pearl weight with *cevvu* values. The verses and computations anticipate and take the reader through different practical situations of measurement, pointing to the practical orientation of the computations. We also find, true to the *Kaṇakkatikāram* tradition of cross-checking computations, the rule of three taking center stage in these verses, mainly in the context of the "*cevvu-mañcāṭi*-pearl count" triad. The first such situation is the simplest one, where we have only one pearl. This is exactly the same as how the MK prescribes it. The second kind of situation arises when we have multiple pearls whose total weight can be more practically measured, compared to the individual weights. For example, 4 pearls that collectively weigh 3 *mañcāṭi*. Verse 2 of the *Kaṇakkatikāram* fragment describes the formula for the calculation of *cevvu* value of pearls in the 20-sized sieve:

> Divide 300 by 20, to give 15. Divide this again by 20, to give ¾. The *cevvu* for one pearl is ¾. To know the *cevvu* for ten pearls in the 20th sieve, multiply the previous *cevvu* value of ¾ by 10, to give 7½. Likewise, use this for the rest of the sieves. Another way of the same calculation: If asked the *cevvu* value for 10 pearls in the 20th sieve: Multiply 300 by 10, to give 3000. Divide this by 20, to give 150. Divide this once again by 20 to give 7½ which is the *cevvu* value. Another way is to ask the count of pearls in the 20th sieve when the *cevvu* value is given as 7½. Multiply 7½ by 20 to give 150. When this is divided by 300, it gives ½ and multiply this by 20 again to give 10, which is the number of pearls (Kaṇakkatikāram 2, in MK text; Prakash and Babu 2024).

[12] *Kaṇakkatikāram* are generic names for treatises of mathematics in Tamil and Malayalam languages that emerged in the south Indian landscape from about the sixteenth century dealing with computational work of the accountant, the teacher and the student. For more on this genre of texts, see (Babu 2022; Wagner and Ashokan, forthcoming).

The above is the same calculation that MK prescribes, which we have already described. The 'multiple ways' of computing such values are simply arithmetically equivalent ways of computing the same thing, but in different practical contexts of what is already known or measured. It is worth noting that, yet again, the computations simply assume the *kaḻañcu cevvu* index of 300 (or equivalently, the *mañcāṭi cevvu* index of ¾), without explaining the reason for such a choice.

This is then followed by a series of examples in the form of word problems and their solutions. For instance,

> Q: If 10 pearls weigh 3 *pā* [*kaḻañcu*], then what would be the *cevvu* value of 5 pearls?
>
> A: Divide 300 by 10, [to get] 30. Multiply this with 3, [to get] 90. Divide this by 10, [to get] 9. Multiply this with 3, [to get] 27. Multiply this with 5 which is 135, the *cevvu* value (Kaṇakkatikāram 3, in MK text; Prakash and Babu 2024).

Following the above hints literally we arrive at the calculation:

$$\frac{300}{10}\times\frac{3}{10}\times 3\times 5\,.$$

But shifting around things just a little, we find that this is a familiar calculation:

$$300\times\left(\frac{3}{10}\right)^2\times 5\,.$$

Since 10 pearls weigh 3 *kaḻañcu*, each one weighs $\frac{3}{10}$ kaḻañcu. Since we are working with kaḻañcu units, the *cevvu* index is 300. Thus, the *cevvu* for each such pearl is

$$300\times\left(\frac{3}{10}\right)^2=27\,.$$

And therefore, the *cevvu* of 5 such pearls is

$$300\times\left(\frac{3}{10}\right)^2\times 5=135\,.$$

Likewise,

> Q: If 10 pearls weigh 3 *pā* [*kaḻañcu*], then how many pearls are there with [total] 54 *cevvu* value?
>
> A: divide 54 by 3, [get] 18. Multiply this with 10, [get] 180. Divide this by 300 which is ⅗ (that is, *araiyē irumā*, 1/2 + 1/10). Multiply this with 10, which gives 6. Divide this by 3, which gives 2, the number of pearls. (Kaṇakkatikāram 3, in MK text; Prakash and Babu 2024).

Following the literal method prescribed, we get $\frac{54}{3}\times 10\times\frac{1}{300}\times 10\times\frac{1}{3}=2$. This is the inverse kind of calculation where the final total *cevvu* value is given, and one is asked to compute the total number of pearls of a given weight. Since 10 pearls weigh 3 *kaḻañcu* in total, the weight of one such pearl is $\frac{3}{10}$ *kaḻañcu*. Thus the *cevvu* for each such pearl is $300\times\left(\frac{3}{10}\right)^2$. If the number of pearls (amounting to the total *cevvu* of 54) is *n*, then we have,

$$300\times\left(\frac{3}{10}\right)^2\times n=54$$

$$\text{That is, } n=\frac{54\times 10\times 10}{300\times 3\times 3}=2.$$

And so on.

The MK also does these same computations, but presents them as prescription rather than descriptions. The poems provide a ready-made chart that allows one to directly reach the final value in such situations, like a ready reckoner (Lines 167–229) (Prakash and Babu, 2024). The remaining lines of the verse here do the same, but in terms of *kaḻañcu* units of weight. Row 3 says, for example, if *arai mā*, that is $\frac{1}{2}\cdot\frac{1}{20}=\frac{1}{40}$ of a pearl weighs 1 *mañcāṭi*, then the total *cevvu* of 1 *mañcāṭi* of such a pearl is 30:

$$\frac{\frac{3}{4}(1)^2}{\frac{1}{40}}=30$$

We can observe various rhythms and patterns in the numbers listed in these verses. The smaller the number of pearls needed to arrive at a weight of 1 *mañcāṭi*, the larger the pearls are, and therefore the higher is their *cevvu* value. The larger the number of pearls on the other hand, the smaller the pearls are individually, making their individual *cevvu* values smaller. The same can be seen in the lines operating with the *kaḻañcu* unit. This produces a certain reciprocity between the two columns of values, one being the inversion of the other – something that must have aided the memorization of the values. We must note that the first and the last entries in the *kaḻañcu* table do not seem to fit into the computation scheme we have described in this essay and we are not sure how to explain this 'aberration'.

Paṭiyaṭi Lakkam *(Henceforth, PL)* This is a section of the *Kaṇakkatikāram* that lays out computations similar to what we have described so far. As an example, we look at *verse 10.*

> If 1 pearl is of 1 *pā* [*kaḻañcu*], the *cevvu* is 300; 2 pearls of 1 *pā* [*kaḻañcu*], *cevvu* is 150; 3 pearls of 1 *pā* [*kaḻañcu*], *cevvu* is 100; 4 pearls of 1 *pā* [*kaḻañcu*], *cevvu* is 75; 5 pearls of 1 *pā* [*kaḻañcu*], *cevvu* is 60; and 6 pearls of 1 *pā* [*kaḻañcu*], *cevvu* is 50 (Kaṇakkatikāram 10, in MK text; Prakash and Babu 2024).
>
> These are the same values we obtain if we follow the MK algorithms. PL also repeats the *cevvu* formula for *mañcāṭi* as well as *kaḻañcu* units: "the procedure to obtain *cevvu*, if the order is of *mañcāṭi*, is square the *mañcāṭi*, reduce it by ¾, divide it by the number of pearls [which] gives the [total] *cevvu*"; "Multiply weight by weight, again multiply it by

300, divide it by the number of pearls gives *cevvu*". The PL then follows the formula up with a series of examples: 1 pearl of 1 *mañcāṭi*, *cevvu* is ¾; 2 pearls of 1 *mañcāṭi*, *cevvu* is ¼ + ⅛; 3 pearls of 1 *mañcāṭi*, *cevvu* is ¼; 4 pearls of 1 *mañcāṭi*, *cevvu* is 3/20 x 3/80; 5 pearls of 1 *mañcāṭi*, *cevvu* is 3/20 (Kaṇakkatikāram 10, in MK text; Prakash and Babu 2024).

And so on.

We only note that the last bit of the original text is damaged, and it is unclear what the *cevvu* for 40 pearls weighing 5 *ka̱lañcu* might be. However, we can make a guess. According to our discussions so far, the *cevvu* should be:

$$300 \times \left(\frac{5}{40}\right)^2 \times 40 = \frac{375}{2} = 187.5$$

Thus, our guess is that the missing numeral in the last bit of the verse is '8'.

In the final sections, the PL lists another collection of examples for the same computation. But this time it does not mention the weight of the pearls. Instead, it mentions the sieve number from which they are obtained and lists the *cevvu*.

For 20 sieve *cevvu* of 1 pearl is ¾; for 30 sieve *cevvu* of 3 pearls is 1; for 40 sieve *cevvu* of 5 pearls is ¾ + (3/20 + 3/80); for 50 sieve *cevvu* of 6 pearls is 5 (Kaṇakkatikāram 11, in MK text; Prakash and Babu 2024).

These can be verified from the tables we have already listed.

Coming back to the MK text as we wrap up, the *Māvuñ cevvum* fragment of the *Muttukaṇakku* explicitly states that (in the context of pearl weights) we ought to think of a *mā* as ⅛ of a *mañcāṭi*:

Convert *mā* in terms of *araikkāl* (⅛), square it, reduce it by ¾ and divide it by the given number of pearls gives *cevvu* (Kaṇakkatikāram 11, in MK text; Prakash and Babu 2024).

Then, as usual, it lists some examples:

1 pearl weighs 4 *mā*, *cevvu* is 3/20 + 3/80; 2 pearls weigh 4 *mā*, *cevvu* is 1/20 + 3/80 + 1/160; 1 pearl weighs 8 *mā*, *cevvu* is ¾; 2 pearls weigh 8 *mā*, *cevvu* is ¼ + ⅛; 1 pearl weighs 16 *mā*, *cevvu* is 3; 1 pearl weighs 160 *mā*, *cevvu* is 300 (Kaṇakkatikāram 11, in MK text; Prakash and Babu 2024).

The computations are the same as we have described so far, and can be easily verified. The final verse of the fragment seems broken at places, cryptic, and quite different from everything we have seen so far. It says:

Convert *mā* into *araikkāl*, add the sum, square 2½ then reduce it by ¾…

For 4 *mā*, cevvu [?] is 3/20 + 3/80; for 4 *mā*, *cevvu* [?] is ¼ [?]; for 5 *mā*, *cevvu* [?] is ¼ + 2/20; 6 *mā* – ¼ + 1/160; 6 *mā* – ½; 7 *mā* – ½ + (1/20 + 1/80); 7 *mā* – ½ + ⅛; 8 *mā* – ¾; 8 *mā* – ¾ + (1/20 + 1/80); 9 *mā* – ¾ + 3/16; 9 *mā* – 1/20; 10 *mā* – ⅛; 10 *mā* – ¼ (Kaṇakkatikāram 12, in MK text; Prakash and Babu 2024).

We are not sure how to make sense of the listed values. The only values that match with what we have seen so far are the *cevvu* for a pearl that weighs 4 *mā* (*cevvu* = 3/20 + 3/80), and for one that weighs 8 *mā* (*cevvu* = ¾).

8.6 In Conclusion: Limitations of the Text and Questions About the Value of Labor

The computational practices we have described above were at the core of the South Indian trade in pearls, enmeshed in the highly speculative and competitive political economies of honor, particularly of royal honor (and thus related to the political authority of the state). Such links go well back in time. For instance, in a Kannada language inscription from 1187 CE Banavur, not far from the medieval market and palace town of Arasikere, one finds mention of activities of the powerful medieval mercantile corporation known as the *Ayyavole 500* that dealt in precious commodities including "pearls like dewdrops". The inscription describes one Chattisetti, a brahmin merchant and member of the mercantile corporation, who was also master of the mint, a courtier of the Hoysala king Viraballala II, and a trader in horses, elephants, pearls, and other valuables (Ali 2010, 186). This trader seems to have gained a lot of fame through such trade, and included many kings as his buyers. In such an economy of prestige, pearls operated as the 'harvest' for the nobility. The Dutch East India Company's factor Floris Blom in 1694 described the pearls fished in the Tamil and Lankan coasts as *"Dit edel zee gewas"* ("this noble harvest from the sea") (Subrahmanyam 1996). According to historian Sanjay Subrahmanyam, despite the use of terms like 'harvest', the specific nature of pearls as precariously dependent on the rhythms of nature for reproduction, and their status as objects of noble prestige, meant that pearl fishing had for long been a privileged sphere for state control, under which the industry was protected from unchecked market competition. The *cevvu* measure (or *chau*, as it is called in the Persian Gulf region (Montigny 2010)) seems to have emerged from a "mutually intelligible system of weights and measures, and a shared currency system" (Bishara and Wint 2021, 57), as part of the attempts to regulate or govern such a highly speculative maritime *bazaar* of honor and prestige items, which was always prone to the excesses of individual, unrestricted greed. This is where the labor of the accountants and measurers became crucial – to maintain and sustain their games of gambling for prestige on the backs of indebted and captured labor. The computational integrity of the games being necessary to organize fairness of the trade for the rich and competitive investors, while at the same time justifying labor obligations to the divers and other workers in the trade.

The wages of the divers were pre-decided as a share of the catch (Steuart 1843, 41; Hornell 1905, 13–15). Keeping the diver indebted in such economies of prestige seems to have been crucial to running the trade itself. Samuel Ostroff discusses in detail how the pearl divers regularly fell into debt traps through usurious and speculative moneylenders, many of whom were also the principal boat renters. While those higher up in the industry hierarchy—such as scribes, accountants, cash keepers and other office servants—seem to have had normal wages in addition to pre-decided shares from the catch, when it came to paying the divers, we find practices of paying wages in unopened oysters—a speculative currency token emerging through the measurement apparatus we have described in this essay. Ostroff notes how the oysters paid to the divers as wages also worked as a local speculative currency—as a form of in-kind payment in exchange for a range of goods and services. The divers could either sell their oysters in the marketplace and exchange them for money, or "try their luck by opening the oysters with the hope of finding a valuable pearl". All this means the diver was paid in what can be described as a 'stochastic currency'. While this might mean windfalls of remittance back to the home village at times, most other times this meant crushing debts (Ostroff 2016, 116–118).

The dangers of pearl-fishing are well-documented. Divers would starve of oxygen, their bodies crushed by the columns of ocean water, pressing down on their respiratory systems and eardrums. British official and traveler Jonathan Fellowes wrote, "The blood sometimes gushes from the nose and ears of the diver after his emersion from the deep" (Ostroff 2016, 89). Nineteenth-century accounts indicate the lack of even basic diving instruments such as nose clips (Ostroff 2016, 90), battling with sharks, "both real and imagined, with the aid of 'shark-charmers'" who stayed ashore to collect a modest fee. The process of putrefaction once the oysters are brought ashore and sold in heaps at auction was marked by a "singular lack of hygiene and overpowering stench", attracting flies, making cholera a frequent occurrence that accompanied pearl-fishing (Thurston 1894, 23). The fishing activity itself is known to have been violently policed. Accounts about the fisheries note how the divers had to remain naked throughout the event, except for "a slip of calico about his loins", so as to prevent them from stealing the precious pearl harvest. The irony is inescapable: the indebted divers had to go naked in order to enable the economy of prestige items meant for bodily adornments of the super rich.

The texts we have described here are understandably silent about all of this, and present to us a face that is arithmetically final and beyond conflict. However, no matter how ironclad the computational logic might seem from the texts, or how brutal the corporal punishments might have been to maintain the computational order and discipline, in reality various subversions hardly seem to have been uncommon. Early nineteenth-century British traveler James Cordiner noted during his visit to the pearl fishery:

> The divers, the boat-men, the persons employed in washing the oysters and sifting the sand, leave no expedient untried to accomplish frauds. Even the peons, employed as a check upon the laborers, have been known to attach a viscous substance to the end of their canes, and extract from washing-troughs valuable pearls, with the very instrument used to punish such delinquencies (Cordiner, Description of Ceylon, 2:59, cited in Ostroff 2016, 119).

We find early colonial records of wage disputes and the divers refusing to work. There is no reason to believe that such labor strikes didn't occur earlier as well. Accounting practices and apparatuses such as the ones normalized through texts like *Muttukaṇakku*, contributed crucially in this way not only to evaluating the pearls being traded, but also, by hiding and devaluing the realities of labor, to the affirmation of social hierarchies and the consequent labor obligations owed to the prestigious classes that ultimately ran the trade.[13]

Given this, despite the impressions embodied in texts like the MK, we believe the colander system should be seen as an active slot machine, rather than a passive instrument simply measuring and recording a pre-inscribed value. Like a casino, it made and ruined people as social beings, thus generating or destroying value. Whether instruments like a sieve are simply a sieve, or a slot machine, depends on what role they play in the political-economic context they are located in. For evaluating this role, we need to move beyond the deceptive, placid face of transparent measurements that a deeply theatrical text such as the *Muttukaṇakku* presents, not unlike the calm ocean surface hiding from view the world of crushing, indebted, precarious, speculative labor plunging underneath the surface. We only see this world when we go beyond the canonized texts themselves, and place the colander in the economic motions of manual labor and social prestige.[14]

Bibliography

Ali, Daud. 2010. Between Market and Court: The Careers of Two Courtier-Merchants in the Twelfth-Century Deccan. *Journal of the Economic and Social History of the Orient* 53:185–211.

Arunachalam, S. 1952. *The History of the Pearl Fishery of the Tamil Coast*. Annamalai Nagar: Annamalai University.

Babu, Senthil D. 2022. *Mathematics and Society: Numbers and Measures in Early Modern South India*. Oxford University Press.

Bishara, F., and H. Wint. 2021. Into the Bazaar: Indian Ocean Vernaculars in the Age of Global Capitalism. *Journal of Global History* 16 (1): 44–64. https://doi.org/10.1017/S174002282000011X.

[13] The prestige economy aspects of the value of commodities and human labor become clearly visible in such trades as that of pearls—a prestige commodity without much 'material' functional or 'production' value in the traditional sense, but whose main socioeconomic value or significance is the display of social prestige and control over the laboring communities that one commands. In this way, this trade marks out two broad types of labor for us: that of the accountant, or weaver of an arithmetic and logic crucial for running the speculative games of prestige and glamor; and that of the indebted, forced laborers, enabling such games of consumption and prestige, while themselves being stripped off their social clothing of self-respect.

[14] That quantitative valuation-oriented texts need not be the only way in which currencies of prestige like beads, tokens or pearls could be abstracted to weave a political economic value system, can be seen from other contexts as well. One example might be the use of beads by the Amerindian Iroquois people across Six Nations in northeastern America, who wove the white and purple shell beads called wampum into belts or strings referred to as "words", as records of collective treaties between the nations. This was a political economy of significant scale—the number of wampum beads in circulation by 1650 has been estimated as high as three million. Such an abstract apparatus of words, languages and mnemonic patterns that enabled the public treaties of collective negotiation is clearly radically different from an arithmetic apparatus that pitted individuals against each other for private possession and individual prestige in an economy of competitive speculation, sovereign powers and bankruptcy. For more, see David Graeber, *Toward an Anthropological Theory of Value: The False Coin of Our Own Dreams*, (New York: Palgrave, 2001), 125.

Cāminātaiyar, U.Vē. 1935. Eṭṭuttokaiyuḷeṭṭāvatākiya puṟanāṉūṟu mūlamum uraiyum. ceṉṉai: lājarṉal accukkūṭam.
Deckla, S. October 2004. *Maritime History of The Pearl Fishery Coast with Special Reference to Thoothukudi*. PhD diss., The Manonmaniam Sundaranar University, Thirunelveli.
Fernando, Tamara. 2022. Seeing Like the Sea: A Multispecies History of the Ceylon Pearl Fishery 1800–1925. *Past & Present* 254 (1): 127–160.
Graeber, David. 2001. *Toward an Anthropological Theory of Value: the False Coin of Our Own Dreams*. New York: Palgrave.
Herdman, W.A. 1903. Report to the government of Ceylon on the pearl oyster fisheries of the Gulf of Mannar . London: The Royal society.
Hornell, James. 1905. *Report to the Government of Madras on the Indian Pearl Fisheries in the Gulf of Mannar*. Madras: Government Press.
Hultzsch, E. (ed. & trans.). 1891. *South Indian Inscriptions: Tamil Inscriptions of Rajaraja, Rajendra Chola and Others in the Rajarajesvara Temple at Tanjavur*. vol. 2, part I. New Delhi: Archaeological survey of India, reprint, 1991.
Hultzsch, E. (ed. & trans.). 1892. *South Indian Inscriptions: Tamil Inscriptions of Rajaraja, Rajendra Chola and Others in the Rajarajesvara Temple at Tanjavur*. vol. 2, part II. Madras: Government Press.
Kamāl, Es. Em. 2017. *Cētupati Maṉṉar Ceppēṭukaḷ*. Ariyalūr: Tiruvaḷḷuvar Patippakam.
Kāri. 2007. In *Kaṇakkatikāram: A Text on Mathematics*, ed. P. Subramaniam and K. Sathyabama. Chennai: Institute of Asian Studies.
Leppai, Kuṭavatti Mukammatu. 1881. *Muttu Cevu Kaṇakku*. Pāḷaiyaṅkōṭṭai Ṭārliṅ Press.
Montigny, Anie. 2010. "Weights & measures related to the oriental pearl trade: the origin of the « chau »". In Diving Memories In Qatar. Forum of Qatar National Day
O'Kane, John. 1972. *The Ship of Sulaiman*. New York: Columbia University Press
Ostroff, Samuel Miles. 2016. *The Beds of Empire: Power and Profit at the Pearl Fisheries of South India and Sri Lanka, C. 1770–1840*. PhD diss., University of Pennsylvania, South Asia Regional Studies.
Prakash, Muthu V., and Gunasekaran Babu. 2024. *Muttukkaṇakkatikāram eṉkiṟa Muttukkaṇakku: Book of Pearl Computations*. Puducherry: Palm Leaf Press.
Prakash, Muthu V. and Gunasekaran Babu. 2025. Kaṇakkatikāram , Pondicherry: Palm leaf press.
Sastri, K. A. Nilakanta. 1972. *The Pandya Kingdom – from the Earliest Times to the Sixteenth Century*. Madras: Swathi Publications.
Sivalingam, S. 1961. The 1958 pearl oyster fishery, Gulf of Mannar. , Bulletin No. II, Ceylon: the Fisheries research station, Department of Fisheries.
Subrahmanyam, Sanjay. 1996. "Noble Harvest From the Sea: Managing the Pearl-Fishery of Mannar, 1500-1925." In *Institutions and Economic Change in South Asia*, edited by Burton Stein and Sanjay Subrahmanyam, 134–72. Oxford University Press.
Stueart, James. 1843. *An Account of the Pearl Fisheries of Ceylon*. Ceylon: Church Mission Press.
Thurston, Edgar. 1894. *Pearl and Chank Fisheries of the Gulf of Mannar*. Madras: Government Press.
Unknown author. 1926. *Eṇ Cuvaṭi: Alakunilai, Catiravāy, Kīḻvāy Ilakkam, Ciṟukuḻi, Peruṅkuḻi, Varuṣappiṟappu*. Ceṉṉai: Coxton Press.
Vane, G. 1887. The Pearl Fisheries of Ceylon. *Journal of the Ceylon Branch of the Royal Asiatic Society*, 10(34), (1888): 14–40.
Wagner, Roy and Arun Asokan. forthcoming. *Kaṇakkatikāram*. Puducherry: Palm leaf Press.
Woolf, Leonard Sidney. 1961. *Growing: An Autobiography of the Years 1904 to 1911*. New York: Harcourt Brace Jovanovich.

Chapter 9
Magic Squares, Mathematical Tricks and Angels in Islamicate Occult Sciences: Perspectives from the Būnian Corpus

Jean-Charles Coulon

Abstract Magic squares were a fundamental element of Islamicate Arabic occult sciences, particularly with the development of the "science of letters, names and magic squares." This science transmitted many of the magic squares developed in mathematical treatises. However, as occultist practitioners were not necessarily high-skilled mathematicians, certain treatises on this science shared techniques and mnemonics describing how to easily build magic squares for practical purposes. In this paper, I propose a reconsideration—from the perspective of history of texts—of the methods of building magic squares attributed to al-Būnī (d. 622/1225 or 630/1232), a major authority in this science. I specifically study a section of his *al-Lumᶜa al-nūrāniyya* and a late apocryphal opuscule entitled *Bughyat al-mushtāq* wrongly attributed to him. This last text also introduces an occult terminology for magic squares.

Keywords Magic squares · *wafq* (pl. *awfāq*) · Arabic occult sciences · Arabic mathematics · al-Būnī · *al-Lumᶜa al-nūrāniyya* · *Bughyat al-mushtāq* · Angels · Spirits

I would like to extend my warmest thanks to the colleagues who helped me write this article, in particular, Ulrich Rebstock for our stimulating discussions about magic squares and his careful reading and thoughtful remarks; Roy Wagner and Babu Senthil for inviting me to submit this piece of research and their benevolent patience and feedbacks; the two anonymous reviewers for their accurate review and examination of the paper and their relevant comments; Korshi Dosoo for his close, careful and valuable reading; Monica Balda-Tillier for her expertise on the lines of poetry; A.O.M., Mahdi Abdeljaouad and René-Vincent Du Grand-Launay for their invaluable help for sharing manuscript references and obtaining copies.

J.-C. Coulon (✉)
IRHT-CNRS, Paris, France
e-mail: jean-charles.coulon@irht.cnrs.fr

Senthil Babu D., R. Wagner (eds.), *Normative Texts and Occupational Practices*, Boston Studies in the Philosophy and History of Science 350,
https://doi.org/10.1007/978-3-032-28419-8_9

9.1 Introduction[1]

"Magic square" is the common translation for the Arabic word *wafq* (pl. *awfāq*), which is defined as follows: "a *magic square* is a square divided into a square number of cells in which natural numbers, all different, are arranged in such a way that the same sum is found in each horizontal row, each vertical row, and each of the two main diagonals. A square with n cells on each side, thus n^2 cells altogether, is said to have the *order n*. The constant sum to be found in each row is called the *magic sum* of this square" (Sesiano 2017, p. 3, 2020, p. 1). The expression "magic square" refers to the fact that magic squares were supposed to possess occult properties and were used as powerful magic seals. However, the Arabic term *wafq* should be understood as a harmonious disposition of numbers (*wafq al-aʿdād*). Discussions of the origins and early developments of Arabic magic squares are beyond the scope of this modest contribution,[2] which is, rather, limited to particular aspects of the relationship between the massive corpus of occult sciences treatises attributed to the Sufi master al-Būnī (d. 622/1225 or 630/1232), also called the *corpus bunianum*,[3] and magic squares. Indeed, this Būnian corpus played a prominent role in the diffusion of ideas of the occult properties and magical uses of magical squares after the seventh/thirteenth century. Yet the prolific historian of magic squares, Jacques Sesiano, considers that al-Būnī "is closely associated with the bad (*sic*) reputation of magic squares" and that in his writings, "magic squares, or even figures that resemble them without having the same characteristics, are used indiscriminately, attributed with the most absurd (*sic*) virtues" (Sesiano 2008, p. 205, see also a similar introduction to al-Būnī in 2004, p. 17). Jacques Sesiano's seminal works have given us a much deeper knowledge of the development of magic squares in Arabic mathematics. In this regard, he mentions al-Būnī's name for certain methods of constructing magic squares (that we will discuss below).

However, we have gained a better knowledge of the Būnian corpus in the past decade. It became clear that the works containing the so-called "figures that resemble [magic squares] without having the same characteristics" are not in the oldest treatises by al-Būnī but are found in later pseudepigraphal works attributed to him. The discrepancy between the magic squares in works can be explained by the complex history of these texts, among which the famous *Shams al-maʿārif al-kubrā* (*The*

[1] Throughout the paper the *IJMES* transliteration system is used for Arabic terms. The dates are given in both Hegirian and Christian eras. Throughout the paper and for the sake of brevity and clarity, manuscripts are referred to by MS followed by institution or collection (as relevant) and shelfmark number; the complete references are given in the final bibliography. Moreover, I have not changed the reading direction of the magic squares reproduced from Arabic treatises, which is from right to left, in order not to invert left and right when quoting Arabic texts. Hence, the "first cell" of a magic square refers to the rightmost cell of the upper row, not the leftmost.

[2] This subject has already been covered in other studies (Ahrens 1916, 1922, 1925; Wiedemann 1918; Cammann 1969a, 1969b; Schuster 1972; Roşu 1989; Sesiano 1998, 2004, 2017, 2019, 2020; Comes and Comes 2009; Descombes 2019; Tolsa 2020; Hallum 2021).

[3] This expression was coined and popularized by Witkam (2007).

Great Sun of Knowledge) attributed to al-Būnī has long been considered to be the most representative, although we now know that it is an apocryphal and late compilation of occult works compared to the oldest works by al-Būnī or attributed to him. Hence, the attribution of non-mathematical magic squares to al-Būnī is, in fact, a widespread misconception coming from the sources attributing these texts to al-Būnī and supported by colonial and orientalist scholarship until recently. Still, it has been challenged and undermined for a decade. As a consequence, to consider al-Būnī the sole author of this heterogeneous corpus of texts with magic squares so different in their methods of construction, their structures and their use is a shortcut that continues these misrepresentations and goes against the current investigations of the Būnian corpus as a composite group of texts that require a new and closer look, including from the perspective of the history of sciences.[4]

Indeed, the "science of magic squares" was considered a part of both the science of mathematics and the science of occult properties. As an example, the famous Ottoman scholar Ṭāshköprüzāde (d. 968/1561) defined the science of magic squares (*ʿilm aʿdād al-wafq*) in his epistemological major work *Miftāḥ al-saʿāda* (Ṭāshköprüzāde 1985, vol. 1, pp. 373–374) thus:

> Magic squares (*al-wafq*) are square-shaped tables (*jadāwil murabbaʿa*) with square cells (*buyūt murabbaʿa*). Numerical numbers (*arqām ʿadadiyya*) are placed in those cells, or letters (*ḥurūf*) instead of numbers, provided that the sides and diagonals of those tables are equal in number and that there are no repeated numbers in those cells.
>
> [Scholars] mentioned that the [perfect] equilibrium of numbers (*iʿtidāl al-aʿdād*) has overflowing properties from the spiritual entity (*rūḥāniyya*) of those numbers or letters. It results in wonderful effects (*āthār ʿajība*) and strange powers (*taṣarrufāt gharība*), provided that appropriate times and eminent hours are chosen. There are many useful books on this science that are well-known among its practitioners. This science is among the branches of numerical science (*ʿilm al-ʿadad*) in terms of calculating numbers and among the branches of the science of properties (*ʿilm al-khawāṣṣ*) in terms of its effects and benefits. You will hear "the science of magic square" (*ʿilm al-wafq*), as well as "the science of letters and fractioning" (*ʿilm al-ḥurūf wa-l-taksīr*)—God Almighty willing—when mentioning "the science of properties" when mentioning the isolated letters of the suras (*muqaṭṭaʿāt al-suwar*).The best books on this art are:
>
> – *Kitāb Shams al-āfāq fī ʿilm al-ḥurūf wa-l-awfāq*,
> – *Kayfiyyat al-ittifāq fī tarkīb al-awfāq*,

[4]As Matthew Melvin-Koushki points out, the Islamicate "positive sciences" (he uses astronomy as an example) are often "studied in strict isolation from their immediate sociopolitical and intellectual contexts" (Melvin-Koushki 2016, p. 53). He links this bias to a "scholarly occultophobia", as he also put it in his introduction to the special issue of *Arabica* "Islamicate Occultism: New Perspectives" (Melvin-Koushki 2017a). Confining and limiting the name of al-Būnī as a historical author to pseudo-magic squares amounts to extending faulty historical biases about the Būnian corpus that historians have already challenged over the last two decades. On the contrary, we think that a closer investigation of the Būnian corpus as a collection of texts by different authors and from historical contexts could lead to a better and sharper understanding of the diffusion of mathematical works on the magic squares in non-mathematicians circles and of how they have been used for different purposes, works and compilations.

> – *Baḥr al-wuqūf fī ʿilm al-awfāq wa-l-ḥurūf.*[5]
> There are many books on this art, more than a hundred tomes, beyond what I have seen and heard. Indeed, this science is a sea without a coast, as it ends with the science of unveiling (*ʿilm al-mukāshafa*), in whose seas the minds of scholars and sages have drowned.

Ṭāshköprüzāde was an encyclopaedist, an "occult-friendly" promoter of sciences.[6] Of course, we can find works exclusively devoted to the mathematical aspects of magic squares and treatises exclusively devoted to their occult properties, but some authors also intend to deal with both aspects in their writings without seeing the slightest contradiction or opposition between these two aspects. Some texts attributed to al-Būnī illustrate this state of mind.

In modern research, al-Būnī's relationship to magic squares is particularly difficult to grasp because of the many later texts attributed to him and, unfortunately, a rather limited knowledge of the history of this highly complex body of texts, many aspects of which still elude scholars. For example, Wilhelm Ahrens published a pioneering article in 1922 entitled "Die 'magischen Quadrate' al-Būnī's,' which followed on from his 1916 article "Studien über die 'magischen Quadrate' der Araber" (Ahrens 1916, 1922). Wilhelm Ahrens' article is based on the *Shams al-maʿārif al-kubrā*, one of the main works attributed to al-Būnī, which we now know to be a much later apocryphal production.[7] Indeed, the prominence of al-Būnī's name and the *Shams al-maʿārif al-kubrā* in the study of the magic squares is misleading and challenging for historians in two ways: firstly, this work is much later than al-Būnī himself and cannot be taken as a source for supposedly seventh/thirteenth-century magic squares; secondly, it overshadows a number of other treatises that were in fact sources for al-Būnī and the corpus of works attributed to him. As Bink Hallum noticed:

> Would-be students of the cultural history of *awfāq* are also often hindered by over-reliance on a few works from the corpus attributed to the Sufi master of letter magic (*ʿilm al-ḥurūf*) Abū l-ʿAbbās Aḥmad al-Būnī (fl. 622/1225), which tend unjustifiably to be granted the archetypal status of *summa* of all previous literature on *awfāq*. This is unfair, not least because Islamicate writings on *awfāq* were produced in several genres and modes, most not represented by the Būnīan corpus. To put it another way, the treatment of *awfāq* in the

[5] The *Kitāb Shams al-āfāq fī ʿilm al-ḥurūf wa-l-awfāq* is a major occult encyclopaedia by the famous Ottoman occultist ʿAbd al-Raḥmān al-Bisṭāmī (d. 858/1454) (on this work, see Gardiner 2017a; Coulon 2016), the *Kayfiyyat al-ittifāq fī tarkīb al-awfāq* a work by the mathematician Abū ʿAbd Allāh Yaʿīsh b. Ibrāhīm al-Umawī (on this author, see below), and the *Baḥr al-wuqūf fī ʿilm al-awfāq wa-l-ḥurūf* an important occult treatise mentioned by ʿAbd al-Raḥmān al-Bisṭāmī in several of his own works and whose authorship he also credited himself. Moreover, the fourth section of the *Shams al-āfāq* also bears the title *Baḥr al-wuqūf fī ʿilm al-awfāq wa-l-ḥurūf.* See, for example, ʿAbd al-Raḥmān al-Bisṭāmī, *Shams al-āfāq fī ʿilm al-ḥurūf wa-l-awfāq*, MS BnF 2689, fol. 170a; MS Carullah 1545, fol. 109b; MS British Museum 7494, fol. 94r. The authorship of the *Baḥr al-wuqūf* is still a matter of debate.

[6] Matthew Melvin-Koushki considers him as a member of a "pro-occultist camp" (Melvin-Koushki 2017b, pp. 128–129).

[7] The magic squares of the *Shams al-maʿārif al-kubrā* have been studied in numerous other works, in which they are attributed to al-Būnī (Nouar and Lebza 2016; Lebza and Nouar 2014; Nouar and Lebza 2016; Nouar 2018, 2022).

> Būnīan corpus lacks context without an appreciation of the centuries-old Islamicate traditions of *awfāq* that precede it. (Hallum 2021, p. 59)

In the present study, we aim to reconsider the construction methods of magic squares and the technical occult terminology supposedly written by al-Būnī in order to situate them in the textual history of Arabic occult sciences. As the history of the Būnian corpus still has many mysteries and grey areas, this cannot be a definitive history of these texts but should rather be considered a modest new stone to be added to the edifice.

9.2 Construction Methods in *al-Lumʿa al-nūrāniyya* and the Intricate History of a Complex Text

Al-Būnī is credited for having presented a method of constructing of magic squares. This method has been discussed by Bernard Carra de Vaux (de Vaux 1948) and Jacques Sesiano (Sesiano 2008, see also 2004). Indeed, Bernard Carra de Vaux used an undated booklet entitled *Sharḥ al-ism al-aʿẓam* (*Commentary of the Greatest Name*) published in Cairo. An edition of the booklet has the date Shaʿbān 1328/ August 1910, but it also has been edited by al-Maṭbaʿa al-Maḥmūdiyya al-Tijāriyya in 1358/1939. Jacques Sesiano, in his book *Les carrés magiques dans les pays islamiques*, refers to Bernard Carra de Vaux's paper concerning al-Būnī's method (Sesiano 2004, p. 17). However, he took up the issue anew in his article "Les carrés magiques d'al-Būnī" (Sesiano 2008), in which he relies on two manuscripts held in Istanbul, namely MS Esat Efendi 3558, fols 67a-73b and Asir Efendi, 169. He notes that "this *risāla* is not a separate treatise: indeed, its beginning tells us that it is an extract from the end of one of al-Būnī's works, the *Lumʿat al-kubrà*" (Sesiano 2008, p. 205). In fact, this observation needs to be refined.

9.2.1 Al-Lumʿa al-nūrāniyya *and the Būnian Corpus*

This method for constructing magic squares does indeed come from al-Būnī's work *al-Lumʿa al-nūrāniyya fī l-awrād al-rabbāniyya* (*The Luminous Glow on the Lordly Prayers*), which is occasionally titled *al-Lumʿa al-nūrāniyya al-kubrā* in some manuscripts[8] and which has been edited as a booklet under the title *Sharḥ al-ism al-aʿẓam*.

Over the past decade, we have gained a better understanding of al-Būnī's life, works, their dissemination, and the history of the corpus attributed to him (Gardiner

[8] See, for example, the manuscript Princeton, Islamic Manuscripts, Garrett no. 1895Y, fols 1b–48b, available at http://arks.princeton.edu/ark:/88435/k643b378k. A label partially covers the title on the title page, but we can clearly read *al-kubrā* after what should probably be *al-Lumʿa al-nūrāniyya*.

2012, 2014, 2017b, 2017c; Coulon 2013, 2016, 2017, 2022). There is still a great deal of work to be done to continue to unravel the mysteries of this complex corpus. Current state-of-the-art recognises as authentic five treatises from the Būnian corpus, identified as its 'historical core': a cosmological work and esoteric commentary of the Arabic letters entitled *Laṭāʾif al-ishārāt fī al-ḥurūf al-ʿuluwiyyāt* (*The Subtleties of the Signs: On the Celestial Letters*); an esoteric commentary of the divine names entitled *ʿAlam al-hudā wa-asrār al-ihtidāʾ fī sharḥ asmāʾ Allāh al-ḥusnā* (*The Banner of Guidance and the Secrets of the Right Path: Commentary on the Beautiful Names of God*); and two manuals of Sufi spiritual exercises and path entitled *Mawāqif al-ghāyāt fī asrār al-riyāḍāt* (*The Stations of the Goals: On the Secrets of the Exercises*) and *Hidāyat al-qāṣidīn wa-nihāyat al-wāṣilīn* (*The Guidance of the Candidates and the End of the Arrivals*) (we take up here the list as we introduced it in Coulon 2022, p. 234; for a more more thorough survey of the "core," see Gardiner 2012, 2014; Coulon 2013, 2017). We may add *al-Lumʿa al-nūrāniyya* to this historical core of works: whilst its authenticity as a work by al-Būnī is a matter of debate, it is one of the earliest treatises attributed to al-Būnī. There is also consensus that the treatise on magic known as *Shams al-maʿārif wa-laṭāʾif al-ʿawārif* is not by al-Būnī but is a later compilation, taking *verbatim* extracts from works in al-Būnī's corpus and borrowing the title of a work by al-Būnī then lost (hence, the title corresponds to a book from the 'core' but does not match the treatise we now know). The later augmented version, entitled *Shams al-maʿārif al-kubrā*, is also apocryphal and only appears in the sixteenth century. *Al-Lumʿa al-nūrāniyya* is one of the first treatises attributed to al-Būnī.

Al-Lumʿa al-nūrāniyya begins with a collection of invocations (*duʿāʾ*) for each hour from the 12 h of Sunday, which closes with a correspondence table between the 12 h of Sunday and the 12 h of every other day and night of the week (usually appearing in the shape of a table, but sometimes it can appear in an entirely written form, as in MS BnF 1225, fol. 13a-18b). The second part of *al-Lumʿa al-nūrāniyya* is also known as *al-Anmāṭ*: a collection of ten groups of divine names (*namaṭ*) roughly gathered according to their meanings and semantics where possible and including descriptions of some of their occult properties and uses. The next part is another collection of invocations for the Night of Destiny (*laylat al-qadr*), the Day of ʿArafa (*yawm ʿarafa*) and the beginning of *hijrī* months. The last section, as a conclusion, offers methods to compose magic squares.

This passage is not the only text of the historical core which discusses magic squares. Indeed, al-Būnī introduces magic squares in his *Laṭāʾif al-ishārāt*, but without any theoretical considerations about their construction. We find in the *Laṭāʾif al-ishārāt* magic squares from order 3 to order 11 (Coulon 2022, p. 243). Some are discussed in sections devoted to letters, usually but not systematically corresponding to their numerical value: the eighth-order magic square is associated with the letter *mīm* whose value is 40 (al-Būnī, *Laṭāʾif al-ishārāt*, MS BnF 2657, fol. 33a-33b), the fifth-order with the letter *hāʾ* whose value is 5 (*ibid.*, fol. 37a-38a), the tenth-order with the letter *yāʾ* whose value is 10 (*ibid.*, fol. 45a-b), the ninth-order with the letter *ṭāʾ* whose value is 9 (*ibid.*, fol. 50b-51a), etc. These mentions of magic squares are then completely different from the treatment of this subject in *al-Lumʿa al-nūrāniyya*.

However, it is meaningful that the only magic squares that we find in the treatises that we can consider as genuinely by al-Būnī (*Laṭāʾif al-ishārāt*) are mathematical magic squares (*i.e.* with integers—either appearing as numbers or letters—from 1 to n^2 with a constant magic sum) and not pseudo-magic squares mimicking magic squares without having their mathematical properties (as in the *Shams al-maʿārif al-kubrā*). Moreover, the magic squares found in the *Laṭāʾif al-ishārāt* are generally different than those of *al-Lumʿa al-nūrāniyya* and in most cases do not even match the methods of constructions presented in it. This could indicate that al-Būnī had other sources on magic squares than those found in the last section of *al-Lumʿa al-nūrāriyya*.

9.2.2 **Al-Lumʿa al-nūrāniyya** *and Magic Squares: A Problematic Chapter*

The last section of *al-Lumʿa al-nūrāniyya* on the construction methods of magic squares raises questions about its relationship to the rest of *al-Lumʿa al-nūrāniyya*. Indeed, the subject of this chapter does not fit in well with the other sections, which are fairly logically structured and epistemologically coherent. Indeed, as stated above, *al-Lumʿa al-nūrāniyya* deals with prayers to recite at any time of the day or night or any specific moment in the year (months and religious celebration) and divine names. Apart from this final section, it does not refer to magic squares and their natural properties, although it does discuss certain techniques of the science of letters such as *taksīr* (lit. "breaking", "crushing"). Nothing in the rest of the treatise refers to any mathematical approach or concern. In addition to this sudden change of theme (suddenly and curiously introduced by the sentence "Let us talk about numerical magic squares"—*fa-l-natakallam ʿalā al-awfāq al-ʿadadiyya*), this final section of *al-Lumʿa al-nūrāniyya* begins with a presentation that suggests that this is an opuscule added to *al-Lumʿa al-nūrāniyya*: in a short introduction, the author states that he limited himself to presenting the best methods of constructing bordered magic squares and that his presentation consists of an introduction, two chapters and a conclusion.[9]

The structure of *al-Lumʿa al-nūrāniyya* changes relatively little from one manuscript to another, with the exception of the final section on magic squares.[10] In fact, two versions of this final section can be found in manuscripts. Four combinations can be found in the manuscripts: this final section is either simply absent, or con-

[9] This kind of formula is common in introductions of treatises to outline their structure. It can also appear at the beginning of sections or chapters, usually where there's a clear structure with subsections (for example, a treatise in four parts, each consisting of several chapters or sections). However, this is not the case in *al-Lumʿa al-nūrāniyya*, where only this final section contains the announcement of a plan.

[10] Of course, we do not include here the problem of odd manuscripts that have been incorrectly bound at some point.

tains the method to construct a magic square as the printed *Sharḥ al-ism al-aʿẓam* (the "zigzag method" described below), or contains another version (the "method of dotting" below), or, in exceptional cases, it contains both versions.[11]

Although most of the manuscripts of *al-Lumʿa al-nūrāniyya* have this final section, this is not true of the oldest identified manuscript of *al-Lumʿa al-nūrāniyya* that has reached us, namely MS Chester Beatty 3168/5, fol. 96a-125a, which dates to 686/1287. This version does not include the final section on the magic squares. It is also the case with a few other manuscripts, such as MS BnF 1225,[12] dated 845/1441-2 or MS Şehid Ali Paşa 2764/9, fols 131b-181a,[13] copied in 976/1568-9. Some manuscripts also do not have this section for the obvious reason that they are incomplete and a part of the text is missing.[14] Of course, in such cases, we cannot suppose that the manuscript never had this section.

This section describing building methods for magic squares focuses on so-called "natural" (*ṭabīʿī*) magic squares, which are magic squares containing successive integers from 1 to n^2. It also resumes the usual distinction in mathematical treatises between so-called odd and even magic squares (that is, magic squares whose order is an odd or even number) while distinguishing among the even ones between oddly even and evenly even. "Oddly even" (*zawj al-fard*) means that the order is double an odd number (e.g. 6, 10, 14, 18, etc.) whereas an "evenly even" (*zawj al-zawj*) is double an even number (e.g. 8, 12, 16, 20, etc.). Another technical term that will be found is the "border" (*ṭawq* in Arabic, lit. "necklace"). As defined by Jacques Sesiano, "a *bordered magic square* is one where removal of the successive borders leaves each time a magic square" (Sesiano 2019, p. 2). This technical vocabulary provides a better understanding of the structure of construction methods.

9.2.3 The "Zigzag Placing"

One version of this section is structured with an introduction (*muqaddima*), two chapters (*bāb*) and a conclusion (*khātima*). The introduction introduces the prerequisites (*sawābiq*) and expected results (*lawāḥiq*) in the magic squares.[15] The two chapters explain how to build even and odd magic squares, respectively. The conclusion explains how to apply this method for non-natural magic squares and magic

[11] This is the case, for example, with MS Mach Yahuda 1879, copied in Tripoli in 834/1430. This manuscript contains a commented version, so several manuscripts may have been used in its preparation and could explain why both versions appears.

[12] N° 58 in Noah Gardiner's list. We refer here to the list of manuscripts he drew up in the appendices of his PhD (Gardiner 2014).

[13] N° 136 in Noah Gardiner's list.

[14] This is the case, for example, of MS Izmir 332 (n° 91 in Noah Gardiner's list).

[15] This introduction explains the technical vocabulary. A fuller presentation of this introduction is beyond the scope of this study.

squares based on names. This is the version discussed and analyzed by Bernard Carra de Vaux and Jacques Sesiano in their above-mentioned works. The latter mentions the method as the "zigzag placing" (*placement en zigzag*) (Sesiano 2004, p. 126), and we use this expression in this paper to identify this method more easily.

This method can be found in many manuscripts from the eighth/fourteenth century onwards, for example, in MS Chester Beatty 4284, fol. 60b-72b,[16] copied in 790/1388, in MS Aya Sofya 2810, fol. 79b-95a,[17] copied in 851/1447-8, in MS Asad 3679, fol. 34a-40b copied in 993/1585, or, but incomplete, in MS Aşir Efendi 169, fol. 55b-61b,[18] dated to the fourteenth century. We must add to these examples a manuscript used by Sesiano: MS Esat Efendi 3558, fol. 67a-73b. This last manuscript deserves special mention: indeed, it is a collection of texts in which *al-Lumʿa al-nūrāniyya* is not copied except for this final section on magic squares, preceded by a title page bearing the inscription *Risāla fī l-awfāq al-ʿadadiyya al-muṭawwaqa* (*Epistle on the numerical bordered magic squares*), and introduced by: "The learned imam Shihāb al-Dīn al-Būnī said at the end of *al-Lumʿa al-kubrā* [...]" (*qāla al-imām al-ʿallāma Shihāb al-Dīn al-Būnī fī ākhir al-Lumʿa al-kubrā*). The text is therefore assumed to be a selected part of al-Būnī's *al-Lumʿa al-nūrāniyya.* The colophon indicates that it was copied in 1152/1739-40 by al-Ḥājj Muṣṭafā Ṣidqī (d. 1183/1769), a famous Ottoman scholar and mathematician who copied a considerable number of manuscripts and owned an important scientific library (on Muṣṭafā Ṣidqī, see Fazlioğlu 2006; see also Charette 2003, p. 34; and on his library, see Bonmariage 2016). In fact, the whole manuscript is a collection of short texts copied between 1149/1736 and 1159/1747, primarily (probably entirely) by Muṣṭafā Ṣidqī (the copyist's name is not indicated for each text, but the handwriting suggests that all texts were copied by the same hand). The manuscript contains another text about fifth-order magic square with an empty middle cell entitled *Sihām al-rabṭ fī l-mukhammas khālī l-wasaṭ* by Muḥammad al-Fāsī (d. 1052/1641) (MS Esat Efendi 3558, fol. 86b-89b) copied the 26th dhū l-ḥijja 1159/9th January 1747.[19]

"Zigzag placement" is an easy way to build magic squares. As the title *Epistle on numerical bordered magic squares* given by Muṣṭafā Ṣidqī to this epistle in his manuscript indicates, this method is based on filling the successive borders of a magic square around a central square. The bordered magic square has a specific property pointed out in the introduction of this section: "the bordered magic square is distinguished from the other magic squares, in that whenever you remove a border from it, you find that the remaining square (*al-murabbaʿa al-bāqiya*) is [itself] a magic square."[20]

---

[16] N° 18 in Noah Gardiner's list.

[17] N° 61 in Noah Gardiner's list.

[18] N° 26 in Noah Gardiner's list.

[19] This text is known in other manuscripts, see for example MS BnF 2732, fol. 1b-4a.

[20] Al-Būnī, *al-Lumʿa al-nūrāniyya*, MS Chester Beatty 4284, fol. 63a; MS Aya Sofya 2810, fol. 82a-b; MS Asad 3679, fol. 34b; MS Aşir Efendi 169, fol. 57b; MS Esat Efendi 3558, fol. 68b:
الشرط الثالث وبه تميّز الوفق المطوّق عن غيره من الأوفاق وهو أن يكون كلّما أزلت عنه طوقا وجدت المربعة الباقية وفقا [...].

"Zigzag Placement" for Even Magic Squares

Al-Lumʿa al-nūrāniyya starts with the even magic squares. The basis of this method is a central fourth-order magic square, in which one places the smallest number (1), in the "cell of the king" (*bayt shāh*), which refers to the piece known as the 'king' in chess (Maciuszak 2003, pp. 94–95). From this cell, the next number (2) is placed in the second row, in the cell corresponding to the move of the knight (*al-faras*) in chess (Maciuszak 2003, p. 97).[21] The next numbers must be placed in the third and fourth rows. Hence, the third number (3) must be placed in the third row also according to the move of knight starting from the first number. Finally, the fourth number (4) must be placed in the fourth row also according to the move of knight, from the previous number. Then, the four next numbers must be placed in the same manner backwards, starting with the fifth number (5) in the right cell of the lower row. We can summarize this step this way (Fig. 9.1):

Then, the same operation must be performed backwards with the last eight numbers of the magic square from the last number to appear in the magic square. If we take the example of the fourth-order magic square (which is not the example given in the text), we must place the integers from 9 to 16. The first to be placed is then 16, in the cell of the king (*bayt shāh*) in the opposite side, then 15, 14 and 13 (Fig. 9.2):

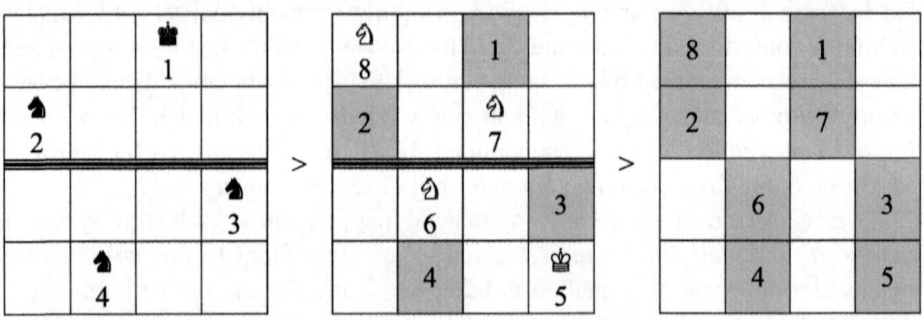

Fig. 9.1 The knight move in the fourth-order magic square (first step)

The text gives the example of the tenth-order magic square: if it is a correct "bordered" magic square, when one removes the external border, the result is an eight-order magic square. If one removes the external border of this eight-order magic square, the result is a sixth-order magic square; and if one removes the external border of this sixth-order magic square, the result is a fourth-order magic square.

[21] The knight's move in chess is often used to describe methods of construction of magic squares in Islamic mathematics. Jacques Sesiano devoted a paper on these methods, but it does not include the exact same method as this one in *al-Lumʿa al-nūrāniyya*, although some are very close (Sesiano 2003b). The closest would be the third method he found in al-Shabrāmallisī's works (see Sesiano 2003b, pp. 12–15, esp. Fig. 9.28).

8	♞ 13	1	
2		7	♞ 14
♞ 15	6		3
	4	♚ 16	5

>

8	13	1	♔ 12
2	♘ 11	7	14
15	6	♘ 10	3
♘ 9	4	16	5

>

8	13	1	12
2	11	7	14
15	6	10	3
9	4	16	5

Fig. 9.2 The knight move in the fourth-order magic square (second step)

Fig. 9.3 The eighth-order magic square by the "zigzag placement" method (first step)

		8	61	1	60		
		2	59	7	62		
		63	6	58	3		
		57	4	64	5		

The example given in the text is the eighth-order magic square, entirely empty of numbers, except the central fourth-order magic square, which is filled with the last eight numbers of this square (that is, from 57 to 64) (Fig. 9.3)[22].

Then, there are two slightly different ways to complete the borders depending on whether they correspond to odd or even internal magic square. For example, an eighth-order magic square has an eighth-order (therefore evenly even) border, which is the external one, and a sixth-order (therefore oddly even) border between the external border and the inside fourth-order magic square.

We will here introduce the method as explained for the eighth-order magic square in the text. The first number to be placed in the border is described in the text as half the mathematical square of the smaller magic square plus 1 (thus always an odd number). It must be placed in the upper right corner cell. In the case of the sixth-order border surrounding the fourth-order central magic square, we will have 9 which is half the square of four plus one ($4^2/2 + 1$); and in the case of the

[22] Al-Būnī, *al-Lumʿa al-nūrāniyya*, MS Chester Beatty 4284, fol. 64b; MS Aya Sofya 2810, fol. 84a; MS Asad 3679, fol. 35b; MS Aşir Efendi 169, fol. 59a; MS Esat Efendi 3558, fol. 69b.

Fig. 9.4 The eighth-order magic square by the "zigzag placement" method (second step)

26							19
	14					9	
		8	61	1	60		
		2	59	7	62		
		63	6	58	3		
		57	4	64	5		

eighth-order border surrounding the sixth-order magic square, we will have 19, which is half the square of six plus one ($6^2/2 + 1$).

Then, the second number to place is the number placed in the upper left corner cell, adding to it the number of cells on the side of the border minus one (thus always be an even number). In the case of the sixth-order border, we will have 14 ($9 + 6 - 1$); and in the case of the eighth-order border, we will have 26 ($19 + 8 - 1$) (see Fig. 9.4).

Then, one must place, in the cell to the left of the lower right corner the odd number that follows the first number placed in the border (hence 11 for the sixth-order border, 21 for the eighth-order border). The following odd number (13 or 23) must be placed in the cell at the left of the cell opposite to the previous one (i.e., the third-rightmost upper cell), and so on, leaving the two last cells that should be filled. This is the so-called "zigzag placement", to use the words of Bernard Carra de Vaux and Jacques Sesiano. In the cell to the left of the last odd number placed this way, one must place the number of the upper right corner plus one (which is $9 + 1 = 10$ for the sixth-order border and $19 + 1 = 20$ for the eight-order border; it will thus always be an even number). Then, one must place the following even number (*i.e.* respectively 12 and 22) in the cell corresponding to the move of the "queen" (*firzān*)[23] in chess, that is diagonally, from the last filled cell (see Fig. 9.5; we highlighted with grey background the cells corresponding to the "zigzag placing").

[23] *Firzān* is usually translated as the "queen" but comes from the Persian word *farzān*, which means "wise, learned" (Steingass 1963, p. 917) and refers here to an advisor or counsellor. This piece's movement was slightly different from our "queen" as it moved one square diagonally (Maciuszak 2003, pp. 95–96).

Fig. 9.5 The eighth-order magic square by the "zigzag placement" method (third step)

26			27		23		19
	14			13		9	
		8	61	1	60		
		2	59	7	62		
		63	6	58	3		
	12	57	4	64	5		
22		10	15		11		
	20	29		25		21	

From there, one must perform the "zigzag placing" method between the left and right sides of the border with the even numbers. However, there are certain specificities: one number that should appear in this series will, in fact, already have been placed in the upper left corner, so it must be skipped,[24] and the "zigzag" is interrupted in the two central cells of the column, which must be filled with the two successive even numbers. The text also states that, in the case of an oddly-even border, the last even number placed will always be on the right side; to the contrary, in the case of an evenly even border, it will be on the left side (see Fig. 9.6).

Half of the cells of the borders are now filled. The text then indicates that one must place in the opposite of every already filled cell a number so that the sum of both equals the total number of cells in the full square plus one (not the border; *i.e.* in our example, the sum of two opposite numbers in every border of the eighth-order magic square is $8^2 + 1 = 65$). The text adds that "the opposite of a corner cell is the diagonally opposite corner (*muqāṭar*)" and the opposite of any other cell corresponds to the move of the rook (*rukh*) in chess (Maciuszak 2003, p. 98).

The text then shows the result in the cases of the eighth-order and tenth-order magic squares (see Figs. 9.7 and 9.8)[25].

[24] For example, in the sixth-order border, 14 is already placed in the upper left corner, so the even number following 12 is 16; in the case of the eight-order border, 26 is already placed so the even number following 24 is 28.

[25] Al-Būnī, *al-Lumʿa al-nūrāniyya*, MS Chester Beatty 4284, fol. 67a; MS Aya Sofya 2810, fol. 87b; MS Asad 3679, fol. 37a; MS Esat Efendi 3558, fol. 70b. The manuscripts usually do not highlight the bordered structure. However, in some cases, some copyists highlighted this structure. For example, in the case of MS Esat Efendi 3558, Muṣṭafā Ṣidqī alternates between black and red ink between each border.

Fig. 9.6 The eighth-order magic square by the "zigzag placement" method (fourth step)

26			27		23		19
34	14			13		9	
		8	61	1	60		32
30		2	59	7	62	18	
28		63	6	58	3	16	
	12	57	4	64	5		24
22		10	15		11		
	20	29		25		21	

Fig. 9.7 The eighth-order magic square by the "zigzag placement" method (final)

26	45	36	27	40	23	44	19
34	14	55	50	13	54	9	31
33	17	8	61	1	60	48	32
30	47	2	59	7	62	18	35
28	49	63	6	58	3	16	37
41	12	57	4	64	5	53	24
22	56	10	15	52	11	51	43
46	20	29	38	25	42	21	39

The "Zigzag Placement" for Odd Magic Squares

Concerning odd magic squares, the first cell to be filled is the central cell with half the mathematical square plus one half ($=n^2/2 + 0.5$), *i.e.* 5 ($=3^2/2 + 0.5$) for the third-order magic square or 13 ($=5^2/2 + 0.5$) for the fifth-order magic square. Then, one must fill a diagonal with the following and preceding numbers. We will take the example of the fifth-order magic square, which must be filled with 5 in the central cell (see Fig. 9.9).

42	67	54	45	58	41	62	37	66	33
49	26	81	72	27	76	23	80	19	52
51	70	14	91	86	13	90	9	31	50
53	69	17	8	97	1	96	84	32	48
46	30	83	2	95	7	98	18	71	55
57	28	85	99	6	94	3	16	73	44
40	77	12	93	4	100	5	89	24	61
63	22	92	10	15	88	11	87	79	38
36	82	20	29	74	25	78	21	75	65
68	34	47	56	43	60	39	64	35	59

Fig. 9.8 The tenth-order magic square by the "zigzag placement" method

Fig. 9.9 The fifth-order magic square by the "zigzag placement" method (first step)

				15
			14	
		13		
	12			
11				

The next step is to determine the "cell of the king" (*bayt shāh*). Indeed, the next number to be placed is the integer following the highest number placed in the diagonal, which is in a corner at the end of the diagonal (here 16 = 15 + 1). It must be placed in the "cell of the king" (*bayt shāh*), which is one of the two cells adjacent to the corner cell of the line and the row of the highest number. There are four possibilities (Fig. 9.10; here in grey, for the rest of our example, we place it next to 15, but three other cells are possible).

Fig. 9.10 The fifth-order magic square by the "zigzag placement" method (second step)

			16	15
			14	
		13		
	12			
11				

Fig. 9.11 The fifth-order magic square by the "zigzag placement" method (third step)

	18		16	15
22			14	
		13		21
20	12			
11		17		19

From this starting point, we can fill the cells in the same border with the successive integers following the zigzag placing until we reach the other end of the diagonal. Then, we start other from the other corner (Fig. 9.11).

Then, the cells left empty on the border must be assigned a number so that the sum of two opposite numbers on the border is equal to the number of cells in the square plus one ($n^2 + 1$; here $5^2 + 1 = 26$). In reality, this amounts to a new zigzag placement with decreasing integers from the smallest number on the diagonal (starting with the cell opposite the "cell of the king"; see Fig. 9.12).

Finally, the six empty cells in the central third-order magic square must be filled with the numbers from 1 to 3 and the three last numbers that must be placed (n^2, $n^2 - 1$ and $n^2 - 2$) (Fig. 9.13).

If we apply this method to the seventh-order magic square, we obtain the following steps. First, we fill a diagonal (see Fig. 9.14).

Then, we proceed the zigzag placing in the external (seventh-order) border with increasing integers from the biggest number of the diagonal, starting with the "cell of the king" (here at the left of the upper right corner cell) and with decreasing integers from the smallest number of the diagonal, starting with the cell at the opposite

7	18	9	16	15
22			14	4
5		13		21
20	12			6
11	8	17	10	19

Fig. 9.12 The fifth-order magic square by the "zigzag placement" method (fourth step)

7	18	9	16	15
22	2	23	14	4
5	25	13	1	21
20	12	3	24	6
11	8	17	10	19

Fig. 9.13 The fifth-order magic square by the "zigzag placement" method (final)

						28
					27	
				26		
			25			
		24				
	23					
22						

Fig. 9.14 The seventh-order magic square by the "zigzag placement" method (first step)

of the "cell of the king" (see Fig. 9.15; increasing numbers with grey background, decreasing numbers in bold).

Then we proceed the same way for the next border (here the fifth-order border), starting with the integers following the highest and preceding the lowest that have

16	33	**18**	31	**20**	29	28
39					27	**11**
12				26		38
37			25			**13**
14		24				36
35	23					**15**
22	**17**	32	**19**	30	**21**	34

Fig. 9.15 The seventh-order magic square by the "zigzag placement" method (second step)

16	33	18	31	20	29	28
39	**7**	42	**9**	40	27	11
12	46			26	**4**	38
37	**5**		25		45	13
14	44	24		48	**6**	36
35	23	**8**	41	**10**	43	15
22	17	32	19	30	21	34

Fig. 9.16 The seventh-order magic square by the "zigzag placement" method (third step)

been placed in the previous border (here 40 and 10, just as the highest and lowest numbers placed in the seventh-order border were 39 and 11) (see Fig. 9.16).

Finally, the central third-order magic square must be filled with the first and last three numbers of the magic square, that is, from 1 to 3 and from 49 (=7^2) to 47 (Fig. 9.17).

Thus we arrive at this seventh-order magic square. The text gives eight examples here: four fifth-order magic squares and four seventh-order magic squares. However, the seventh-order magic squares introduced in the text are based on a slightly

Fig. 9.17 The seventh-order magic square by the "zigzag placement" method (final)

16	33	18	31	20	29	28
39	7	42	9	40	27	11
12	46	2	47	26	4	38
37	5	49	25	1	45	13
14	44	24	3	48	6	36
35	23	8	41	10	43	15
22	17	32	19	30	21	34

Fig. 9.18 Another seventh-order magic square by the "zigzag placement" method (first step)

6	**43**	8	**41**	10	**39**	28
49					27	1
2				26		**48**
47			25			3
4		24				**46**
45	23					5
22	7	**42**	9	**40**	11	**44**

different method. Indeed, the "zigzag placement" starts with 1, which is placed in the cell next to a corner cell in a row or column containing the highest number of the initial diagonal. In order to obtain the same sum from opposite cells, the other "zig-zag placing" must be filled with the integers preceding the highest number of the magic square (n^2) (see Fig. 9.18 for this first step; the example here is the first of the seventh-order magic squares given in the text as an example).

We proceed the same way for the border of the fifth-order magic square inside the seventh-order (see Fig. 9.19) before completing the whole magic square with the central third-order magic square (Fig. 9.20).

Fig. 9.19 Another seventh-order magic square by the "zigzag placement" method (second step)

6	43	8	41	10	39	28
49	15	**34**	17	**32**	27	1
2	**38**			26	12	48
47	13		25		**37**	3
4	**36**	24			14	46
45	23	16	**33**	18	**35**	5
22	7	42	9	40	11	44

Fig. 9.20 Another seventh-order magic square by the "zigzag placement" method (final)

6	43	8	41	10	39	28
49	15	34	17	32	27	1
2	38	20	29	26	12	48
47	13	31	25	19	37	3
4	36	24	21	30	14	46
45	23	16	33	18	35	5
22	7	42	9	40	11	44

Jacques Sesiano has explained the mathematical foundations behind this last method. As he notes, "the placement taught by the author is perfectly regular: a method, as well as being applicable to all borders, must also be memorable so that it can be filled in without calculation" (Sesiano 2008, p. 206).

This version of the last section of *al-Lumʿa al-nūrāniyya* is followed by considerations to build magic squares based on the numerical value of a name. Two elements can be modified in the construction of the magic square: the smallest number in the magic square and the difference between the numbers of the magic square (*tafāḍul*), it being understood that, in the case of a "natural" magic square, this difference (as well as the smallest number) is 1 by default. Two examples are given which may also be found in a later treatise attributed to al-Būnī, namely *al-Uṣūl wa-l-ḍawābiṭ al-muḥakkama* (al-Būnī 1951, p. 30), whose earliest identified manuscript is dated to the tenth/sixteenth century (Gardiner 2012, p. 134; Coulon 2013, vol. 1, pp. 552–556).

Although the section is centered on methods to build natural magic squares, we can see in it an opening towards more practical aspects linked to the occult properties of magic squares.

9.2.4 The "Method of Dotting"

The other version of this part of the text may also be a later addition, as it begins with the eulogic formula (*ḥamdala*) and the *ammā baʿda*, which usually marks the beginning of a text. It introduces two other methods to build odd and even magic squares, but their order is inverted in some manuscripts. These methods contain magic squares with dots to point the cells that must be filled first; hence, Jacques Sesiano refers to it as the "method of dotting" (Sesiano 2019, pp. 45–47). Jacques Sesiano does not associate this method with al-Būnī.[26]

In al-Būnī's *al-Lumʿa al-nūrāniyya*, this method may be found in manuscripts from the eighth/fourteenth century, contemporary with the "zigzag placement." Based on the manuscript evidence we have today, we cannot sustain the anteriority of one method over another. We can find this method, among others, in MS Asad 18,554, fol. 27b-31b, copied in 773/1372, MS Aya Sofya 2803/1,[27] fol. 63b-69b, dated fourteenth century, MS Carullah 2083/1, fol. 98a-101b,[28] copied 893/1487-8 or MS Laleli 1594/2, fol. 47a-72a, here 69a-72a,[29] copied 818/1415-16.[30]

[26] In this paper, Jacques Sesiano edits, translates and discusses a text found in a manuscript (MS Fatih 3439, fol. 178a-182b) copied in 587/1191 and whose author claims he relies on Ibn al-Haytham (d. 430/1039), al-Asfizārī (d. 480/1087) and al-Isfarāyinī (d. early eleventh c.) (Sesiano 2003a, pp. 144–152 (discussion), 166–172 (transl.) and 184–189 (Arabic text)). Ibn al-Haytham's text is now lost, and we can only hypothesize that it could potentially have been the source of this method of construction in *al-Lumʿa al-nūrāniyya*, but it would need further investigations.

[27] N° 28 in Noah Gardiner's list.

[28] N° 77 in Noah Gardiner's list. Noah Gardiner indicates that *al-Lumʿa al-nūrāniyya* is in fol. 1b-31 of the manuscript. In fact, it seems that the manuscript is incorrectly bound and some other parts of *al-Lumʿa al-nūrāniyya* can be found elsewhere, including the magic-squares part.

[29] N° 35 in Noah Gardiner's list.

[30] We can also add to this list MS Aya Sofya 1870/6, fol. 44a-84a, here fol. 81b-83b, copied in 828/1424-5 (n° 43 in Noah Gardiner's list) and MS Fatih 2803, fifteenth century, fol. 41a-45b (n° 106 in Noah Gardiner's list). In these manuscripts, magic squares do not appear, but the text of the last section matches the "method of dotting" version with a few differences. In the case of the latter, we can even see that there are blanks in some places, which may indicate that the copyist let some space to add the magic squares subsequently. As Ulrich Rebstock pointed out to me, this phenomenon "can often be observed in mathematical manuscripts where—for one reason or another—all numbers are missing. Obviously the second hand of the specialist missed the chance to fill the lacunae" (personal correspondence, the 20th June 2024).

This version is no less important than the previous one. Indeed, ʿAbd al-Raḥmān al-Bisṭāmī (d. 858/1454) comments on this "method of dotting" (not on the "zigzag placement") in his lengthy commentary on *al-Lumʿa al-nūrāniyya* entitled *Rashḥ adhwāq al-ḥikma al-rabbāniyya fī sharḥ awfāq al-Lumʿa al-nūrāniyya*[31] he wrote in 842/1439.[32] We also find in his earlier and shorter commentary of *al-Lumʿa al-nūrāniyya* entitled *Kashf al-asrār al-rabbāniyya fī sharḥ al-Lumʿa al-nūrāniyya* and written in 822/1419[33] that he based this commentary on a text containing the "dotting," not the "zigzag" method, as we can see in MS Carullah 1560, fol. 116b-123b, copied in 953/1547 and MS Hekimojlu 540, fol. 57b-60a, copied in 1002/1594 (*Rashḥ adhwāq al-ḥikma al-rabbāniyya fī sharḥ awfāq al-Lumʿa al-nūrāniyya* appears to be an augmented and more elaborate version of the *Kashf al-asrār al-rabbāniyya*).[34] Therefore, this is the version that ʿAbd al-Raḥmān al-Bisṭāmī most likely knew and taught.[35]

The manuscripts may have a slightly different order: after the introduction, some manuscripts start with even magic squares, others with the odd magic squares, but, except for this inversion, it is the same method. However, there is a gap in this method: the text says that it presents a method for even and odd magic squares. However, in the section on even magic squares, the method presented only concerns evenly even magic squares. The method, based on the fourth-order square, cannot be applied to oddly even magic squares.

[31] For example, in MS Kadizame Mehmed 333, fol. 144a-150b, copied in 914/1509; or in MS Carullah 1543/2, fol. 143b-151a, copied in 953/1547; this last one is nº 105 in Noah Gardiner's list, but he considered this manuscript to be *al-Lumʿa al-nūrāniyya* although it is ʿAbd al-Raḥmān al-Bisṭāmī's commentary.

[32] This date of composition Monday 27th dhū l-qaʿda 842/11th May 1439 appears in the text itself. See ʿAbd al-Raḥmān al-Bisṭāmī, *Rashḥ adhwāq al-ḥikma al-rabbāniyya fī sharḥ awfāq al-Lumʿa al- nūrāniyya*, MS Kadizade Mehmed 333, fol. 122. This year of composition is also confirmed in ʿAbd al-Raḥmān al-Bisṭāmī's intellectual auto-bio-bibliography *Durrat tāj al-rasāʾil wa-ghurrat minhāj al-wasāʾil*. See ʿAbd al-Raḥmān al-Bisṭāmī, *Durrat tāj al-rasāʾil wa-ghurrat minhāj al-wasāʾil*, MS Nuruosmaniye 4905, fol. 37a.

[33] There are also some differences in the title, which is *Kashf al-asrār al-zamāniyya fī sharḥ al-Lumʿa al-nūrāniyya* (not *al-rabbāniyya* as usual) in MS Carullah 1560, fol. 7a. The date of composition is not exactly the same as in the *Durrat tāj al-rasāʾil*, in which it appears in the year 816, and not 822.

[34] See also, for example, ʿAbd al-Raḥmān al-Bisṭāmī, *Sharḥ al-Lumʿa al-nūrāniyya wa-l-barqa al-raḥmāniyya*, MS ULiège 5406, fol. 102a-104b.

[35] In the *Durrat tāj al-rasāʾil*, in several years, ʿAbd al-Raḥmān al-Bisṭāmī states that he gave reading certificates (*qaraʾa ʿalayya*) for *al-Lumʿa al-nūrāniyya*. It seems that it is the work he taught the most in his career, and his own production reflects this preeminence, as he wrote several commentaries of *al-Lumʿa al-nūrāniyya* and its parts throughout his active years.

•			•
	•	•	
	•	•	
•			•

>

4			1
	7	6	
	11	10	
16			13

>

4	14	15	1
9	7	6	12
5	11	10	8
16	2	3	13

Fig. 9.21 The fourth-order magic square by the "method of dotting"

The "Method of Dotting" for [Evenly] Even Magic Squares

The method for the fourth-order magic square is child's play. One must place dots in the cells of the two diagonals of the magic square (left-hand square in Fig. 9.21), according to the move of the *firzān* in chess as the text states (see above Sect. 9.2.3.1). These cells are then filled with the number corresponding to their order in the square (middle square in Fig. 9.21).[36] Then, the operation is simply repeated backwards with the empty squares (right-hand square in Fig. 9.21).

This method can be found in earlier treatises, for example for the fourth-order magic square in the *Maqāla fī wujūd ʿillat al-ʿadad al-mutaḥābba wa-l-ashkāl al-murabbaʿa dhawāt al-sutūḥ al-ʿadadiyya wa-khawāṣṣi-hā min jihat awḍāʿ khawāṣṣ al-ajrām al-falakiyya* by Abū l-Walīd al-Mālaqī (d. *c.* 530/1135) (on al-Mālaqī, see Hallum 2021, pp. 99–102).[37] Not all the manuscripts of al-Mālaqī's epistle contain the same amount of text. The more developed version, containing this method of construction, can be found in MS Majlis Šūrā-ye Millī 31395, p. 50.[38] Although al-Mālaqī does not apply the method to other magic squares, it would represent one of its earliest mentions if this section of the text is authentic.[39]

[36] The squares containing dots are the first, fourth, sixth, seventh, tenth, eleventh, thirteenth and sixteenth. Simply put 1 in the first square, 4 in the fourth, 6 in the sixth, 7 in the seventh, 10 in the tenth, 11 in the eleventh, 13 in the thirteenth and 16 in the sixteenth.

[37] I am particularly indebted to Ulrich Rebstock for drawing my attention to this very important treatise by al-Mālaqī. My use of the text here owes much to our stimulating exchanges. I would like to thank him warmly and sincerely for this.

[38] It cannot be found in the manuscript used by Bink Hallum in his paper, MS Staatsbibliothek 98. Doubt may be cast on the authenticity of these parts due to the fact that the treatise, in both the Berlin and Tunis manuscripts, presents the magic squares from the third-order to the tenth-order in relation to the planets, and then, in the part of the Tunis manuscript that is not in the Berlin manuscript, presents the magic squares from the third-order to the tenth-order once again, this time with their numerical correspondences and corresponding spirits (among many other themes, such as the method of constructing the 4×4 magic square, the use of letters and numerical calculations, etc.). This form of repetition, while not impossible in a treatise, nevertheless raises questions about the authenticity of this part and whether it might not be a later addition.

[39] Since much of the literature on magic squares has yet to be discovered and studied, it is quite possible that new sources discovered in the future will shed new light on the history of this method.

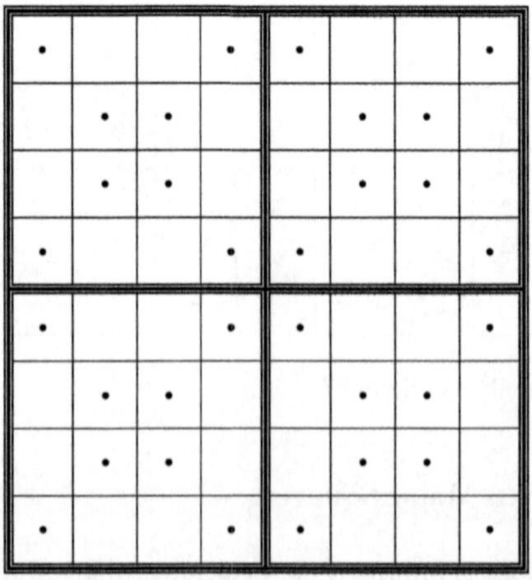

>

8	58	59	5	4	62	63	1
49	15	14	52	53	11	10	56
41	23	22	44	45	19	18	48
32	34	35	29	28	38	39	25
40	26	27	37	36	30	31	33
17	47	46	20	21	43	42	24
9	55	54	12	13	51	50	16
64	2	3	61	60	6	7	57

Fig. 9.22 The eighth-order magic square by the "method of dotting"

Pseudo-Būnian works seem to have particularly valued this easy method as we can find it in pseudo-Būnī's *Bughyat al-mushtāq* (see below) or in pseudo-Būnī's *al-Uṣūl wa-l-ḍawābiṭ al-muḥakkama* (al-Būnī 1951, p. 30), a text we cannot trace back before the tenth/sixteenth century (Gardiner 2012, p. 134; Coulon 2013, vol. 1, pp. 552–556).

Concerning the eighth-order magic square, the operation is essentially the same, as we can divide the eighth-order magic square into four fourth-order magic squares to place the dots (left-hand square in Fig. 9.22), then successively fill the cells with the corresponding number in the dotted cells, then backwards in the blank cells (right-hand square in Fig. 9.22):

From a mathematical perspective, this method can be found, for example, in al-Būzjānī's (d. 388/998) (Sesiano 2017, pp. 214–216 (transl.) and 342–344 (Ar.)) and in a sixth/twelfth-century compilation in a manuscript dated 587/1191 (Sesiano 2003a, pp. 144–145, 166–168 (transl.) and 184–185 (Ar.), see also 2004, pp. 43–45), and a similar method is also contained in the anonymous eleventh-century *Iʿdād fī wafq al-aʿdād* (*Numeration on the Harmony of Numbers*) (Sesiano 1996, pp. 41–42 (transl.) and 198–196 (Ar.)).

Therefore, the "method of dotting" is particularly easy to apply to evenly even magic squares, and this may why it is reproduced in al-Mālaqī's manuscript or also given in the late apocrypha *Bughyat al-mushtāq*.

The "Method of Dotting" for Odd Magic Squares

As for the odd magic squares, the dots are not put in the diagonals but in the vertical and horizontal rows that cross in the central cell. The central cell is that at which the two diagonals intersect: it is a specificity of odd magic squares (even magic squares

	•	
•	•	•
	•	

>

	1	
3	5	7
	9	

>

8	1	6
3	5	7
4	9	2

Fig. 9.23 The third-order magic square by the "method of dotting"

have four central cells) and is described in the text as being "like the heart in the human body" (*ka-l-qalb fī jasad al-insān*).[40] This central cell contains half the number of cells plus one, *i.e.* $(n^2 + 1)/2$. However, contrary to the evenly even magic squares, the dotted cells are not the two main diagonals, but the cells constituting a smaller "diagonal square" whose sides intersect the middle of the sides of the magic square (left-hand square in Fig. 9.23). These cells will receive the odd numbers of the magic square.

The text explains that we must first place the odd numbers of the square, *i.e.* 1, 3, 5, 7 and 9. One must start with one at the upper end of the central column. The next odd number, 3, is placed in the cell corresponding to the move of the *firzān*. The next number to be placed is 5 in the central cell. Then, we move to the cell next to the central one in the right, where we place 7. Then, 9 is placed on the cell corresponding to the move of the *firzān* from the cell of the 7. In fact, we fill each dotted diagonal with the successive odd numbers (middle square in Fig. 9.23). Finally, the even numbers (*i.e.* 2, 4, 6 and 8) must be placed in the empty cell in the increasing order from the right to the left and from the bottom to the top (right-hand square in Fig. 9.23). The three steps are depicted in the text (as in Fig. 9.23).

Al-Lumʿa al-nūrāniyya then states that this method can be applied to the other odd magic squares, such as the fifth-order and seventh-order magic squares, but does not give examples for them. We can here see an example with the fifth-order magic square (Fig. 9.24).

The text in *al-Lumʿa al-nūrāniyya* is not very clear about the even numbers to be placed, but, in fact, it also follows the "move of the *firzān*" in the empty triangles, starting with the upper cell of the right bottom triangle, and going on successively with the upper cells of the left bottom, the upper right and the upper left triangles (2 → 4, then 6 → 8, for the two upper cells of the lower triangles; then 10 → 12, 14 → 16, 18 → 20 and 22 → 24 for the four upper cells of the upper triangles).

This method of building odd magic squares is reported by al-Kharaqī (d. 533/1138-9) in his *Talkhīṣ fī l-ʿadad al-wafq* (*Epitome on the Harmonious Number*). He attributes this method to al-Isfizārī (d. 510/1116) (Sesiano 2004, p. 34). We also find it in the anonymous eleventh-century *Iʿdād fī wafq al-aʿdād* (Sesiano 1996, pp. 34–39 (transl.) and 201–198 (Ar.)), as well as in treatises by later authors such

[40] Al-Būnī, *al-Lumʿa al-nūrāniyya*, MS Asad 18,554, fol. 29a; MS Aya Sofya 2803/1, fol. 66a; MS Carullah 2083/1, fol. 100b; MS Laleli 1594/2, fol. 71a.

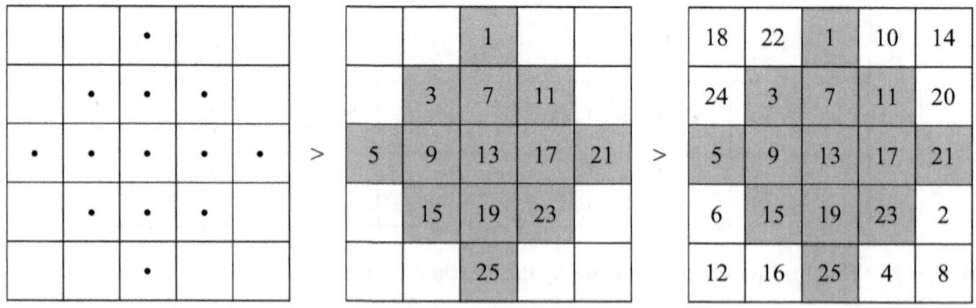

Fig. 9.24 The fifth-order magic square by the "method of dotting"

as al-Shabrāmallisī (d. 1087/1677) and al-Kishnāwī (d. 1154/1741) (Sesiano 2004, p. 34). However, it seems that it is not associated with the "method of dotting." It is probable that the dotting of the cells for adding numbers is there as a means to draw a link between the two methods.

As a conclusion for this particular section of *al-Lumʿa al-nūrāniyya*, both methods aim to provide an easy method to build magic squares without complicated calculations. Jacques Sesiano states that, after the major mathematical works on the magic square by Abū l-Wafāʾ al-Būzjānī, ʿAlī b. Aḥmad al-Anṭākī (d. 376/987) or that of Ibn al-Haytham (d. *ca* 430/1040) among others,

> General studies did not stop after the 11th century, although they tended to consist in finding other, simpler or more elegant methods of constructing magic squares. The treatises also became more numerous. This meant, however, that they less demanded of readers, the main concern being to teach methods without going into the mathematical background. […] (Sesiano 2020, p. 6)

Indeed, both versions of the last section of *al-Lumʿa al-nūrāniyya* focus on the practical aspect of easily constructing a magic square, and do not go into mathematical demonstration. Of course, that does not necessarily mean that the author of *al-Lumʿa al-nūrāniyya* (if we consider any of the two versions of the last section as genuine) or later practitioners were not able to perform the calculations to check the validity of their magic squares: these methods must be understood as mnemonic devices for constructing the magic squares as easily as possible, a kind of standard framework to be adapted according to the goal. It also seems that both versions introduce methods that have been drawn from other works.

9.3 A Late Apocrypha: The *Bughyat al-mushtāq fī maʿrifat waḍʿ al-awfāq*

There is another work attributed to al-Būnī which introduces a construction method for magic squares: *Bughyat al-mushtāq fī maʿrifat waḍʿ al-awfāq*. This short epistle has been edited by Muṣṭafā al-Bābī al-Ḥalabī in Cairo in *Manbaʿ uṣūl al-ḥikma*, a collection of four treatises (falsely) attributed to al-Būnī. It has not yet been possible to identify the manuscript used by the publisher. The *Bughyat al-mushtāq* never appears in any bibliographical lists of works attributed to al-Būnī, and his authorship seems to be unfounded, although can be explained by reference to its content with regards to al-Būnī's reputation in occult sciences.

We find some mentions of similar titles in bibliographical works or manuscripts. Thus, in *Kashf al-ẓunūn ʿan asāmī l-kutub wa-l-funūn*, Ḥājjī Khalīfa mentions a *Bulghat al-mushtāq fī ʿilm al-awfāq* by Muḥammad b. ʿAlī b. Aḥmad al-Fāriqī (Ḥājjī Khalīfa 1941, vol. 1, col. 202). It has not been possible to find a manuscript containing this precise title, but a manuscript bearing the title *Bulghat al-mushtāq min muntakhab al-awfāq*, copied in Rabīʿ II 967/January 1560 can be found in MS National Library of Turkey 7015.[41] Without entering into a detailed analysis of the content, which would go far beyond the objective of this article, the table of contents shows the similarities between the two texts:

> Section One: Introductions that must be mentioned in its matter, the properties of the magic square and its benefit
>
> Section Two: The Amicable and Hateful Numbers and the description of the operation in each description
>
> Section Three: The times in which the operation must be done in order to succeed, the explanation of its benefit, and the purpose of each magic square
>
> Section Four: The description of deducing the names of the assistants (*al-aʿwān*) from the magic square to swear an oath by it and for the power (*li-l-taṣrīf*) in the operation
>
> Section Five: The description of what each magic square is fumigated with at the time of [its] writing until the operation is completed
>
> Section Six: The benefits of magic squares that have been experienced in prose and verse
>
> Section Seven: The magic squares that have been experienced according to the sun especially and its presence in the Zodiac[42]

However, even if it deals with similar subjects, the *Bulghat al-mushtāq* is completely different from the pseudo-Būnī's *Bughyat al-mushtāq*, whether at the level of its organization or its content itself, and the former cannot be considered as the source of the latter.

Later, Ismāʿīl Bāshā al-Bābānī (d. 1920), in his supplement to Ḥājjī Khalīfa's work *Kitāb Īḍāḥ al-maknūn fī l-dhayl ʿalā Kashf al-ẓunūn*, mentions a *Bughyat al-mushtāq fī asrār al-ḥurūf wa-l-awfāq* by Muḥammad b. Muḥammad b. Sannār

---

[41] I deeply thank AOM for making this manuscript known to me. The title *Bulghat al-mushtāq min muntakhab al-awfāq* appears in the text of the manuscript (fol. 2a) but the title page has *al-Muntakhab min Bulghat al-mushtāq fī manāfiʿ al-awfāq*.

[42] Muḥammad b. ʿAlī b. Aḥmad al-Fāriqī (?), *Bulghat al-mushtāq min muntakhab al-awfāq*, MS National Library of Turkey 7015, fol. 2b:

الفصل الأوّل في مقدّمات يجب ذكرها في أمرها خواصّ الوفق ومنفعته
الفصل الثاني في الأعداد المتحابّة والمتباغضة وصفة ما يعمل في كلّ صفة
الفصل الثالث في الأوقات التي يجب العمل فيها لينجح وبيان منفعته وما القصد من كلّ وفق
الفصل الرابع في صفة استخراج أسماء الأعوان من الوفق ليقسم به عليه وللتصريف في الأعمال
الفصل الخامس في صفة ما يبخّر به كلّ وفق في وقت الكتابة حتّى يتمّ العمل
الفصل السادس فيما وقعت التجربة عليه من منافع الأوفاق نشرًا ونظمًا
الفصل السابع فيما جرّب من الأوفاق بحسب الشمس خاصّة وكونها في البروج

al-Aḥmadī al-Shāfiʿī held in the khedival library (*min kutub al-khidīwiyya*) (Ismāʿīl Bāshā al-Bābānī 1971, vol. 1, col. 190) (Brockelmann 2016-2019, S2, 1053). However, an examination of the manuscript (MS Baladiyya 1686 B[43]) shows us that it was not the source of our pseudo-Būnian *Bughyat al-mushtāq*, although there are some similarities that we will discuss below.

Despite the vagueness and grey areas surrounding the history of this treatise, we can nevertheless find elements of it in other works. As we have no information that would allow us to date it, it is still difficult to establish the precise relationship between these different texts, but pointing out these parallels is a first step towards a better understanding of the history of this text.

9.3.1 *Structure and Topics of the* Bughyat al-mushtāq*: Magic Squares, Spirits and Astrology*

The first part of the text introduces construction methods for magic squares. It is divided into three sections: respectively odd magic squares, evenly even magic squares, and oddly even magic squares. These methods are, in fact, mnemonic techniques to build a magic square easily (or, at least, supposedly easily). The first section on odd magic squares gives poetry verses or "rules" (*qāʿida*), which are a succession of words or letters to build these magic squares. The second section on evenly even magic squares explains the "method of dotting" exposed in some manuscripts of *al-Lumʿa al-nūrāniyya* (which may also explain the erroneous attribution of this treatise to al-Būnī). This part of the treatise is purely descriptive, explaining how to construct magic squares: the question of their properties, uses, correspondences, etc., is not addressed at all here.

The second part of the *Bughyat al-mushtāq* contains a first section about the definitions of the "foundations" (*uṣūl*) of magic squares. We find here technical terms which correspond to arithmetical operations based on the numbers of the magic squares. They are used later in the treatise, in the third section of the third part, to deduce the names of angels and spirits linked to the magic square. This section is followed by two other sections about numerical technical operations for divine names and Qurʾānic verses, namely *taksīr* (lit. 'fractioning') and *takʿīb* (lit. 'cubation').

The last part of the work introduces the "operative" aspects of magic squares. Indeed, the first section deals with astrological correspondences: firstly, the correspondence between auspicious and inauspicious moments and the zodiacal signs and planets are presented; secondly, each magic square is associated with a planet and a field of application. The second section is devoted to the Balance of Letters, a correspondence system inspired by the Jābirian corpus and at the heart of the "science of letters": it draws correspondences between the Arabic letters, the four

[43] I would like to extend my warmest thanks to René-Vincent Du Grand-Launay for his invaluable help in obtaining a copy of this manuscript.

elements or qualities, and seven degrees. Then, the third section expounds on how to deduce the names of higher and lower spirits (*rūḥāniyya ʿulwiyya* and *rūḥāniyya sufliyya*) associated with magic squares based on the technical terms in order to invoke them in specific rituals involving incense fumigations (*bakhūr*) and oaths (*qasam*).

Finally, the conclusion, claiming the authority of the theologian al-Ghazālī (d. 505/1111), gives general indications on the spiritual retreats and recitations to be made in association with the magic squares. It is not surprising here that al-Ghāzālī's authority is summoned insofar as he was apocryphally attributed teachings on the third-order magic square, sometimes referred to as *muthallath al-Ghazālī* (Coulon 2017, pp. 187–191; Hallum 2021, pp. 102–104).

9.3.2 Mnemonic Rules for Magic Squares

The pseudo-Būnī *Bughyat al-mushtāq* introduces a number of mnemonic techniques to easily build magic squares.

Odd Magic Squares in Pseudo-Būnī's *Bughyat al-mushtāq*

The first part is dedicated to odd magic squares and introduces for the third-order magic square a "method" (*ṭarīqa*) entitled "B.Ṭ.D—Z.H.J—W.Ā.Ḥ" (al-Būnī 1951, p. 57), which is essentially the letters corresponding to the numerical value of the integers of the natural third-order magic square (2.9.4—7.5.3—6.1.8). Learning this sequence appears to be fundamental to building higher odd magic squares.

Indeed, the next odd magic square construction techniques build upon bordered magic squares centred on a third-order magic square. The author introduces the method with the fifth-order magic square (al-Būnī 1951, p. 51):

> As for the fifth-order magic square (*al-mukhammas*), place the 1 in the cell of the *kāf*,[44] the 2 in the cell of the *ʿayn*, the 3 in the cell of the *dhāl*, and so on until the 8 in the cell of the *bāʾ*, following the method (*ṭarīqa*):
>
> Every time you have seen someone good with your own eyes, they have [only] imagined it.
>
> His passion came to strike me by [throwing me into] a desert without water
> (*Kullumā ʿāyantu Dhā l-Ḥasan Khālu-hu/Jāʾa Yarmīnī Hawā-hu Bi-l-falāt*)[45]

---

[44] By *bayt al-kāf*, we should understand the eleventh cell of the square: the *kāf* is worth 20 according to the *abjad* system, but it is the eleventh letter in ascending numerical order. Following this system, we have then *ʿayn* for the 16th cell, *dhāl* for the 25 cell and *bāʾ* for the second cell.

[45] The first hemistich is based on an opposition between seeing with one's own eyes (*ʿāyana*) and imagining forms (*khāla*). We would like to thank Monica Balda-Tillier for her help in understanding this line of poetry.

Fig. 9.25 The fifth-order magic square in the *Bughyat al-mushtāq* (first step)

7			8	
6				
				1
				2
3	4			

Fig. 9.26 The fifth-order magic square in the *Bughyat al-mushtāq* (final)

7	22	5	8	23
6	12	17	10	20
25	11	13	15	1
24	16	9	14	2
3	4	21	18	19

> [At the beginning of] each word [in this verse, there is] a letter, which is the cell in which it is placed. This is how it is represented (see Fig. 9.25)[46].

To put it more clearly, the line of poetry is a mnemonic device: the first letter of each word indicates the box in which to place the corresponding number (*i.e.* 1 for the first word, 2 for the second, etc). The text goes on:

> The completion rule (*qāʿidat al-takmīl*) is "T.Sh.W.Th—D.Ā.R.S". Place the 9 in the cell of the 1 of the third-order magic square (*al-muthallath*) which is in its middle, the 10 in its second [cell] [and so on] until the last, in which there will be 17. Then place the 18 in the cell opposite the most important of the eight numbers taken from the [completion] rule, which is 8, [place] the 19 opposite the 7, the 20 opposite the 6, and so on until opposite the 1, which is the *mighlāq* (lit. "the lock"),[47] provided that the opposite of the side is a side, and the opposite of the diagonal a diagonal. [The magic square] will thus be completed according to the rule of completion which has been mentioned, and this is what its representation will be (see Fig. 9.26).

In the previous paragraph, the poetry verse indicated how to place the eight first integers in the magic square. Here, the next nine numbers (*i.e.* from 9 to 17) must be placed in the same order as from 1 to 9 in the central third-order magic square.

[46] In the text, only the border is represented, no cell is demarcated in the central third-order magic square at this stage. I have restored the lines here with dotted lines to better highlight the cells to allow the reader to count them.

[47] This term refers to the highest number in the square, as explained below.

Fig. 9.27 The fifth-order magic square in the *Bughyat al-mushtāq* (structure)

7	*22*	**5**	**8**	*23*
6	12	17	10	*20*
25	11	13	15	**1**
24	16	9	14	**2**
3	**4**	*21*	*18*	*19*

It assumes that the practitioner must know by heart the basic third-order magic square. Finally, the "completion rule" (*qāʿidat al-takmīl*) "T.Sh.W.Th—D.Ā.R.S" indicates where to place the eight last integers in the magic square (that is from 18 to 25). Indeed, *tāʾ* (T) is the 22nd letter according to the abjad system, so 18 must be placed in the 22nd cell. *Shīn* (Sh) is the 21st letter, so 19 must be placed in the 21st cell; *wāw* (W) is the 6th letter/cell, where 20 must be placed; *thāʾ* (Th) is the 23rd letter/cell for 21; *dāl* (D) is the 4th letter/cell for 22; *alif* (Ā) is the 1st letter/cell for 23; *rāʾ* (R) is the 20th letter/cell for 24; and *sīn* (S) is the 15th letter/cell for 25. We provide (Fig. 9.27) a representation of the structure of the final magic square with the numbers corresponding to the mnemonic verse in bold characters, the central third-order magic square in grey, and the numbers from the "completion rule" in italics. As we can see, the mnemonic techniques concern the border of the magic square.

Evenly Even Magic Squares in Pseudo-Būnī's *Bughyat al-mushtāq*

The method of constructing evenly even magic squares in this treatise is basically the "Method of Dotting." The author gives the example of the fourth-order magic square whose diagonals must be filled with dots, which corresponds to the move of the *firzān* in chess. As explained above (see Sect. 9.2.4), from there, one simply fills in the dotted cells with the number corresponding to their order in the magic square, then repeats the same operation starting from the end with the undotted cells (al-Būnī 1951, p. 59).

The author then takes the eighth-order magic square (*al-muthamman*) as an example and explains that it can be divided into four fourth-order magic squares. The dots must be placed in each sub-fourth-order magic square as in the basic fourth-order magic square, and the integers must be placed following the same rule (al-Būnī 1951, p. 59).

This method is by far the simplest way of creating magic squares. It requires no mathematical knowledge, apart from knowing how to count from 1 to n^2!

Oddly Even Magic Squares in Pseudo-Būnī's *Bughyat al-mushtāq*

The construction method for oddly even magic squares is by far the trickiest. The author here gives two examples, which are the sixth-order and the tenth-order magic squares. The author seems less at ease proposing a simple technique for constructing these magic squares. He uses primarily mnemonic devices, which are far less easy to remember than those for odd-numbered squares.

For example, for the sixth-order magic square, the author explains that (al-Būnī 1951, pp. 59–60):

> As for the sixth-order magic square (*al-musaddas*), place the 1 in the first cell of the first row, the 2 in the second cell of the sixth row, the 3 in the sixth cell of the fifth row, the 4 in the sixth cell of the first row, the 5 under the 1, the 6 and the 7 after the 2, the 8 above the 3, the 9 under the 5, the 10 before the 4, according to the rule (*qāʿida*) of this cell—[rule that will follow]—that is to say that the first letter of the word [corresponds] to the number of the cells (*abyāt*) and the second to the number of the rows, which is:
>
> Did he go away confessing while she denied him proximity,
> Wizened limbs and his eyelids were stricken with intense fever[48]
> (*a-ahjara bawḥā wa-hiya al-qarb abʿā—jawāriḥ dūnī wa-daqqa ajfān-hā hāmā*)
>
> Then place the 11 in the "key"[49] of the fourth-order magic square (*miftāḥ al-murabbaʿ*) that there is inside, according to the [following] rule: "Ā.Z.L.N—S.Ṭ.W.D—Y.ʿ.J.H—Ḥ.B M.K": the 12 in the second [cell] and so on, one by one, until its "lock" (*mighlāq*). Then place the 27 in the cell opposite the largest number in the border, which is the 10, then the 28 opposite the 9 and so on, like what went before in the odd [magic squares] (*fī fard al-fard*), until the magic square is complete. This is the description of its arrangement—God knows best (see Fig. 9.28)[50].

As in the case of odd magic squares, the process of distributing the integers from 1 to 36 in the square is divided into three parts: first, the integers from 1 to 10 must be distributed in the outer border, then the numbers from 11 to 26 in the central fourth-order magic square, then the numbers from 27 to 36 are distributed in the empty border cells.

To understand the line of poetry, we need to take the first two letters of each of the ten words and deduce in which column and row to put the number corresponding to the order of the word in the line (see Table 9.1; the letters between parentheses are the letter equivalent of the number).

---

[48] I have amended here the edition that introduced the poetry verse this way: *wa-hiya 11*, followed by the verse. However, by comparing the square given as an example with the mnemonic corresponding to this verse, the verse should begin with two *alif*, easily confused in manuscripts with twice the number 1, which is why I have changed here *hajara* to *a-ahjara*. The first hemistich is built on parallelism and opposition between *hajara* and *qaraba* and between *bawḥā* and *abʿā*. The verse uses the topoi of love poetry. I would like to thank Monica Balda-Tillier for her help in understanding this line of poetry.

[49] The "key" is the smallest number in the magic square, as explained below (Sect. 9.3.3).

[50] I have corrected the magic square in the edition. Here are the changes: the edition has 23 instead of 26; 39 instead of 29; 24 instead of 34; 32 instead of 36; and 23 instead of 33.

Fig. 9.28 The sixth-order magic square in the *Bughyat al-mushtāq* (final)

4	10	30	31	35	1
32	18	21	24	11	5
28	23	12	17	22	9
8	13	26	19	16	29
3	20	15	14	25	34
36	27	7	6	2	33

Table 9.1 Analysis of the mnemonic verse/rule to build the sixth-order magic square

Number to be placed	Column	Row	Word	Cell number
1	1 (Ā)	1 (Ā)	***A**-**A**hjara*	1st
2	2 (B)	6 (W)	***BaW**ḥā*	32nd
3	6 (W)	5 (H)	***Wa**-**H**iya*	30th
4	6 (W)	1 (Ā)	*al-qarb*[a]	6th
5	1 (Ā)	2 (B)	***AB**ʿā*	7th
6	3 (J)	6 (W)	***JaW**āriḥ*	33rd
7	4 (D)	6 (W)	***DW**nī*	34th
8	6 (W)	4 (D)	***Wa**-**D**aqqa*	24th
9	1 (Ā)	3 (J)	***AJ**fān-hā*	13th
10	5 (H)	1 (Ā)	***HĀ**mā*	5th

[a]This term does not fit here. We should probably have *wa-l-qarb* because we should have a word beginning with a *wāw* and an *alif*

The second step is to fill the central fourth-order magic square with the 16 next numbers (from 11 to 26). Contrary to case of odd magic squares, such as the fifth-order, the placement of the numbers does not follow the structure of the basic fourth-order magic square, and the author gives a "rule" for filling it: "Ā.Z.L.N—S.Ṭ.W.D—Y.ʿ.J.H—Ḥ.B M.K." Each of these 16 letters gives the position of the corresponding number in the fourth-order magic square. Indeed, the first letter (1, *i.e.* 11) is Ā, whose number equivalent is 1, so 11 must be placed in the first cell. The second letter (for 12) is Z, whose equivalent is 7, so 12 must be placed in the seventh cell; the third letter (for 13) is L, whose equivalent is 12, so 13 must be placed in the 12th cell; and so on. The last number in this central fourth-order magic square is 26 which must be placed in the cell corresponding to *kāf* (K = 11th letter/cell). All that remains is to place the numbers from 27 to 36 in the last empty cells in the border. Each number from 27 to 36 must be placed in the opposite cell in the border of the numbers 10 to 1 (27 opposite 10, 28 opposite 9, 29 opposite 8 and so on until 36 in the square diametrically opposite 1). As we can see, the method is not too difficult,

4	10	11	14	88	89	94	95	99	1
96	22	30	73	32	70	27	81	19	5
92	77	36	42	62	63	67	33	24	9
86	21	64	50	53	56	43	37	80	15
83	72	60	55	44	49	54	41	29	18
17	75	40	45	58	51	48	61	26	84
16	25	35	52	47	46	57	66	76	85
8	23	68	59	39	38	34	65	78	93
3	82	71	28	69	31	74	20	79	98
100	91	90	87	13	12	7	6	2	97

Fig. 9.29 The (corrupted) tenth-order magic square in the *Bughyat al-mushtāq*

but one needs to know by heart the two rules for the ten first cells in the border and the 16 cells in the central fourth-order magic square.

The second example given is the tenth-order magic square and, in this case, we are faced with a major difficulty: the magic square shown as an example is incorrect in the edition and, despite my best efforts, I have not succeeded in correcting it yet.[51] What is more, the rules set out do not working properly, and some are certainly corrupted.[52] We can, of course, accept that the text originally offered a method that worked, but that errors in successive copies may have led to a text that

[51] Of course, a skilled mathematician could succeed in that task, and I do not assume that the magic square had never been correct in the history of the text. On the contrary, I think that the successive mistakes in the copying process led to such a situation. However, if the rules given in the text partially lead to a result, I have not succeeded in obtaining a perfect magic square following them. I won't reproduce the magic square given in the edition here because it is highly corrupted (with some doublets or missing integers, for example). The closest result I have been able to produce is in Fig. 9.29. However, the second and ninth lines and columns do not match the magic sum. I submit an amended version in Fig. 9.30.

Although the second and ninth columns do not have the exact magic sum (504 and 506 instead of 505), it is the closest result I have been able to provide, in trying to follow the instructions as closely as possible (but, even in this case, I had to break certain rules to get closer to this result).

[52] Some mistakes can easily be corrected but, as the magic square given in the example is incorrect, some are particularly tricky to restore. Such a mathematical-philological discussion would go far beyond the scope of this paper.

4	10	11	14	88	89	94	95	99	1
96	72	30	73	32	70	27	81	19	5
92	77	36	42	62	63	67	33	24	9
86	21	64	50	53	56	43	37	80	15
83	26	60	55	44	49	54	41	75	18
17	79	40	45	58	51	48	61	22	84
16	25	35	52	47	46	57	66	76	85
8	23	68	59	39	38	34	65	78	93
3	82	71	28	69	31	74	20	29	98
100	91	90	87	13	12	7	6	2	97

Fig. 9.30 The (corrupted) tenth-order magic square in the *Bughyat al-mushtāq* (amended version)

is too corrupted to be properly understood and used. In this case, it also means that the generations of copyists did not accurately understand the text. In any case, the method follows the same steps, successively filling the border with half the numbers supposed to be placed in it according to a "rule" (*qāʿida*) until the central fourth-order magic square. Hence, the external border of the tenth-order magic square must be filled with the numbers from 1 to 18,[53] the border of the eighth-order magic square from 19 to 32 and the border of the sixth-order magic square from 33 to 42. Then, after filling the central fourth-order magic square with the numbers from 43 to 58, all that is left to do is fill in the blanks in the successive borders, by placing the numbers in ascending order in the square opposite the numbers already entered in descending order. Hence, following this rule, the empty cells of the border of the sixth-order magic must be filled with the numbers from 59 to 68, the border of the eighth-order from 69 to 82 and the tenth-order from 83 to 100.

However, this last example shows the limits of this approach: the rules to be remembered become too complex to facilitate the construction of magic squares. The corruption of the text in this particular example shows that the difficulty is too great compared with the methods for odd and evenly even magic squares.

[53] The text of the edition reads 8 here, which is obviously a typo.

9.3.3 Angels and Magic Squares: The Elaboration of an "Occult" Technical Terminology of Magic Squares

The second part of the *Bughyat al-mushtāq* defines the eight "foundations" (*uṣūl*) of the (even and odd) magic squares:

The *miftāḥ* (lit. "the key") is the smallest number we put into the magic square, and the *mighlāq* (lit. "the lock") is its greatest.

The *ʿadl* (lit. "equality"[54]) is the sum (*majmūʿ*) of the *miftāḥ* and the *mighlāq*.

The *aṣl* (lit. "root, base, foundation"[55]) is the remainder of the magic square (*isqāṭ al-wafq*), which is [also] called *al-ṭarḥ* (lit. "the subtraction"[56]) and is the result of the product of the square root of the *shakl* (lit. "the form") by its half, after subtracting 1 from it (*al-ḥāṣil min ḍarb tarbīʿ al-shakl fī nuṣfi-hi baʿd ṭarḥ wāḥid min-hu*).

The *wafq*, which is [also] called *al-ḍilʿ* (lit. "the row"), is the result of multiplying the square root of the *shakl* by its half after adding 1 (*al-ḥāṣil min ḍarb tarbīʿ al-shakl fī nuṣfi-hi baʿd ziyādat wāḥid ʿalay-hi*).

The *misāḥa* (lit. "measure"[57]) is the result of the product of the *wafq* by the *shakl*,[58] or the sum of the rows, in length and height (*majmūʿ al-aḍlāl ṭūlan wa-ʿarḍan*).

The *ḍābiṭ* (lit. "general rule") is the sum of the *misāḥa* and the *wafq*.

The *ghāya* (lit. "goal, extremity") is double (*ḍiʿf*) the *misāḥa* and double the *wafq* or the sum of the rows in length, height and diagonal (*quṭr*[an]).

It then gives the example of the third-order magic square with the values of the eight 'foundations' (see Table 9.2, not in the *Bughya*: we added the formulae corresponding to these 'foundations' in the case of 'natural' magic squares).

Table 9.2 The eight foundations of the magic squares and the example of the third-order magic square

miftāḥ	1	1
mighlāq	9	n^2
ʿadl	10	$1 + n^2$
aṣl (or *ṭarḥ*)	12	$n \times ((n^2 - 1)/2)$
wafq (or *ḍilʿ*)	15	$n \times ((n^2 + 1)/2)$
misāḥa	45	$[n \times ((n^2 + 1)/2)] \times n$
ḍābiṭ	60	$[n \times ((n^2 + 1)/2)] + [n \times ((n^2 + 1)/2)] \times n$
ghāya	120	$[n \times ((n^2 + 1)/2)] \times n \times 2 + n \times ((n^2 + 1)/2)$

[54] This is a highly polysemous word in Arabic. It can also mean "measure" in some contexts.

[55] It is another highly polysemous word in Arabic.

[56] This word can also refer to the act of advancing a piece in chess, which, of course, recalls the fact that chess pieces are mentioned in the methods of constructing magic squares that we have seen.

[57] The *misāḥa* is the output of a measurement and *ʿilm al-misāḥa* is the geometry. *Misāḥa* also became one of the terms to be calculated (added, multiplied…). I thank one of the anonymous reviewers for this last piece of information.

[58] If we take up here the vocabulary elements previously described and compare them with the result announced in the following paragraph, there seems to be an error: *tarbīʿ* seems to be missing before *shakl* as in the two previous calculations insofar as the *misāḥa* announced is 45 and the *wafq* is 15. So here we should have the product of 3 and 15 and not 9 and 15.

Table 9.3 The *ṭarḥ* of the magic squares from the third- to the tenth-order magic squares

Third-order	12
Fourth-order	30
Fifth-order	60
Sixth-order	105
Seventh-order	168
Eighth-order	252
Ninth-order	360
Tenth-order	495

Finally, the *Bughyat al-mushtāq* provides the *ṭarḥ* (or *aṣl*) for the magic squares from the third-order to the tenth-order (summarised in Table 9.3).

These technical terms seem to appear in occult literature devoted to magic squares but not in the mathematical texts about them. Although the origins and the elaboration of this vocabulary remain quite shadowy, some elements which can be traced earlier tend to confirm that this terminology is intrinsically linked to the "occult" uses of the magic squares and not their mathematical aspects.

Among the possible earlier sources of these occult mathematical developments are the *Kanz al-asrār wa-dhakhāʾir al-abrār* attributed to Hermes through a recension by Thābit b. Qurra, and the aforementioned *Maqāla* by Abū l-Walīd al-Mālaqī.

Although al-Mālaqī seems primarily interested in the occult properties of magic squares, he nevertheless stresses the importance of calculation for all operations involving them:

> When you have a number and you want to insert it into the third-order magic square to deduce the kings (*al-mulūk*) by whom you are sworn to operate according to the number of the magic square, which is 45,[59] and to deduce the incense from it, and to deduce the zodiacal sign from it, and to deduce from it the mansion in which the moon is, then examine your operation, then perform skillfully and correctly its calculation according to the [rule of the] Great *abjad* values (*bi-l-jummal al-kabīr*).[60] If the number is correct, your operation is correct, and your need is satisfied.
>
> Know that this science (*al-ʿilm*) does not make mistakes, and the reason for its error is only due to the student's (*al-ṭālib*) lack of calculation. If the student is poor in calculation, then his operation [can be] right [as well as] wrong. If he is an arithmetician (*ḥāsib^{an}*) and astronomer (*falakiyy^{an}*) who knows the four seasons and the division of the [lunar] mansions and Zodiacal signs according to the four natures and knows the mixing (*mumāzaja*) of the letters with each other, then his operation is correct and will never be wrong, as long

[59] The sum of all the numbers in a third-order magic square is 45.

[60] *Al-jummal al-kabīr* is an expression for an operation converting the name of an individual into its numerical value. The difference between the small (*al-ṣaghīr*) and the great (*al-kabīr*) *jummal* is that the "small" does not take into account tens and hundreds (*alif*, *yāʾ* and *qāf* both count for 1), but the "great" differentiates tens and hundreds (*alif*, *yāʾ* and *qāf* respectively count for 1, 10 and 100).

> as the [celestial] sphere (*al-falak*) will continue its movement on its own, with the permission of God […].[61]

This call for a sound knowledge of calculation, arithmetic and astronomy for occult practices is in accordance with the technical nature of some methods.[62] Al-Mālaqī explains that to deduce the higher "kings" (*al-mulūk*), the first is deduced from the cell of the "key" (*bayt al-miftāḥ*), the next three from the magic square's "pillars" (*arkān al-wafq*) and the fifth entirely from the diagonal (*min al-quṭr kulli-hi*), adding that the fifth rules these four higher kings (*al-arbaʿa al-ʿulwiyya*). The "terrestrial kings" (*al-mulūk al-arḍiyya*) can be deduced from other cells: the first from the cell following the "key", etc.

The text goes on to explain that, in order to deduce the names of the corresponding entities, one must subtract 51 from the number, convert the remainder into letters and add the suffix *-āʾīl* (whose numerical value is 51) or subtract 319 from the number and add the suffix *-ṭīsh* (whose numerical value is 319) to the remainder after it being converted into letters. This is the operation described in *Bughyat al-mushtāq* to deduce the names of the higher (*ʿulwī*) and terrestrial (*arḍī*) "kings" (*malik*) respectively. The ambiguity of the term *mlk*, which can be read *malak* (angel) or *malik* (king), can be tricky here: an angel is supposed to be a celestial being and can hardly be qualified as "terrestrial" (*arḍī*). The plural used here, even for the higher ones, is *mulūk* which is the usual plural for *malik*, not *malak*. However, the theophoric suffix *-āʾīl* obviously makes us think of angels. As we will see, the terms used to describe these entities can slightly change from one source to another. The method to deduce the names by subtracting 51 or 319 and adding a suffix to the letters corresponding to the result of the subtraction is roughly the same as in the *Bughyat al-mushtāq*, except in al-Mālaqī's text, the calculation is based on the numbers in the cells of the magic square, although the *Bughyat al-mushtāq* bases the calculations on the "foundations" calculated from specific numbers in the magic square. We can see here that al-Mālaqī's epistle does not use such an elaborate terminology, but if these passages are authentic, this text could be one of the first steps in this development.

As for the *Kanz al-asrār wa-dhakhāʾir al-abrār*, it seems to be an influential text on the occult aspects of magic squares and it also introduces similar techniques. Ḥājjī Khalīfa describes the text as (Ḥājjī Khalīfa 1941, vol. 2, cols. 1512–1513; see also Bonmariage and Moureau 2016, p. 4, fn. 16):

---

[61] MS Majlis Šūrā-ye Millī 31395, p. 57.

فإذا كان معك عدد وأردت الدخول به في وفق المثلث لإخراج الملوك الذي تقسم بهم على العمل عدد الوفق وهو ٤٥ ولإخراج البخور منه ولإخراج البرج منه ولإخراج المنزلة منه الذي فيها القمر فانظر في عملك ثمّ أحكم حسابه بالجمل الكبير حكما صحيحا فإذا صحّ عملك وقضيت حاجتك.

واعلم أنّ هذا العلم ليس يخطئ وما يكون سبب خطأه إلّا من قلّة حساب الطالب فإذا كان الطالب قليل الحساب فإن عمله يصيب ويخطئ وإذا كان حاسبًا فلكيًا عالمًا بالفصول الأربع وبقسمة المنازل والبروج على الطبائع الأربع ويعلم ممازجة الحروف بعضها مع بعض فإن عملك يصحّ ولا يخطئ أبدًا ما دام الفلك حركته عماله وذلك بإذن الله [...].

[62] This technical nature has been highlighted for example by Ulrich Rebstock: "Solche Techniken erfordern zwar auf der Seite des Produzenten eine profunde Kenntnis der mathematischen Regeln. Sie erlauben aber auf der anderen Seite eine außerordentlich flexible Anpassung des Talismans an die persönlichen Bedürfnisse des Empfängers. Die Rechenkunst avanciert hier zum Messdiener der Religion!" (Rebstock 2019, p. 231).

> *Kanz al-asrār wa-dhakhāʾir al-abrār* (*The Book of Secrets and Treasures of the Righteous*) [in the science of letters and magic squares][63] by Hermes (*Hirmis al-harāmisa*), [which is his fifth book on the science of letters].[64] It is a great book, among the foundations of this art. It is from [this book] that the *shaykh* Abū ʿAbd Allāh Yaʿīsh b. Ibrāhīm al-Umawī[65] extracted the *Kitāb al-Istinṭāqāt* (*The Book of the Articulated Languages*).[66] It was commented in Arabic language[67] by Teukros of Babylon (*Tinkalūshā al-Bābilī*), as well as by Thābit b. Qurra al-Ḥarrānī [d. 288/901] and Ḥunayn b. Isḥāq al-ʿIbādī[68] [d. 260/873]. It is a great book and a foundation in the science of magic squares and letters.

Hence, the dating of this text is very problematic and the current state-of-the-art does not allow us to provide a sound periodization.[69] However, we can read in it the beginning of the use of specific numbers to deduce names of entities linked to the magic squares. We do not find in this treatise the developed terminology (*miftāḥ*, *mighlāq*, etc.) as it is clearly expressed in the *Bughyat al-mushtāq*, but we can observe similarities:

> The first name is deduced from the number in the centre (*wasaṭ*) of the magic square if it is an odd [magic square],[70] and, if it is an even [magic square], from half the equalizing number in that magic square.
>
> The second name [is deduced] from the last number located into that magic square.
>
> The third name is deduced from the 'equalizing' number (*al-ʿadad al-ʿadl*) in that magic square, which is the sum of the smallest number in it and the largest number in it.
>
> The fourth name [is deduced] from the number located on the side of that magic square.
>
> The fifth name [is deduced] from the total number by which this magic square is established.
>
> An example of this in a 3 by 3 square: so the first name [is deduced] from the pivot (*quṭb*) of the magic square and the middle cell, which is 5; the second name [is deduced] from the last number located in this square, which is 9; the third name [is deduced] from the number (*sic*),[71] and that is if you add the smallest number in it to the largest number in it, then in this example it will be 10; the fourth name [is deduced] from the number located in

[63] This does not appear in Gustav Flügel's edition.

[64] This likewise does not appear in Gustav Flügel's edition.

[65] The date of this arithmetician is problematic: although Ḥājjī Khalīfa gives 895/1489-90 as his death date, a manuscript of his *Marāsim al-intisāb fī ʿilm al-ḥisāb* contains a teaching licence by him dated 774/1373 to the copyist of the manuscript, who wrote in the colophon that he copied the text in this same year 774/1373 (Saidan 2008).

[66] The term *naṭq* or *istinṭāq* is used in later texts to designate words derived from the conversion of numbers from the 'foundations' of magic squares. Unfortunately, I have not been able to consult a manuscript of this text, but it is possible that the title refers to this terminology.

[67] According to Flügel's edition Teukros provided an "Arabic commentary" (*sharḥ*[an] *ʿarabiyy*[an]), but in the 1941 edition he provided a "strange commentary" (*sharḥ*[an] *gharīb*[an]).

[68] In the 1941 edition, we read al-Qabāwī instead of the usual *nisba* al-ʿIbādī.

[69] The *terminus post quem* would be the second half of the third/ninth century to be concordant with Thābit b. Qurra or Ḥunayn b. Isḥāq's putative authorship, and the *terminus ante quem* would be the end of the eighth/fourteenth century as Abū ʿAbd Allāh Yaʿīsh b. Ibrāhīm al-Umawī is said to have drawn from this text.

[70] In this case, the *wasaṭ* equals $(n^2 + 1)/2$.

[71] The text has *al-ʿadad*, but we should read *al-ʿadad al-ʿadl*, as above.

> one side, which is 15[72]; and the fifth name [is deduced] from the total number located in the magic square, which in this example is 45.[73] Understand this.[74]

We can draw parallels between these five numbers and the eight foundations in the *Bughyat al-mushtāq*, with the exception of the first name.[75] The numbers linked to the second, third, fourth and fifth names respectively correspond to the *mighlāq*, the *ʿadl*, the *wafq* and the *ghāya*.

The text goes on to explain that, in order to deduce the names of the corresponding entities by subtracting 51 or 319 from the number, one must convert the remainder into letters and add the suffix *-āʾīl* or *-ṭīsh*. A notable and singular difference in the *Kanz al-asrār* is that names (*asmāʾ*) in *-āʾīl* are described here as "spiritual" (*rūḥānī*) while names in *-ṭīsh* as "demonic" (*shayṭānī*).

The influence of the *Kanz al-asrār* in the development of this terminology can be seen for example in the already mentioned *Bulghat al-mushtāq*, which gives the same definitions of the five names and indicates that the terminology would have been transmitted from the Greek (*nuqila ʿan al-Yūnān*), and, later in the section, that it comes from the *Kanz al-asrār wa-dhakhāʾir al-abrār* by Hermes (*Hirmis al-harāmisa*) through the commentary (*tafsīr*) by Thābit b. Qurra.[76] This influence is further visible in the huge occult sciences treatise *Kashf al-asrār al-makhfiya fī ʿilm al-ajrām al-samāwiyya wa-l-ruqūm al-ḥarfiyya* by ʿUmar b. Masʿūd b. Sāʿid al-Mundhirī al-Sulayfī (d. 1160/1747) (on this author, see Wright 2018, pp. 101–105). Al-Mundhirī devotes a chapter on the deduction of the angels (not kings) of the magic squares (*bāb fī maʿrifat istikhrāj amlāk al-awfāq*) which he claims to have drawn from the *Kanz al-asrār* attributed to Hermes (al-Mundhirī 2003, vol. 2, pp. 283–285). However, he introduces the same eight terms of the *Bughyat al-mushtaq* with almost the same wording, with the exception of the *aṣl*, which he says is the product of the *mighlāq* and the *ghāya* (and which therefore comes logically after the *ghāya*). He gives the example of the third-order and fourth-order magic squares. Then, he explains how to deduce the seven names for the magic square (using the

---

[72] *I.e.* the "magic sum".

[73] *I.e.* the sum of all the numbers in the magic square.

[74] (Pseudo-)Hermes and Thābit b. Qurra, *Kanz al-asrār wa-dhakhīrat al-abrār*, in MS Garrett 673Y, fol. 11a-b:

والاسم الأول يستخرج من جملة ما في وسط الوفق من الأعداد إن كان فردًا وإن كان زوجًا فمن نصف العدد العدل في ذلك الوفق
الاسم الثاني من آخر عدد يقع في ذلك الوفق
الاسم الثالث يستخرج منه العدد العدل في ذلك الوفق وهو جميع أقل عدد فيه وأكثر عدد فيه
الاسم الرابع من العدد الواقع في ضلع ذلك الوفق
الاسم الخامس من جملة العدد الذي يحق به ذلك الوفق
مثال ذلك في مربّع ٣ في ٣ فكان الاسم الأوّل من قطب الوفق والبيت الأوسط فكان ٥ والاسم الثاني من آخر عدد يقع في هذا المربّع وهو ٩ والاسم الثالث من العدد وذلك إذا جمعت أقلّ عدد فيه وأكثر عدد فيه فيكون على هذا المثال ١٠ والاسم الرابع من العدد الواقع في الضلع الواحد وهو ١٥ والاسم الخامس من جملة العدد الواقع في الوفق وهي في هذا المثال ٤٥ فافهم ذلك.

[75] We can hypothesise that the deduction of the first name is abandoned insofar as it depends on the even or odd nature of the magic square, which makes its deduction slightly more complex.

[76] Al-Fāriqī (?), *Bulghat al-mushtāq min muntakhab al-awfāq*, MS National Library of Turkey 7015, fol. 22a-24b.

numbers except the *aṣl*) with the suffix *-āʾīl* for "spiritual" names and *-ṭīsh* for "demonic" names.

The method of deducing the names of the higher kings (*al-mulūk al-ʿulwiyya*) and the lower assistants (*al-aʿwān al-sufliyya*) from the eight foundations of magic squares is also presented in the anonymous *Supplement* (*dhayl*) to the *Tadhkirat ūlī l-albāb wa-l-jāmiʿ li-l-ʿajab al-ʿujāb* by the *littérateur* and physician Dāwūd al-Anṭākī's (d. 1008/1599). This *Supplement* is supposed to be authored by a disciple of Dāwūd al-Anṭākī, which sets it at the beginning of the eleventh/seventeenth century at the latest. In addition to the theory, the disciple gives a specific example, based on the sentence "Ṣadr al-Dīn asks for supply/pay" (*Ṣadr al-Dīn yaṭlubu rizq*), which equals 747 once the numerical equivalents of the letters are added together (Dāwūd al-Anṭākī 1877, vol. 3, pp. 128–129, vol. 3, p. 81) (to be compared to MS Laleli 1618, fol. 447b-448a). The author then provides the corresponding third-order magic square (see Fig. 9.31).

The process of obtaining such a magic square is simple, although the author does not explain it. Firstly, you divide the number of the sentence (747) by the order of the magic square (3): the result is 249. Then, you take a third-order "natural" magic square whose central cell is 5, and simply replace 5 with 249, and add the difference between 5 and 249 (=244) to each number of the "natural" magic square. Of course, this method works in this case because 747 can be divided by 3, but Dāwūd al-Anṭākī's disciple does not say anything about the cases in which the number cannot be divided by 3. However, such a magic square is reminiscent of a similar kind of third-order magic square based on a divine name. For example, a third-order magic square based on the divine name *Kāfī* (lit. "The Sufficient"), whose numerical value is 111 (K = 20, Ā = 1, F = 80, Y/Ī = 10), can be obtained by adding 106 (=111–5) to each cell of the "natural" third-order magic square (see Fig. 9.32).

Fig. 9.31 The "Ṣadr al-Dīn asks for supply" magic square

248	253	246
247	249	251
252	245	250

4	9	2
3	5	7
8	1	6

> + 106 >

110	115	108
109	كافي	113
114	107	112

Fig. 9.32 The *Kāfī* magic square

Such a magic square can be found, for example, in the *Shams al-āfāq* by ʿAbd al-Raḥmān al-Bisṭāmī, without any explanation about how it was built.[77] The use of this kind of technique is also explained in the pseudo-Būnī's *Bughyat al-mushtāq* in the second section of the second part. The main difference is that this last method can be applied to any word in the central cell, either it can be divided by 3 or not.

The specificity of Dāwūd al-Anṭākī's disciple's example—which also constitutes its value and interest—is that the numerical value of the sentence is the magic sum, and not only the value of the central cell (this could not work for a sentence whose numerical equivalent cannot be divided by three). On the contrary, the method for divine names is that the numerical equivalent of the name is the value of the central cell, not the magic sum (in this case, we can produce a magic square, whatever the name's numerical value).

This kind of method is elaborated in a number of treatises: for example, it is introduced in an anonymous treatise held in a manuscript copied in 1085/1675 studied by Jacques Sesiano who hypothesizes that "the original text might go back to the 13th or 14th century" (Sesiano 2022, p. 93). This text presents as examples instructions for how to build a third-order magic square whose magic sum equals the numerical value of the divine name Malik (lit. "King"), that is 90 (the text specifies that it has an integral third), but also fifth-order and fourth-order squares based on the same name, and how to build fourth-order and fifth-order magic squares based on the name Jalīl (lit. "Great") whose numerical value is 73 (Sesiano 2022, pp. 119–120 (transl.) and 144–145 (Ar.)). Such a method can also be observed in objects: for example, the British Museum holds a brass seal[78] containing a third-order magic square (Fig. 9.33) surrounded by the name of four angels (Jibrāʾīl, Mīkāʾīl, Isrāfīl and ʿAzrāʾīl).

The magic sum is 1185, which is the numerical value of the invocation "O Greatest Name [of God]" (*yā l-ism al-aʿẓam*) (Porter et al. 2017, p. 540). Also 1185 divided by 3 gives 395, so all you have to do is add 390 (=395 − 5) to each number in the basic third-order magic square to obtain this magic square.

Fig. 9.33 The Greatest Name's invocation magic square

394	399	392
393	395	397
398	391	396

[77] ʿAbd al-Raḥmān al-Bisṭāmī, *Shams al-āfāq fī ʿilm al-ḥurūf wa-l-awfāq*, MS BnF 2689, fol. 143b; MS Carullah 1545, fol. 87b; MS British Museum 7494, fol. 74a.

[78] The British Museum 1893,0215.3.

Table 9.4 The eight foundations of the "Ṣadr al-Dīn asks for supply" magic square

ʿadl	*aṣl*	*mighlāq*	*miftāḥ*
498	1,511,928	253	245
ghāya	*ḍābiṭ*	*misāḥa*	*wafq*
5976	2988	2241	747

Dāwūd al-Anṭākī's disciple also goes further in his example and indicates the corresponding eight foundations for his example (see Table 9.4)[79].

Finally, the disciple shows to his reader the steps to deduce the names of the entities of the magic square (see Table 9.5).

The numbers and names in this table drawn up by the disciple of Dāwūd al-Anṭākī are for illustrative purposes only (or, it could be applied in the only case of a man named Ṣadr al-Dīn asking for supply or money) and represent a practical example of the rules set out in the previous paragraph of the text. The copying errors can, therefore, be rectified thanks to the explanations given, but also reflect the fact that not all copyists ensured the accuracy of numbers and names by systematically checking calculations and equivalences between numbers and letters. Moreover, we have not found evidence that such calculations were actually used in the process of making talismans. Indeed, most talismans use commonly known angels either from the Islamic Quranic, exegetic or occult traditions,[80] but we rarely find completely unknown angel names in talismans that would result from a specific calculation process.

Finally, we must add that, although it is a different text than the pseudo-Būnī's *Bughyat al-mushtāq*, the homonymous *Bughyat al-mushtāq fī asrār al-ḥurūf wa-l-awfāq* by Muḥammad b. Muḥammad b. Sannār al-Aḥmadī al-Shāfiʿī held in the khedival library also contains the method to deduce the names of higher and lower angels (*amlāk*) based on the eight foundations (MS Baladiyya 1686 B, fol. 3a) defined as in al-Mundhirī's *Kashf al-asrār al-makhfiya* with the *aṣl* being defined as the product between the *ghāya* and the *mighlāq* (MS Baladiyya 1686 B, fol. 3b, cf. (al-Mundhirī 2003, vol. 2, p. 283)). In another place, the treatise gives the *ṭarḥ* of the "natural" magic squares, as in the pseudo-Būnī's *Bughyat al-mushtāq*, *i.e.* 12 for the third-order magic square, 30 for the fourth-order, 60 for the fifth-order, 105 for the sixth-order, 168 for the seventh-order, 163 for the ninth-order, 495 for the tenth-order (MS Baladiyya 1686 B, fol. 2a).

[79] We have here corrected the number corresponding to the *aṣl*: in the edition, we find 5768, but in MS Laleli 1618 we find 1,511,968 in this table and 1,511,927 in the next table. If we calculate the *aṣl* as the product of the *mighlāq* and the *ghāya*, we find 5976 × 253 = 1,511,928.

[80] Quranic tradition refers to the angel names actually in the Quran such as Jibrīl/Jibrāʾīl, Isrāfīl, etc.; exegetic tradition refers to the angel names from hadith or exegetic literature such as ʿIzrāʾīl/ʿAzrāʾīl, which is the name given to the "angel of death" (*malak al-mawt*) in the Quran; and occult tradition refers to the angel names from occult literature, such as the seven angels associated in the *Shams al-maʿārif* with the seven letters absent from the first sura of the Quran, the 7 days or the seven planets, for example Rūqiyāʾīl, Samsamāʾīl, etc.

Table 9.5 The higher and lower spirits of the eight foundations of the "Ṣadr al-Dīn asks for supply" magic square

Foundations (*uṣūl*)	Number (ʿ*adad*)	Remainder (*al-bāqī*)	Letter equivalent (*naṭq*)	Higher [entity] (ʿ*ulwiyya*)	Remainder (*al-bāqī*)	Letter equivalent (*naṭq*)	Lower [entity] (*sufliyya*)
miftāḥ	245	194	Qṣd	Qṣdāʾīl	286[a]	Rfw	Rfwaṭīsh
mighlāq	253	202	Rb	Rbāʾīl	294[b]	Rṣd	Rṣdaṭīsh
aṣl[c]	1,511,928	1,511,778	Ghghthghyā Ghḍʿz	Ghghthghyā Ghḍʿzāʾīl	1,511,609	Ghghthghyā Ghkhṭ	Ghghthghyā Ghghkhṭṭīsh
ʿ*adl*	498[d]	447	Tmz	Tmzāʾīl	179[e]	Qʿṭ	Qʿṭṭīsh
wafq	747	696	Khṣw	Khṣwāʾīl	428	Tkḥ	Tkḥṭīsh
misāḥa	2241	2190[f]	Bghqṣ	Bghqṣāʾīl	1922	Ghẓkb	Ghẓkbṭīsh[g]
ḍābiṭ	2988[h]	2937	Bghẓlz	Bghẓlzāʾīl	2669	Bghkhsṭ	Bghkhsṭṭīsh
ghāya	5976	5925	Hghẓkh	Hghẓkhāʾīl	5657	Hghkhnz	Hghkhnzṭīsh[i]

[a]MS Laleli 1618, fol. 448a has 782 and the editions have 786. I have corrected here the number as it should be 286 (=245 + 360 − 319). The number 286 corresponds to the letters rfw, which appears both in the manuscript and the editions, which tends to indicate that the number has been mistakenly copied at some point (and that the mistake has not been corrected)

[b]MS Laleli 1618, fol. 448a has 214, which is a mistake

[c]This line seems to be completely faulty in the edition. I find

aṣl	235	1370	Gghthghyā	Ghḍzāʾīl	180	Ghghthghyā	Ghkhṭṭīsh

MS Laleli 1618, fol. 448a also have some erroneous information. Indeed, the *aṣl* appears here as 1,511,927 instead of 1,511,928. The remainder to deduce the celestial name appears as 1,511,778 instead of 1,511,877 and the remainder for the inferior name appears as 151,160 which should be read 1,511,609. The remainder to deduce the celestial name is Bāʿ Ghṣʿr Ghghsʿ in the manuscript instead of Ghghthghyā Ghḍʿz. The lower name is Ghghthghyā Ghghkhṭṭīsh in the manuscript, instead of Ghghthghyā Ghkhṭṭīsh (the name contains an additional *ghayn* compared to the corresponding remainder in the previous cell)

[d]MS Laleli 1618, fol. 448a has 489 here instead of 498 (the ʿ*adl* is correctly 498 in the previous table in the manuscript)

[e]MS Laleli 1618, fol. 448a has 189 here instead of 179

[f]MS Laleli 1618, fol. 448a has 219 here instead of 2190

[g]MS Laleli 1618, fol. 448a has Ghkhlbṭīsh instead of Ghẓkbṭīsh. There is no diacritical dot on the *ẓāʾ* (ظ) and under the *bāʾ* (ب) in the remainder, hence the possible reason why *ẓāʾ* appears as *ṭāʾ* (ط) in the corresponding name

[h]MS Laleli 1618, fol. 448a has 2977 instead of 2988 (but the *ḍābiṭ* is correctly 2988 in the previous table in the same manuscript)

[i]MS Laleli 1618, fol. 448a has Hghrṭīsh instead of Hghkhnzṭīsh (although the correct remainder for this name is Hghkhnz in the previous cell, but the corresponding number appears as 8568 instead of 5657). There are also mistakes for the celestial spirit in the manuscript: the number is 50215 instead of 5925, the letters Hghṭkh instead of Hghẓkh (the *ẓāʾ* does not have the dot) and the name Hghẓlhyāʾīl instead of Hghẓkhāʾīl

As we can see, this method to deduce the names of lower and higher entities tracing back to al-Mālaqī's epistle and pseudo-Hermes' widespread *Kanz al-asrār*.

9.3.4 The Balance of Letters

Pseudo-Būnī's *Bughyat al-mushtāq* also introduces calculations based on the Balance of Letters, which is summarized in a table (which is directly introduced as a table in the epistle) (see Table 9.6).

Although Jābir b. Ḥayyān is not mentioned, the system of equivalence of the Arabic letters with the four elements and qualities and seven degrees comes from the Jābirian Science of Balance (*ʿilm al-mīzān*). Interestingly, the third-order magic square is mentioned in the *Kitāb al-Mawāzīn al-ṣaghīr* attributed to Jābir b. Ḥayyān and devoted to the topic of the Balances (Coulon 2013, vol. 1, pp. 241–242; Hallum 2021, pp. 78–80). As Paul Kraus explained,

> The science of the Balance (*ʿilm al-mīzān*) aims to reduce all the data of human knowledge to a system of quantity and measurement, thus giving them the character of an exact science. It applies primarily to the three kingdoms of the sublunary world, subjecting their changes, generation and corruption, to mathematical laws. It is also used to measure the distances and movements of the stars, and even relates to the hypostases of the spiritual world. In short, "all things fall under (the principle of) the Balance" and thus become comprehensible to man. (Kraus 1942, vol. 2, p. 187; and for an overview about the Balance of letters, see pp. 223-236)

The Balance table can be found in al-Būnī's *Laṭāʾif al-ishārāt*, in which he offered this table of correspondence (al-Būnī, *Laṭāʾif al-ishārāt*, MS BnF 2657, fol. 13a). Hence, it is not surprising to find a discussion about the Balance of Letters in an epistle of pseudo-Būnī. However, the link between magic squares and the Balance of Letters can also be found in other sources, for example, in al-Mālaqī's *maqāla*, in which we find the table and the picture of the four balances (esp. p. 50–53). We also find the table and four Balances in the homonymous *Bughyat al-mushtāq* by Muḥammad b. Muḥammad b. Sannār al-Aḥmadī al-Shāfiʿī (MS Baladiyya 1686 B, fol. 4b). In fact, al-Mālaqī's *maqāla* and the two *Bughyat al-mushtāq* introduce two different representations of the Balance of Letters, the first as a table, as in al-Būnī's

Table 9.6 The Balance of letters in the *Bughyat al-mushtāq*

	Watery	Airy	Earthy	Fiery
Rank (*martaba*)	D	J	B	Ā
Degree (*daraja*)	Ḥ	Z	W	H
Minute (*daqīqa*)	L	K	Y	Ṭ
Second (*thāniyya*)	ʿ	S	N	M
Third (*thālitha*)	R	Q	Ṣ	F
Fourth (*rābiʿa*)	Kh	Th	T	Sh
Fifth (*khāmisa*)	Gh	Ẓ	Ḍ	Dh

Laṭāʾif al-ishārāt, the second as four Balances each one corresponding to an element, with three letters on each plate of the Balances and a seventh letter at the top center of the Balances. This last representation corresponds more visually to the idea of a Balance that "balances" the letters.

9.4 Final Remarks on the Talismanic Use of Mathematical Magic Squares

A question naturally arises from these texts: were/are these methods for easy construction of natural magic squares actually used by practitioners or talisman makers? It is difficult for a historian to answer such a question. Indeed, when we find a magic square, we can in some cases determine which method was used to compose it by observing the arrangement of the numbers. On the other hand, unless we have the scribe's drafts, we cannot determine whether he applied the method himself or simply used a model.

Talismans often contain magic squares. Ursula Hammed states "[s]o far, no magical squares have been attested on papyrus, but they appear on paper documents, both handwritten and block-printed. There is even one example of magical squares on an Arabic ostracon" (Hammed 2022, p. 223).[81] Many of magic squares in Arabic talismans are not mathematical magic squares in the sense that they do not contain integers from 1 to n^2, distributed so as to obtain the same sum in all the lines horizontally, vertically and diagonally. Indeed, many are built from a divine name or a Quranic verse (Hammed 2022, p. 223), and do not require the same composition methods.

What is more, the numerical magic square most commonly found in talismanic objects is the third-order magic square, the most basic, the one that any practitioner worthy of the name should know by heart. It is known by the nickname *budūḥ*, a name made up of the four letters equivalent to the four even numbers distributed around the four corners of this square. It is also nicknamed the *muthallath* (third-order magic square) of al-Ghazālī (d. 505/1111). This famous theologian, in fact, refers to this magic square in one of his treatises. However, as mentioned above, a whole subsequent literature made al-Ghazālī a repository of occult knowledge of this magic square, in a line going back to the Greek Sages and Āṣaf b. Barakhiyā, the vizier of Solomon. For example, in his *Enigmatic Charms*, the only magic square found by Karl R. Schaefer in a block-printed talisman in his corpus is the third-order (Schaefer 2006, pp. 76–79, plate 8).

The fourth-order magic square is also quite common in talismans. It seems that the surviving examples often match the result of the zigzag or dotting methods introduced in both versions of *al-Lumʿa al-nūrāniyya*. For example, a talisman in

[81] Karl Schaefer in the same volume also states that "Magic squares containing numbers, on the other hand, are quite common in the block printed amulets" (Schaefer 2022, p. 199).

the collection of Almohad plaques in the Numismatic Cabinet of the Museo Arqueológico Nacional in Madrid features a magic square that corresponds perfectly to the "dotting method" (Rodríguez Lorente and Kind 1991).

The question of the method used by practitioners or their models is, in fact, particularly relevant when it comes to high-level magic squares and/or when part of the magic square is missing and must be reconstructed. For example, a block-printed talisman held in the Beinecke Library at Yale[82] contains a 13 × 13 magic square, much of which is missing due to the poor condition of the support. For his analysis of this talisman, Mark Muehlhaeusler needed to go back to the mathematical treatises studied by Jacques Sesiano in order to reconstruct the missing parts of the square (Muehlhaeusler 2010, pp. 608–610).

In fact, many studies describing talismanic objects containing magic squares only describe them: what is the magic sum or the central number, and does it match the numerical value of a divine name or Quranic verse? Is the sum identical in each line, column and diagonal? In the case of the mathematical natural magic square, what is the order of the magic square, and with which planet is it associated? But historians—including the author of these lines—rarely wonder what kind of method was used to build the magic square in texts and talismans. The presentation of these methods in texts attributed to al-Būnī invites us to reconsider the way that magic squares are usually tackled in academic studies.

9.5 Conclusion

The history and elaboration of Būnian texts on the mathematical aspects of magic squares retain many mysteries. The appendix to *al-Lumʿa al-nūrāniyya* on construction methods exists in two different versions, and we cannot tell at this stage which one was the first in the manuscripts. In any case, it seems that the "method of dotting" was particularly common as this is the method discussed by ʿAbd al-Raḥmān al-Bisṭāmī in his commentaries on *al-Lumʿa al-nūrāniyya*. This section also explains why some magic square treatises have been attributed to al-Būnī, such as the *Manbaʿ*'s *Bughyat al-mushtāq*.

In any case, their purpose was to teach how to easily build magic squares. However, later pseudo-Būnian works introducing construction methods also exposed mathematical techniques and more elaborated vocabulary to enhance the esoteric and occult properties of magic squares, far beyond those of astrological correspondences. Hence, a mathematical angelology emerged and flourished in the science of magic squares. Although some modern scholarship has been tempted to reject the occult aspects of magic squares outside the scientific field, one must bear in mind that the occult properties of magic squares did not prevent the mathematical study of the squares; to the contrary, it served as a powerful motivation to deepen

[82] P. CTYBR INV. 2016.

their study, and authors such as al-Mālaqī insist on the importance of mathematical skills to use magic squares.

I would like to give another, famous example of the magic squares as a motivation to study and develop mathematics. The famous Ottoman scholar Mollā Lüṭfī al-Tūqātī (d. 900/1495), wrote a certain *Risālat Taḍʿīf al-madhbaḥ* (*The Treatise on the Doubling of the Altar*) exposing the "Delian problem" as developed by the mathematician and philosopher Theon of Smyrna (d. *ca* 135): in this story, the Delians face a huge epidemic and are informed by an oracle that they have to double their cubic altar to make the disease end. They consult Plato who explains that the plague was sent to them as a test, as they had neglected geometry and could not find the solution without a sound knowledge of this science.[83] In Mollā Lüṭfī al-Tūqātī's treatise, geometry is finally replaced by the science of magic squares. He states that "the science of magic squares is the first science that God made to exist, and since [this science exists] the prophets, the saints and the sages have never ceased to transmit it as an inheritance from one master to another" (*ʿilm al-wafq awwal ʿilm awjada-hu Allāh taʿālā bi-nafsi-hi wa-lam yazul baʿda dhālika yatawārathu al-anbiyāʾ ʿalay-him al-salām wa-l-awliyāʾ wa-l-ḥukamāʾ kābiran ʿan kābir*) (Mollā Lütfī al-Maqtūl 1940, p. 54 (French transl.), 17 (Arabic text)). Hence, the study of these mathematical harmonious arrangements is not only a kind of difficult mathematical mind game, but a means to gain access to a powerfully effective divine knowledge.

References

Manuscripts

Al-Bisṭāmī, ʿAbd al-Raḥmān. *Durrat tāj al-rasāʾil wa-ghurrat minhāj al-wasāʾil.* MS Istanbul, Nuruosmaniye 4905.

Al-Bisṭāmī, ʿAbd al-Raḥmān. *Kashf al-asrār al-rabbāniyya fī sharḥ al-Lumʿa al-nūrāniyya.* MS Istanbul, Süleymaniye Kütüphanesi, Carullah 1560.

Al-Bisṭāmī, ʿAbd al-Raḥmān. MS Istanbul, Süleymaniye Kütüphanesi, Hekimojlu 540.

Al-Bisṭāmī, ʿAbd al-Raḥmān. *Rashḥ adhwāq al-ḥikma al-rabbāniyya fī sharḥ awfāq al-Lumʿa al-nūrāniyya.* MS Istanbul, Süleymaniye Kütüphanesi, Kadizame Mehmed 333.

Al-Bisṭāmī, ʿAbd al-Raḥmān. MS Istanbul, Süleymaniye Kütüphanesi, Carullah 1543/2.

Al-Bisṭāmī, ʿAbd al-Raḥmān. *Sharḥ al-Lumʿa al-nūrāniyya wa-l-barqa al-raḥmāniyya.* MS Liège, ULiège, Ms5406.

Al-Bisṭāmī, ʿAbd al-Raḥmān. *Shams al-āfāq fī ʿilm al-ḥurūf wa-l-awfāq.* MS Paris, BnF, Arabe 2689.

Al-Bisṭāmī, ʿAbd al-Raḥmān. MS Istanbul, Carullah 1545.

Al-Bisṭāmī, ʿAbd al-Raḥmān. MS London, British Museum, 7494.

Al-Būnī. *Laṭāʾif al-ishārāt fī al-ḥurūf al-ʿuluwiyyāt.* MS Paris, BnF, Arabe 2657.

[83] Edition and French translation in Mollā Lütfī l-Maqtūl, *La Duplication de l'autel (Platon et le problème de Délos)*, transl. H. Corbin and A. A. Adıvar, Paris 1940.

Al-Būnī. *al-Lumʿa al-nūrāniyya*. MS Damascus, al-Asad National Library, 3679.
Al-Būnī. MS Damascus, al-Asad National Library, 18554.
Al-Būnī. MS Dublin, Chester Beatty, Ar. 3168/5, fol. 96a-125a.
Al-Būnī. MS Dublin, Chester Beatty, Ar. 4284.
Al-Būnī. MS Istanbul, Süleymaniye Kütüphanesi, Aşir Efendi 169.
Al-Būnī. MS Istanbul, Süleymaniye Kütüphanesi, Aya Sofya 1870/6, fol. 44a-84a.
Al-Būnī. MS Istanbul, Süleymaniye Kütüphanesi Aya Sofya 2803/1.
Al-Būnī. MS Istanbul, Süleymaniye Kütüphanesi, Aya Sofya 2810.
Al-Būnī. MS Istanbul, Süleymaniye Kütüphanesi, Carullah 2083/1.
Al-Būnī. MS Istanbul, Süleymaniye Kütüphanesi, Esat Efendi 3558.
Al-Būnī. MS Istanbul, Süleymaniye Kütüphanesi, Fatih 2803.
Al-Būnī. MS Istanbul, Süleymaniye Kütüphanesi, Izmir 332.
Al-Būnī. MS Istanbul, Süleymaniye Kütüphanesi, Laleli 1594/2, fol. 47a-72a.
Al-Būnī. MS Istanbul, Süleymaniye Kütüphanesi, Şehid Ali Paşa 2764/9, fols 131b-181a.
Al-Būnī. MS Paris, BnF, Arabe 1225.
Al-Būnī. MS Princeton, Islamic Manuscripts, Garrett no. 1895Y, fols 1b-48b, Available at http://arks.princeton.edu/ark:/88435/k643b378k.
Al-Būnī. MS Princeton, Mach Yahuda 1879.
al-Anṭākī, Dāwūd. *Dhayl tadhkirat ūlī l-albāb wa-l-jāmiʿ li-l-ʿajab al-ʿujāb*. MS Istanbul, Süleymaniye Kütüphanesi, Laleli 1618.
Al-Fāriqī, Muḥammad b. ʿAlī b. Aḥmad. *Bulghat al-mushtāq min muntakhab al-awfāq*. MS Istanbul, National Library of Turkey, 7015.
Al-Fāsī, Muḥammad. *Sihām al-rabṭ fī l-mukhammas khālī l-wasaṭ*. MS Paris, BnF, Arabe 2732, fol. 1b-4a.
(Pseudo-)Hermes, & Qurra, Thābit b. *Kanz al-asrār wa-dhakhīrat al-abrār*. MS Princeton, Princeton University Library, Garrett 673Y. Available at: http://arks.princeton.edu/ark:/88435/wh246s20m.
Al-Mālaqī, Abū l-Walīd. *Maqāla fī wujūd ʿillat al-ʿadad al-mutaḥābba wa-l-ashkāl al-murabbaʿa dhawāt al-sutūḥ al-ʿadadiyya wa-khawāṣṣi-hā min jihat awḍāʿ khawāṣṣ al-ajrām al-falakiyya*. MS Tehran, Ketābḫāne Majlis Šūrā-ye Millī, 31395.
Al-Mālaqī, Abū l-Walīd. MS Berlin, Staatsbibliothek, Ms. or. quart. 98.
Sannār al-Aḥmadī al-Shāfiʿī, Muḥammad b. Muḥammad b. *Bughyat al-mushtāq fī asrār al-ḥurūf wa-l-awfāq*. MS Alexandria, al-Maktaba al-baladiyya, 1686 B.

Edited Sources

(Pseudo-)al-Būnī, Abū l-ʿAbbās Aḥmad b. ʿAlī. 1951. *Manbaʿ uṣūl al-ḥikma*. Beirut: al-Maktaba al-thaqāfiyya.
al-Anṭākī, Dāwūd. 1877. *Tadhkirat ūlī al-albāb wa-l-jāmiʿ li-l-ʿajab al-ʿujāb*. Cairo: al-Maṭbaʿa al-wahbiyya, 1294/1877-1878.
Khalīfa, Ḥājjī. 1941. *Kashf al-ẓunūn ʿan asāmī l-kutub wa-l-funūn*. Beirut: Dār iḥyāʾ al-turāth al-ʿarabī, Maarif Matbaasi.
al-Bābānī, Ismāʿīl Bāshā. 1971. *Īḍāḥ al-maknūn fī l-dhayl ʿalā Kashf al-ẓunūn ʿan asāmī l-kutub wa-l-funūn*. Istanbul: Aufl. 2. Milli Ejitim Basımevi.
al-Maqtūl, Mollā Lütfī. 1940. *La Duplication de l'autel (Platon et le problème de Délos)*, ed. Mehmed Şerefeddin Yaltkaya, transl. Henry Corbin and Abdülhak Adnan Adıvar. Paris: De Boccard.
Al-Mundhirī, ʿUmar b. Masʿūd b. Saʿīd. 2003. *Kashf al-asrār al-makhfiya fī ʿilm al-ajrām al-samāwiyya wa-l-ruqūm al-ḥarfiyya*. Beirut: Muʾassasat al-Aʿlamī li-l-maṭbūʿāt.
Ṭāshköprüzāde, Aḥmad b. Muḥammad. 1985. *Miftāḥ al-saʿāda wa-miṣbāḥ al-siyāda fī mawḍūʿāt al-ʿulūm*. Beirut: Dār al-kutub al-ʿilmiyya.

Studies

Ahrens, Wilhelm. 1916. Studien über die «magischen Quadrate» der Araber. *Der Islam* 7 (3): 186–248.
Ahrens, Wilhelm. 1922. Die «magischen Quadrate» al-Būnī's. *Der Islam* 12 (3–4): 157–177.
Ahrens, Wilhelm. 1925. Nochmals die «magischen Quadrate». *Der Islam* 14:104–110.
Bonmariage, Cécile. 2016. La bibliothèque de Mustafa Sıdkı: Contribution à l'étude d'une collection de manuscrits au XVIIIe siècle. *Arabica* 63 (5): 533–556.
Bonmariage, Cécile, and Sébastien Moureau, eds. 2016. *Le Cercle des lettres de l'alphabet Dā'irat al-aḥruf al-abjadiyya: un traité pratique de magie des lettres attribué à Hermès*. Leiden/Boston: Brill.
Brockelmann, Carl. 2016-2019. *History of the Arabic Written Tradition*. Transl. Joep Lameer. Leiden/Boston: Brill.
Cammann, Schuyler. 1969a. Islamic and Indian Magic Squares: Part I. *History of Religions* 8 (3–4): 180–209.
Cammann, Schuyler. 1969b. Islamic and Indian Magic Squares: Part II. *History of Religions* 8 (3–4): 271–299.
Carra de Vaux, Bernard. 1948. Une solution arabe du problème des carrés magiques. *Revue d'histoire des sciences et de leurs applications* 1 (3): 206–212.
Charette, François. 2003. *Mathematical Instrumentation in Fourteenth-Century Egypt and Syria: The Illustrated Treatise of Najm al-Dīn al-Miṣrī*. Leiden/Boston: Brill.
Comes, Mercè, and Rosa Comes. 2009. Los cuadrados mágicos matemáticos en al-Andalus. El tratado de Azarquiel. *Al-Qanṭara* 30 (1): 137–169.
Coulon, Jean-Charles. 2013. *La magie islamique et le «corpus bunianum» au Moyen Âge*." PhD thesis. Université Paris-Sorbonne.
Coulon, Jean-Charles. 2016. Building al-Būnī's Legend: The Figure of al-Būnī Through ʿAbd al-Raḥmān al-Bisṭāmī's *Shams al-āfāq*. *Journal of Sufi Studies* 5 (1): 1–26.
Coulon, Jean-Charles. 2017. *La magie en terre d'islam au Moyen Âge*. Paris: CTHS.
Coulon, Jean-Charles. 2022. Amulets and Talismans in the Earliest Works of the Corpus Bunianum. In *Amulets and Talismans of the Middle East and North Africa in Context: Transmission, Efficacy and Collections*, ed. Marcela A. Garcia Probert and Petra M. Sijpesteijn, 230–252. Leiden/Boston: Brill.
Descombes, René. 2019. Les carrés magiques planétaires d'Agrippa. In *Bibnum Textes fondateurs de la science analysés par les scientifiques d'aujourd'hui*.
Fazlioğlu, İhsan. 2006. Mustafa Sıdkı (ö. 1183/1769). Osmanlı matematik ve astronomi âlimi, şair. In *Türkiye Diyanet Vakfı İslâm Ansiklopedisi*, vol. 31, 356–357.
Gardiner, Noah. 2012. Forbidden Knowledge? Notes on the Production, Transmission, and Reception of the Major Works of Aḥmad al-Būnī. *Journal of Arabic and Islamic Studies* 12:81–143.
Gardiner, Noah. 2014. *Esotericism in a Manuscript Culture: Ahmad al-Buni and His Readers Through the Mamluk Period*. PhD thesis. University of Michigan.
Gardiner, Noah. 2017a. The Occultist Encyclopedism of ʿAbd al-Raḥmān al-Bisṭāmī. *Mamlūk Studies Review* 20:3–38.
Gardiner, Noah. 2017b. Stars and Saints: The Esotericist Astrology of the Sufi Occultist Aḥmad al-Būnī. *Magic, Ritual, and Witchcraft* 12:39–65.
Gardiner, Noah. 2017c. Esotericist Reading Communities and the Early Circulation of the Sufi Occultist Aḥmad al-Būnī's Works. *Arabica* 64 (3–4): 405–441.
Hallum, Bink. 2021. New Light on Early Arabic *Awfāq* Literature. In *Islamicate Occult Sciences in Theory and Practice*, ed. Liana Saif, Francesca Leoni, Matthew Melvin-Koushki, and Farouk Yahya, 57–161. Leiden/Boston: Brill.
Hammed, Ursula. 2022. Arabic Magical Texts in Original Documents: A papyrologist Answers Five Questions You Always Wanted to Ask. In *Amulets and Talismans of the Middle East and*

North Africa in Context: Transmission, Efficacy and Collections, ed. Marcela A. Garcia Probert and Petra M. Sijpesteijn, 275–290. Leiden/Boston: Brill.

Kraus, Paul. 1942. *Jābir ibn Ḥayyān: Contribution à l'histoire des idées scientifiques dans l'Islam*. Cairo: Imprimerie de l'institut français d'archéologie orientale.

Lebza, Amel, and Ahmed Nouar. 2014. Sur l'arithmologie des Asmāʾ Allāh al-Ḥusnā dans le 16e chapitre du «Shams al-Maʿārif» d'Aḥmad al-Būnī. In *Actes du XIe colloque maghrébin sur l'histoire des mathématiques arabes. Alger, 26-28 octobre 2013*, 277–301. Alger: Éditions Al-Khalduniya.

Maciuszak, Kinga. 2003. Persian Checkmate - 'The King is Oppressed' on the Origin of the Chessmen's Names. *Studia Etymologica Cracoviensia* 8:91–101.

Melvin-Koushki, Matthew. 2016. Of Islamic Grammatology: Ibn Turka's Lettrist Metaphysics of Light. *al-ʿUṣūr al-Wusṭā* 24:42–113.

Melvin-Koushki, Matthew. 2017a. Introduction: De-Orienting the Study of Islamicate Occultism. *Arabica* 64 (3–4): 287–295.

Melvin-Koushki, Matthew. 2017b. Powers of One: The Mathematicalization of the Occult Sciences in the High Persianate Tradition. *Intellectual History of the Islamicate World* 5 (1): 127–199.

Muehlhaeusler, Mark. 2010. Math and Magic: A Block-Printed Wafq Amulet from the Beinecke Library at Yale. *Journal of the American Oriental Society* 130 (4): 607–618.

Nouar, Ahmed. 2022. Sur les carrés magiques d'ordre 3 dans le 16e chapitre du *Shams al-maʿārif wa-laṭāʾif al-ʿawārif* d'Aḥmad al-Būnī (m. 1225). In *Actes du XIVe colloque maghrébin sur l'histoire des mathématiques arabes. Sousse, 6-8 mai 2022*, ed. Hmida Hedfi and Mahdi Abdeljaouad, 245–256. Tunis: COMHISMA.

Nouar, Ahmed, and Amel Lebza. 2016. Les carrés magiques dans le 16e chapitre du «Shams al-Maʿārif» d'Ahmad al-Būnī. In *Actes du XIIe colloque maghrébin sur l'histoire des mathématiques arabes. Marrakech 26-28 mai 2016*, ed. Ezzaim Laabid, 164–185. Marrakech: Université Cadi Ayyad.

Nouar, Ahmed. 2018. Les carrés magiques d'ordre 5 dans le 16e chapitre du *Shams al-maʿārif* d'Aḥmad al-Būnī. In *Actes du XIIIe colloque maghrébin sur l'histoire des mathématiques arabes. Tunis, 30-31 mars et 1er avril 2018*, ed. Hmida Hedfi and Mahdi Abdeljaouad, 247–266. Tunis: Graphika Imprimerie.

Porter, Venetia, Liana Saif, and Emilie Savage-Smith. 2017. Medieval Islamic Amulets, Talismans, and Magic. In *A Companion to Islamic Art and Architecture. Volume I: From the Prophet to the Mongols*, ed. Finbarr Barry Flood and Gülru Necipoğlu, 521–557. Hoboken: G. John Wiley & Sons, Inc.

Rebstock, Ulrich. 2019. Islamische Zahlenunterwelten. In *Zahlen- und Buchstabensysteme im Dienste religiöser Bildung*, ed. Laura V. Schimmelpfennig and Reinhard Gregor Kratz, 225–245. Tübingen: Mohr Siebeck.

Rodríguez Lorente, Juan José, and Hans Dieter Kind. 1991. Un amuleto arábigo con un cuadrado mágico en el monetario del Museo Arqueológico Nacional de Madrid. *al-Qanṭara* 12 (2): 401–413.

Roşu, Arion. 1989. Les carrés magiques indiens et l'histoire des idées en Asie. *Zeitschrift der Deutschen Morgenländischen Gesellschaft* 139:120–158.

Saidan, Ahmad Salim. 2008. Al-Umawī, Abū ʿAbdallāh Yaʿīsh Ibn Ibrāhīm Ibn Yūsuf Ibn Simāk al-Andalusī. In *Complete Dictionary of Scientific Biography*, ed. Charles Coulston Gillispie. Detroit: Charles Scribner's Sons.

Schaefer, Karl R. 2006. *Enigmatic Charms: Medieval Arabic Block Printed Amulets in American and European Libraries and Museums*. Leiden/Boston: Brill.

Schaefer, Karl R. 2022. The Material Nature of Block Printed Amulets: What Makes Them Amulets? In *Amulets and Talismans of the Middle East and North Africa in Context: Transmission, Efficacy and Collections*, ed. Marcela A. Garcia Probert and Petra M. Sijpesteijn, 180–208. Leiden/Boston: Brill.

Schuster, H.-S. 1972. Magische Quadrate im islamischen Bereich. Ihre Entlehnung ins Abendland im Mittelalter sowie ihre Vorstufen. *Der Islam* 49 (1): 1–84.

Sesiano, Jacques. 1996. *Un traité médiéval sur les carrés magiques: de l'arrangement harmonieux des nombres*. Lausanne: Presses polytechniques et universitaires romandes.
Sesiano, Jacques. 1998. Les carrés magiques de Manuel Moschopoulos. *Archive for History of Exact Sciences* 53 (3): 377–397.
Sesiano, Jacques. 2003a. Une compilation arabe du XIIe siècle sur quelques propriétés des nombres naturels. *Sciamus* 4:137–189.
Sesiano, Jacques. 2003b. Construction of Magic Squares Using the Knight's Move in Islamic Mathematics. *Archive for History of Exact Sciences* 58 (1): 1–20.
Sesiano, Jacques. 2004. *Les carrés magiques dans les pays islamiques*. Lausanne: Presses polytechniques et universitaires romandes.
Sesiano, Jacques. 2008. Les carrés magiques d'al-Būnī. In *Actes du IXe colloque maghrébin sur l'histoire des mathématiques arabes. Tipaza, 12-14 mai 2007*, 205–218. Alger: Imprimerie Fasciné.
Sesiano, Jacques. 2017. *Magic Squares in the Tenth Century: Two Arabic Treatises by Anṭākī and Būzjānī*. Cham: Springer.
Sesiano, Jacques. 2019. *Magic Squares: Their History and Construction from Ancient Times to AD 1600*. Cham: Springer International Publishing.
Sesiano, Jacques. 2020. *An Ancient Greek Treatise on Magic Squares*. Stuttgart: Franz Steiner Verlag.
Sesiano, Jacques. 2022. A Treatise on the Use of Magic Squares. *Suhayl* 19:89–150.
Steingass, Francis Joseph. 1963. *A Comprehensive Persian-English Dictionary*. 5th ed. London: Routledge & Kegan Paul Limited.
Tolsa, Cristian. 2020. The Earliest Arabic Magic Squares. *Suhayl* 18:7–24.
Wiedemann, Eilhard. 1918. Zu den magischen Quadraten. *Der Islam* 8:94–97.
Witkam, Jan Just. 2007. Gazing at the Sun: Remarks on the Egyptian Magician al-Būnī and his Work. In *O ye Gentlemen: Arabic Studies on Science and Literary Culture in Honour of Remke Kruk*, ed. Arnoud Vrolijk and Jan P. Hogendijk, 183–199. Leiden/Boston: Brill.
Wright, Zachary. 2018. Secrets on the Muhammadan Way: Transmission of the Esoteric Sciences in 18th Century Scholarly Networks. *Islamic Africa* 9 (1): 77–105.

Chapter 10
Two Sciences, Two Methods for Handling Fractions: Arithmetic and Inheritance Shares in Nineteenth-Century West Africa

Alexis Trouillot

Abstract The objective of this chapter is to contextualize the links between manuals on arithmetic (*ḥisāb*) and manuals dealing with inheritance shares (*farāʾiḍ*) in nineteenth-century West Africa by highlighting some key differences between the two disciplines. To do so, I take the case study of texts written by various authors found in the library of a West Saharan scholar, Shaykh Sīdiyyā al-Kabīr (d. 1868), and the inheritance-share manual that he himself wrote.

I show that the way numbers are represented and conceptualized differs significantly across disciplines. Whereas *ḥisāb* puts emphasis on the representation of numbers using numerals, *farāʾiḍ* ostensibly has to do with a practical activity of inheritance-share calculation and uses number-words. Whereas *ḥisāb* considers fractions as mathematical objects that can be summed or multiplied, *farāʾiḍ* obfuscates manipulations of fractions through a process called *taṣḥīḥ*, which could be translated as "transformation into integers".

I posit that this difference in the conceptualization of fractions in *ḥisāb* and *farāʾiḍ* is less due to a strict opposition between the theoretical and the practical and more due to the fact that two traditions regarding fractions, each with their own historical trajectory, were made to coexist.

Keywords History of mathematics · Islamic law · Science in Africa · West Africa

A. Trouillot (✉)
Center for African Studies, University Mohammed VI Polytechnic, Benguerir, Morocco
e-mail: alexis.trouillot@um6p.ma

D. Senthil Babu, R. Wagner (eds.), *Normative Texts and Occupational Practices*, Boston Studies in the Philosophy and History of Science 350,
https://doi.org/10.1007/978-3-032-28419-8_10

10.1 Introduction

Historians of science often emphasize the link between arithmetical texts in Arabic and the Muslim inheritance system.[1] After all, the first incidence of written fractions in Arabic is the Qurʾān.[2] This is further confirmed by some primary sources: *farāʾiḍ* (the science of inheritance shares) and *ḥisāb* (accounting, arithmetic) are traditionally considered as disciplines neighboring each other. Some authors, like the Andalusian Qalaṣādī (d. 1486), produced texts in both disciplines (Souissi 1997, 477), while some other authors, like Al-Khwārizmī (c. 780–850), used *farāʾiḍ* as a source of exercises. Laabid Ezzaim has explored how Andalusian mathematicians conceived the link between *ḥisāb* and *farāʾiḍ* (Laabib 2006). Ulrich Rebstock tried to analyze the question in a bigger geographical scope (Rebstock 2009) and Marc Moyon has also shown how geometry could intervene in succession to divide land between successors (Moyon 2011, 532–533).

The objective of this chapter is to contextualize the links between mathematical manuals and manuals dealing with inheritance shares in nineteenth-century West Africa by highlighting some key differences. To do so, I will take the case study of the texts from the library of a West Saharan scholar, Shaykh Sīdiyyā al-Kabīr (d. 1868). Shaykh Sīdiyyā was the head of a *zāwiya*, a religious institution linked to Sufism that, in the region now called Mauritania, also took up the teaching prerogatives of a *madrasa* in subjects like law, grammar or mathematics (Pettigrew 2007, 64–71). He was a highly influential scholar and jurist who acted as a mediator in local conflicts through his knowledge of Islamic jurisprudence. Local individuals and groups recognized themselves as Muslims and followed the precepts of Islamic law under the school (*madhhab*) of Mālik b. Anas (d. 795), called the Mālikī school. Thus, it was crucial to have individuals well-versed in the principles of Islamic Law, as well as in discussions on how to adapt it to local issues.[3] While Sīdiyyā was active in the nineteenth century and wrote and used some texts from this period, many of his references come from other parts of the Muslim world going as far back as the twelfth century.[4] All the references below to manuscripts or edited volumes in Arabic are texts found in his family library.[5]

In the case of the science of inheritance shares, I will mainly focus on the text he wrote himself: the *Sharḥ bāb al-tarika min mukhtaṣar Khalīl b. Isḥāq* ("Commentary on the Chapter of Inheritance of the Handbook of Khalīl"). It is Sīdiyyā al-Kabīr's

[1] See for example Berggren (2016) and Rashed (2007, 24–30).

[2] For a general history of the notion of fraction in Arabic mathematics see (Djebbar 1992).

[3] Warscheid (2017) for debates and adaptations see, for example: Warscheid (2018) or Hall (2013).

[4] On Islam in precolonial West Africa and its trade connections to the rest of the Muslim world, see Lydon (2012). On manuscript culture and literacy in West Africa, see Brigaglia and Nobili (2017). On the books and subjects studied by local scholars, see Stewart and Hall (2011). On the intellectual production of a scholar active at the end eighteenth century and beginning of the nineteenth, see Marcus-Sells (2022).

[5] On the archives used for this paper, see Stewart (1991, 349–58). On the life of Sīdiyyā, see Stewart (1973).

commentary of the *Mukhtaṣar Khalīl,* the most widely used handbook on the Maliki School of Islamic law, which was the one most followed in West Africa at the time.[6]

As for arithmetic, Sīdiyyā's library contains the *Kashf al-asrār ʿan ʿilm ḥurūf al-ghubār* ("The Unveiling of the Secrets of the Science of Numerals"), a short treatise written by the Andalusian ʿAlī b. Muḥammad al-Qalaṣādī al-Bisṭāmī (d. 858/1454) on the methods of *ḥisāb* focusing on the use of numerals (al-Qalaṣādī 1988a, 1988b). This text was commented on by another West Saharan author, Anbūya al-Walātī (d. 1843–4), in his *Nuzhat al-albāb fī ʿilm al-ḥisāb* ("The Promenade of the Mind in the Sciences of Accounting"). While this text was not in Sīdiyyā's library, it provides helpful insight into how numeracy was taught.

Looking at these texts, I will show that the way numbers are represented and conceptualized is quite different from one discipline to the other. Whereas *ḥisāb* puts emphasis on the representation of numbers using numerals, *farāʾiḍ* uses number-words. *Ḥisāb* sees fractions as mathematical objects that can be summed or multiplied, *farāʾiḍ* seems intent on obfuscating manipulations of fractions through a process called *tasḥīḥ* that could be translated to "transformation into integers".

I posit that this difference in the conceptualization of fractions in *ḥisāb* and inheritance division is less due to a strict opposition between the theoretical and the practical and more due to the fact that two traditions regarding fractions, each with their own historical trajectory, were made to coexist.

10.2 The Science of Inheritance Shares

Farāʾiḍ are the rules that determine which heir necessarily inherits from the departed and in which amount. It is not the whole of Islamic inheritance (*mīrāth*), but rather the science of dividing inheritance shares. At the very basis of *farāʾiḍ* is a short passage of the *sūrat al-nisāʾ* in the Qurʾān:

> For men there is a share in what their parents and close relatives leave, and
>
> for women there is a share in what their parents and close relatives leave—whether it is little or much. These are obligatory shares.
>
> […]
>
> Allah commands you regarding your children: the share of the male will be twice that of the female. If you leave only two or more females, their share is two-thirds of the estate. But if there is only one female, her share will be one-half. Each parent is entitled to one-sixth if you leave offspring. But if you are childless and your parents are the only heirs, then your mother will receive one-third. But if you leave siblings, then your mother will receive one-sixth—after the fulfilment of bequests and debts. Be fair to your parents and children, as you do not fully know who is more beneficial to you. This is an obligation from Allah. Surely Allah is All-Knowing, All-Wise.
>
> You will inherit half of what your wives leave if they are childless. But if they have children, then your share is one-fourth of the estate—after the fulfilment of bequests and debts. And your wives will inherit one-fourth of what you leave if you are childless. But if

[6] On the *mukhtaṣar*, see Fadel (1996).

> you have children, then your wives will receive one-eighth of your estate—after the fulfilment of bequests and debts. And if a man or a woman leaves neither parents nor children but only a brother or a sister from their mother's side, they will each inherit one-sixth, but if they are more than one, they all will share one-third of the estate—after the fulfilment of bequests and debts without harm to the heirs. This is a commandment from Allah. And Allah is All-Knowing, Most Forbearing.
>
> These entitlements are the limits set by Allah. Whoever obeys Allah and His Messenger will be admitted into Gardens under which rivers flow, to stay there forever. That is the ultimate triumph!. (Webb et al. 2016, Q4:7. Q4:11–13)

This passage explicitly defines shares for certain heirs based on their relationship to the deceased: daughters, parents, wives and uterine siblings.[7] Such shares, prescribed by God, are called *furūḍ* (sing. *farḍ*, lit. "obligations", hereafter "prescribed shares"), hence the name of the discipline of *farāʾiḍ*. Its main goals are to calculate the value of the *furūḍ* all while staying consistent with an agnatic system of transmission (*ʿaṣaba*) in which the closest male relatives inherit the estate of the departed,[8] and to find mathematical solutions in cases where the shares exceed or fail to exhaust the estate and elaborate new problems in the name of the totality of *fiqh*.

In his commentary of the *Mukhtaṣar*, Sīdiyyā describes a *farḍ* as follows:

> the prescribed share (*farḍ*) [in simple] language (*fī lugha*) is the separation or the cut (*al-qiṭʿ wa-l-ḥazz*). And in more technical language (*iṣṭilāḥān*) it is the portion (*naṣīb*) corresponding to the heir in the Sharia without adding anything except by *radd* (proportional return) for the one accepting it and without subtracting anything except by increase (*ʿawl*). ("Sharḥ bāb al-tarika", 3)

Notable in this definition is the absence of the term "fraction" (*kasr*). He then follows the structure of the *Mukhtaṣar Khalīl* by categorizing the heirs depending on what the associated share is.

- Half $\left(\frac{1}{2}\right)$ is for the husband, when the deceased has no descendants, or for the daughter, when she has no brother, or the daughter of the son, when she has no brother, or aunt or the germane or consanguine sister, when she has no brothers.
- Except for the husband, if the above heirs are more than one, they share two-thirds $\left(\frac{2}{3}\right)$ under the same conditions as above.
- A third $\left(\frac{1}{3}\right)$ is to be shared between the mother and the uterine siblings of the deceased.
- A sixth $\left(\frac{1}{6}\right)$ is due to the uterine siblings, when they are competing with a descendant, for the mother, when she is competing with germane or consanguine siblings, or for the non-uterine sisters or the daughters of the son, when they are

[7] The inheritance system differentiates between germane siblings, meaning siblings that share both parents; consanguine siblings that share only a father and uterine siblings that share only a mother.

[8] Who exactly the closest male relatives were in preislamic time is still subject to debate, see Powers (1993, 15–17). Some argue for a patriarchal structure and others for a fratriarchal structure.

competing with exactly one daughter or germane sister, so as to "complete the two thirds".

- A fourth $\left(\frac{1}{4}\right)$ is due to a husband, when competing with descendants, and for the wives, however many they are, when not competing with descendants.
- An eighth $\left(\frac{1}{8}\right)$ is due to the wives, however many they are, when they are competing with descendants.

Similarly, he defines the *ʿaṣab*, saying:

> It is the one who (inherits the totality of the money) if he is alone or the rest after the prescribed share ("Sharḥ bāb al-tarika", 10).

Following this definition, I will translate *ʿaṣab* as residuary heir in the rest of the text when dealing with calculations.[9] Much like the prescribed heirs, the residuary heirs influence one another's shares but do so more extensively. Whereas two prescribed heirs of different categories (*ṣinf*) can alter what the other receives without exclusion (for example: the mother and the siblings), only one category can inherit the remainder.

The residuary heirs are, in order:

- The sons. They share the remainder equally between them. If daughters are present with them, they make their sisters into residuary heirs in a process called *taʿṣīb* and the estate is shared between all of the siblings with the males taking twice what the females take. They cannot be excluded from the inheritance by the presence of another relative.
- The sons of the sons. They likewise make their sisters into residuary heirs. They are excluded by the sons.
- The father. He is excluded by the descendants.
- The male ascendant of the father, however far removed.
- The germane brothers.
- The consanguine brothers.
- The uncle and his descendants, however far removed.
- The manumitter (*walāʾ*).

A residuary heir in this system does not necessarily inherit more than the prescribed heir. In fact, the residuary heir may not even receive any property, as the remainder after prescribed shares might well be zero. A fundamental difference between the Maliki school and the others on matters of inheritance is that the *bayt al-māl*, a form of public treasury supposed to help the whole community, is the last residuary heir, inheriting only if there is no other residuary heir, whereas in other schools it gets

[9] This is consistent with existing literature on the subject. It is used for example in (Layish 1983) or (Coulson 1971) and makes the understanding of the calculations easier.

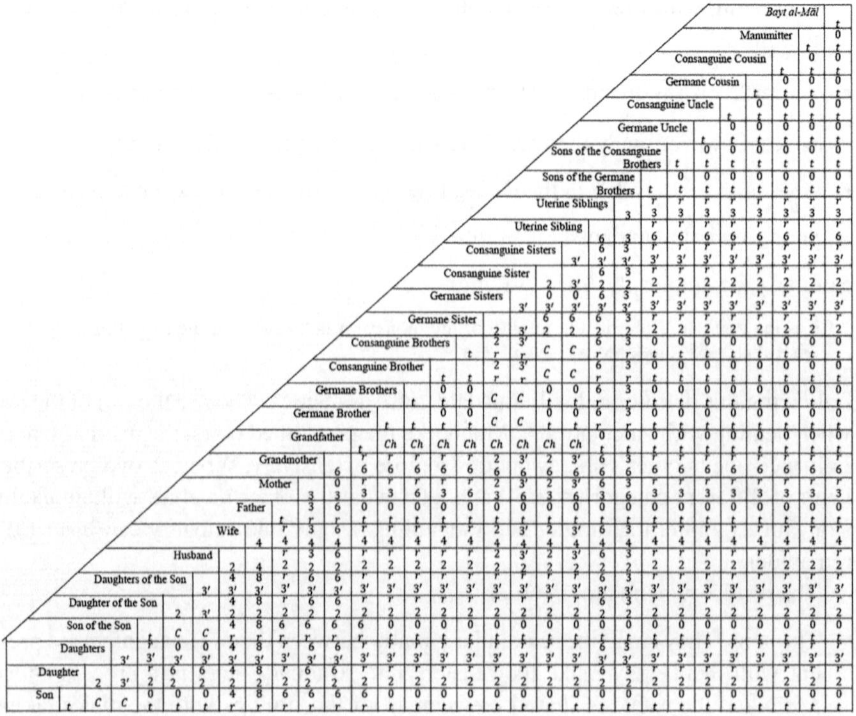

Fig. 10.1 Table summarizing the rules of farāʾiḍ

money only if no heir is present at all.[10] Hence, while the estate is always fully spent in the Maliki school, the *bayt al-māl* never explicitly appears in problems of *farāʾiḍ*. For example, if someone dies leaving only a daughter, the texts will state that the daughter inherits half of the fortune, leaving implicit the fact that the *bayt al-māl* takes the other half.[11] As such, while the estates in the problems to follow might appear under-exhausted, they never are.

The following table (Fig. 10.1) summarizes the repartition of shares between competing heirs discussed above. To find the shares of two competing categories of heirs, go down the table diagonally until you find the first heir and go to the first column on their right. Then go down that column until you are on the same line as the second heir. The number atop the cell is the due of the first heir and the number below the due of the second. As for the letters, they represent:

[10] While the *bayt al-māl* is also accepted as an heir in the Shafiʿi school, it poses the condition that it should be "properly administered" to be considered an heir.

[11] Whereas it would, in other schools, inherit nothing.

t—the totality: said heir takes all the inheritance
r—the remainder
0, 2, 3, 4, 6, 8—respectively 0, $\frac{1}{2}, \frac{1}{3}, \frac{1}{4}, \frac{1}{6}, \frac{1}{8}$
3′—$\frac{2}{3}$
Ch—Intervene in the convoluted case of the grandfather. He has a choice of how he wants the inheritance to be shared

C—The two categories share among themselves, a man inheriting twice the share of a woman.

For example, if we want to know what the grandmother inherits when in competition with the wife, we start from "*bayt al-māl*", go down diagonally to "grandmother", go to the right on the cell with "6" and then down this column until we are on the line lead by "wife". We read in the cell "$\frac{6}{4}$", so the grandmother gets $\frac{1}{6}$ and the mother $\frac{1}{4}$ of the total estate.

From the previous description, it should be obvious that the process of inheritance-sharing must reckon with fractions since its main objects are fractions expressed in the Qurʾān. How? Among the texts dealing with inheritance sharing from Sīdiyyā's library that are not commentary of the *Mukhtaṣar*, only the *Kāfī fī l-fiqh* of Andalusian Maliki jurist Ibn ʿAbd al-Barr (d. 1070) gives some details on calculation in the last section of its chapter on *farāʾiḍ*: "chapter of problems in inheritance shares" (*bāb masāʾil min al-farāʾiḍ*). Conversely, the *Risāla* of Ibn Abī Zayd al-Qayrawānī (923/4-996)—a text for students which became the standard legal manual of the Maliki school in the Islamic West, including the Saharan West—while mentioning inheritance shares, talks very little about calculations. It only exposes in turn the share of sons and daughters, spouses, parents, uterine, germane and consanguine siblings, grandfathers, and grandmothers, and rules on priority of inheritance and those who are forbidden from inheriting. That is not the case for all of the treatises, and when they are dealing with number manipulations, the treatises on inheritance shares in our corpus present a unified picture: they use both the same type of method and the same vocabulary, most of which is introduced by the *Mukhtaṣar* itself.

10.3 Solving a Case

10.3.1 A Short Mathematical Prelude

For the sake of clarity, let us first describe the mathematical procedure described in the treatises studied in anachronistic terms. If two heirs are to receive respectively $\frac{a}{b}$ and $\frac{c}{d}$ of the total estate, where a, b, c, and d are integers; the goal is to find e, the lowest common multiple (LCM) of b and d so that the two fractions have the same denominator.

Hence, if we name e_1 and e_2 the integers so that $e = e_1 \times b = e_2 \times d$ then $\frac{a}{b} = \frac{a \times e_1}{e}$ and $\frac{c}{d} = \frac{b \times e_2}{e}$.

The core of most problems is to find e. Then the estate to be divided can be defined as having a value of e so as to eliminate the need for fractions: the heirs no longer receive $\frac{a}{b}$ and $\frac{c}{d}$ of an indetermined estate but $a \times e_1$ and $c \times e_2$ of an estate now fixed to be e. The fractions, while still present since the *furūḍ* in the *surat al-nisāʾ* are fractions, are no longer a type of number we have to reckon with: everything is expressed in integers.

To find e, e_1 and e_2 we have to see how b and d divide one another. In treatises on *farāʾiḍ*, four cases are presented

1. $b = d$. In this case their LCM is themselves: the case is already solved! It is called "equality" (*tamāthul*).
2. b is a factor of d: b "overlaps" (*tadākhul*) with d. We see this by performing a Euclidean division of d by b. If there is a null remainder, they overlap (d is divisible by b). In this case the bigger number, d, is the LCM.
3. b and d are coprime: they "disagree" (*tabāyun*). The Euclidean division of the biggest by the smallest yields a remainder of 1. In this case, $e = b \times d$.

b and d have a common divisor: they are said to "be in harmony" (*tawāfuq*). This is the case when the Euclidian division of the biggest by the smallest yields a remainder that is neither 1 nor 0. In this case their LCM is $b \times \frac{d}{\mathrm{GCD}(b,d)} = d \times \frac{b}{\mathrm{GCD}(b,d)}$.

For example, if a woman dies leaving a husband and a daughter then the husband is entitled to $\frac{1}{4}$ of the estate and the daughter to $\frac{1}{2}$. So we have to compare 2 and 4.

4 is a multiple of 2 since $4 = 2 \times 2$ so there is overlap.

Hence, both fractions should have 4 as their denominator.

This is already the case for the husband. For it to be the case for the daughter, we have to multiply the daughter's numerator and denominator by 2. So the daughter's share becomes $\frac{1}{2} = \frac{1 \times 2}{2 \times 2} = \frac{2}{4}$.

Hence, the result can be reformulated in this way: assuming that the estate to divide is 4, the husband gets 1 and the daughter 2. In other words, this procedure serves the mathematical purpose of making the fractions commensurable but also the stylistic purpose of eliminating them from the *farāʾiḍ* texts in the Sīdiyyā library.

Since the LCM of three numbers a, b, c is equal to LCM (LCM(a, b), c), we can apply the above methods for as many heirs as we want.

Let's take a case with a mother, a wife and three daughters: the mother takes $\frac{1}{6}$ of the estate, the wife $\frac{1}{8}$ and the daughters share $\frac{2}{3}$.

Now, we have to compare the denominators 6, 8, and 3.

6 and 8 are both divisible by 2 so their *LCM* is $6 \times 4 = 3 \times 8 = 24$.

We now compare 24 and 3.

3 is a factor of 24 since $3 \times 8 = 24$. Hence, their *LCM* is 24. Therefore $e = 24$.

We will now consider how the steps we just saw were formulated in Sīdiyyā's library.

10.3.2 The Aṣl al-Mas'ala

The first step is the determination of the *aṣl al-mas'ala* (pl. *uṣūl* lit. "the foundation of the problem"). Sīdiyyā describes an *aṣl* as "the number from which the shares of each type of heir result" (*al-ʿadad allatī yakhraj min-hā sihām*).

He further elaborates:

> So the half from two and the fourth from four and the eighth from eight and the third or the two third from three and the sixth from six.

In other words, though Sīdiyyā does not use the term *mukhraj* ("denominator"), if there is only one prescribed share in the problem, I show that the *aṣl* (pl. *uṣūl*) refers to the denominator of the share.

> The fourth with the third or the sixth from twelve. And the eighth and the third and the sixth or the third. And the aim of this is the two thirds since the third and the eighth cannot be together. Because the eighth truly is for the wife with descendants (*furūʿ*) and the third truly is for the mother or her children without descendants—from twenty-four.[12]

The number 12 is given without explanation as the quantity to use when $\frac{1}{4}$ and either $\frac{1}{3}$ or $\frac{1}{6}$ are present in the problem. Similarly, if $\frac{1}{8}$ and either $\frac{2}{3}$ or $\frac{1}{6}$ are present, the *aṣl* is 24.[13] This passage makes clear that these *uṣūl* are not computed but are to be known. The number of prescribed fractions used in the calculation of inheritance shares is finite; therefore, it follows that the number of their shared common denominators is also finite. These denominators are the *uṣūl*. While it is not quite made explicit in the treatise of Sīdiyyā, we will see shortly that he is cognizant of this fact.

---

[12] The original text reads:

فالنصف من اثنين والربع من أربعة والثمن من ثمانية والثلث أو الثلثان من ثلاثة والسدس من ستة.
والربع والثلث أو السدس من أثنى عشر. والثمن والثلث والسدس أو الثلث -والمراد به الثلثان إذ لا يجتمع الثلث والثمن لأن الثمن إنما هو للزوجة بفرع والثلث إنما هو للأم أو ولديها بلا فرع- من أربعة وعشرين.

("Sharḥ bāb al-tarika", 16).

[13] As the biggest of the *uṣūl*, 24 is a common multiple of any combination of denominator of *furūḍ*. Such a constatation led to the use of the *qīrāṭ*, a unit designating a 24th of a quantity. See Hunwick (1999, 103), Souissi (1993, 23), and Meriwether (1999, 162).

10.3.3 The Case of the Grandfather

The heir involving the most convoluted calculations is the grandfather. While the texts in Sīdiyyā's library do not explicitly note any particular difficulties, Coulson quotes the following saying, attributed to the caliph ʿUmar (c. 582/583–644): "If anyone is attracted by the prospect of rushing headlong into the depths of hell-fire, let him attempt to adjudicate a competition between the grandfather and the collaterals." (Coulson 1971, 79).

The rules start out easily enough: if siblings are not present, the grandfather inherits as would the father, on the condition of his absence. If the father is present, the grandfather inherits nothing. Complexities arise when the grandfather is in competition with siblings, be they uterine or non-uterine. The proposed solution is to have the grandfather inherit in three different ways when in competition with non-uterine siblings, depending on what would net him more money: either he receives $\frac{1}{6}$ of the estate as a prescribed heir, or he is considered as an agnate on the same level of the siblings (*muqāsama*), or he inherits $\frac{1}{3}$ of what is left after the prescribed shares are allotted (ibn Anas 2004, 380). In competition with uterine siblings, he bars them from inheritance except in the extremely convoluted case of the *akdariyya*.

The case of the grandfather comes with Sīdiyyā's adoption of an innovation only present in one text from his library, a commentary of the *Mukhtaṣar* by Muḥammad al-Khirshī (d. 1690), (al-Khirshī and ʿAlī al-ʿAdawī 1900): to the seven *uṣūl al-masʾala* he adds two others, 18 and 36. As opposed to the seven more canonical *uṣūl*, the calculation process is detailed:

> The first is the *aṣl* of the sixth and the third of the rest like the mother and the grandfather and three brothers. The optimum for the grandfather is a third of the rest after the sixth of the mother. Then multiply the denominator (*maqām*) of the third by the denominator of the sixth, [you get] eighteen.
>
> And the second is the *aṣl* of the sixth and the fourth and the third of the rest, like the mother and the wife and the grandfather and three siblings. The optimum for the grandfather is a third of the rest after the sixth of the mother and the fourth of the wife. So multiply the denominator of the third by the denominator of the fourth and the sixth, [you get] thirty-six.[14]

This is the first and only time that Sīdiyyā hints at the fact that the objects manipulated are, in fact, fractions in the mathematical sense with a numerator and a denominator. Viewed in this light, the calculations are straightforward: we want to obtain, respectively $\frac{1}{3}$ of $\frac{5}{6}$ and $\frac{1}{3}$ of $\frac{7}{12}$ so we put those fractions on the same

[14] The original text reads:

فالأول أصل السدس وثلث الباقي كأم وجد وثلاثة إخوة فاكثر. الأفضل للجد ثلث الباقي بعد سدس الأم فتضرب مقام الثلث في مقام السدس بثمانية عشر.

والثاني أصل السدس والربع وثلث الباقي كأم وزوجة وجد وثلاثة إخوة فاكثر. الأفضل للجد ثلث الباقي بعد سدس الأم وربع الزوجة. فتضرب مقام الثلث في مقام الربع والسدس بستة وثلاثين.

("Sharḥ bāb al-tarika", 15).

denominator. It also confirms how the *uṣūl* are computed, since 12, the *aṣl* of the fourth and the sixth, is here their *maqām*, meaning the lowest common multiple of their denominator.

These two new *uṣūl* are introduced by Sīdiyyā to be known by the reader so that they would have less difficulty calculating the cases involving a grandfather by minimizing their interactions with fractions. But at the same time, they reveal to us that he did see the *uṣūl* as objects resulting from calculations made possible by seeing the *furūḍ* as fractions.

10.3.4 Integers as a Result

If we go back to the case of a mother, a wife and three daughters, we have seen that the mother takes 4, the wife 3 and the daughters share 16 from an estate of 24.

While we had stopped there, the problem is in fact not yet entirely solved: while $e = 24$ allows us to express what each type of heir gets using an integer—what is called *sihām* in *farāʾiḍ* texts—if we want to know what one individual daughter inherits we would have to use fractions and write that she receives $\frac{16}{3}$ of 24. To avoid expressing the number as an improper fraction, we now want to compare the primality of the share that the whole category of heir inherits to the number of heirs in this category n.

Hence we have to compare 16 and 3. They are coprime, so we multiply all the quantities involved by 3. The solution to the problem is that the mother inherits 12, the wife 9 and the daughter 48 from an estate of 72.

This step of multiplying the numbers is called the *taṣḥīḥ*. It makes clear the place of fractions in the science of inheritance shares. The *taṣḥīḥ* is the process of making fractions disappear from *farāʾiḍ*, both as objects to be directly manipulated and as solutions. Hence, while some authors have translated "*taṣḥīḥ*" as "correction" or "verification" based on the root *ṣ-ḥ-ḥ*, I propose a more accurate translation: "integerization," or the conversion of fractions into integers. Indeed, what this process does is to take fractions and convert operations on them into operations on integers. Hence its name should be understood as derived from *ʿadad ṣaḥīḥ*, which is the term for integers. A text of the library, the *al-Tāj wa-l-iklīl li-Mukhtaṣar Khalīl* of al-Ḥaṭṭāb (1497–1547), makes this explicit:

> The purpose of the scholars of inheritance shares (*faraḍiyīn*) for the making out of this result [the *aṣl al-masʾala*] (*hādhihi al-makhārij*) was double: one of them is the division of the shares of each type of heir (*sihām*) into integers (*aʿdād ṣaḥāḥ*) without fractions (*kasr*). And the second to ask the *smallest number to "integerize" by* (*taṣaḥḥu fī-hi*) so that they rely on it (*yuʿawwalūn ʿaly-hi*). (al-Jundī et al. 2007, 596)

It is also this last step that introduces the need for calculation for the practitioner since, for the previous steps, most problems can be solved by knowing the list of the *uṣūl*.

Here, the practitioners are concerned only with coprimality of numbers, even for complex cases.[15] According to *farāʾiḍ*, two numbers can be equal, overlapping, in harmony or in disagreement. Sīdiyyā explains the latter three:

> The overlap—and it is called proportionality (*tanāsub*)—if one exhausts the other (*bafanā aḥad-humā al-ākhar*).
>
> […]
>
> If it remains from the biggest something, then if it remains from it one then disagreement like two with five.
>
> And if not then between them the harmony. It is called the communion (*al-mushāraka*) according to the last number of the exhaustion like four and six. [16]

The treatises studied, even when they propose a method of resolution of a case and its solution, are relatively discreet in terms of how exactly the computation is to take place. As we have seen, the *uṣūl* are known quantities deriving from the prescribed shares intervening in the case. Hence, it would be possible for a practitioner to progress far into the resolution of a case with very minimal calculations.

The lack of fractions in treatises on inheritance share is also made possible because they are removed from practice, as their solution is to manipulate the estate to be divided. This is made possible by the fact that treatises on inheritance shares are never concerned with dividing an actual estate. In theory, changing the estate e to be divided by whatever quantities needed is a fine solution, but it is maladapted to actual cases where the estates to be divided are set. The manuals studied are silent on how to translate the methods they use to practical problem solving. However, some documents attest that this repartition of shares was not merely theoretical in nineteenth-century West Africa.

In 1966, Mervyn Hiskett edited a manuscript that he determined to be from the latter half of the nineteenth century detailing the possessions of an official of the Kano court (present-day Nigeria), converting them into cowries mats and then dividing it among his children.[17] The estate is valued at 599 mats. From this is deducted "the fee" (*ijāra*) and it is then divided among two sons and three daughters. Only the results are presented:

> And the grand total is five hundred and ninety-nine mats. And the deceased has two male heirs and three females. The names of the two males are Muḥammad and ʿAbd al-Qādir; the

---

[15] For example, cases where an intersex is present (*khunthā*). See Rebstock (2007).

[16] The original text reads:

التداخل ويسمي التناسب إن بفني أحدهما الأخر.

[…]

فإن بقي من الأكثر شيء فإن بقي منه واحد فمتباين كالاثنين مع الخمسة.

وإلا فبينهما الموافقة وتسمي المشاركة بنسبة المفرد للعدد المفني أخرا كالأربعة والستة

("Sharḥ bāb al-tarika", 22).

[17] Hiskett (1966a). Hiskett uses this text to evaluate the prices of various goods. Copies of the *Mukhtaṣar* and of the *Risāla* are both parts of the estate.

> names of the females are Raqiyah, Rābiʿah, Khadījah. Then deduct the fee: sixty-nine mats, and the remainder is five hundred and thirty-one mats to be divided among two male (heirs) and three daughters. Each one of the males inherits one hundred and fifty-four mats and two thousand, two hundred and eighty-six cowries, and each one of the three daughters inherits seventy-seven mats, and one thousand, one hundred and forty-three cowries. (Hiskett 1966a, 141)

As Hiskett notes himself, those results do not add up (Hiskett 1966b, 365). The announced grand total 599, when subtracted by the announced fee of 69, makes 530 and not 531. Then the sum of all the parts of the heirs adds up to 539 and some. So the estate is exceeded. If we were to do the calculations we would obtain, taking 530 as a result of the subtraction, 150 and some for each son and 75 and some for each daughter. It is hard to ascertain from where this inconsistency stems from, but it seems that 150 for each son was the result in the manuscript before a marginal insertion.[18]

The second document is a translation by John Hunwick of a manuscript from the Aḥmed Baba institute (Mali) detailing the division of the estate of a woman named Nāna ʿĀʾisha, a great-great-great-granddaughter of the famous Aḥmed Baba (1556–1627). The estate is valued at 81.5 gold *mithqāl* (mq.) after funeral expenses, debts and payment of the witnesses. The money is then divided between her daughter, her brother and her two sisters. "The daughter was apportioned half of it, which is 40.75 mq." and:

> Then the second half was divided among the brother and the two aforementioned sisters. The brother was apportioned 20.25 mq. and each of the daughters [sic. It should be "sisters"] was apportioned 10 mq. and 3 *qīrāṭs*. The aforementioned brother took half of the house at its value and the slave called al-Manṣūr and 10 mq. for himself, and for his two sisters was the value of half the aforementioned house, 22.5 mq. and the slave at 18 mq. (Hunwick 1999, 103)

Hunwick's translation makes it sound like the house was divided into three halves valued 22.5 mq. and gives an impossible result, namely: 40.5 mq. for the daughter, 32.5 and an enslaved man for the brother and 40.5 mq. for the two sisters. It adds up to 113.5 mq. plus an enslaved man which is quite a bit more than the original estate. There is also the problem that the enslaved person, al-Manṣūr, is not valued. Without access to the original manuscript, it is impossible to be sure from where the problem originates, but a possibility is that one should understand that "the aforementioned brother took *half of* half of the house at its value and *of* the slave called al-Manṣūr and 10 mq. for himself, and for his two sisters was the value *of half* of half of the aforementioned house, *valued* 22.5 mq. and *of* the slave *al-Manṣūr valued* at 18 mq." This would solve some but not all of the problems: the repartition would now be closer to the correct result with 40.5 to the daughter, 30.25 to the brother and the sisters sharing 20.25. This result is almost correct and adds up to 91 mq. It would in fact add up to 81 were it not for the 10 *mithqāl* that the brother gets, the only

[18] Hiskett (1966a, 141). The mistake on the subtraction of the fee could be the result of another marginal insertion (Hiskett 1966a, 140).

liquidity in this estate. With the value of the house and the enslaved man already adding up to 81 mq., the role of those 10 *mithqāl* is unclear.

In these two examples, the methods used to calculate the division of shares is not made explicit, but the results approach the value dictated by Islamic Law. We can infer that a method was used once a case was solved to apply the result to divide an actual estate.

10.4 The Treatises on Arithmetic

The only manual on arithmetic available in the Sīdiyyā library is the *Kashf al-asrār ʿan ʿilm ḥurūf al-ghubār* of Qalaṣādī, itself a summary version of the *Tabṣirat al-mubtadiʾ bi-qalam al-hindī* both written by the same author. It is a manual dealing with the use of numerals in calculations for integers, fractions and roots ("Kashf al-asrār"). The text was widely circulated in the region as was its commentary by a local nineteenth-century author.[19] We will now compare the techniques made available to Sīdiyyā in these texts with what we have seen of the treatment of fractions in the science of inheritance shares.

10.4.1 Invisibles Numerals

So far, while describing the calculations, we have left unsaid an important detail. Only number-words are used in the *farāʾiḍ* treatises consulted, not numerals. The arithmetic (*ḥisāb*) treatises available locally, in contrast, use numerals. In fact, a local commentary of Qalaṣādī's *Kashf al-asrār ʿan ʿilm al-ghubār*, the *Nuzhat al-albāb* by Anbūya b. al-Imām al-Walātī (d. 1260/1843-4) adds a section about writing numerals. The introduction of the *Kashf* presents the glyphs for the numerals, then gives examples of how to write some quantities. While al-Walātī largely reproduces the text, he starts by describing how to actually trace each numeral by making reference to letters in the Arabic alphabet whenever possible, in effect adding to the text his own lengthy introduction:

> An *alif* for one, meaning that the shape of the one is like the shape of the *alif*, like so: 1. ("Nuzhat al-albāb", 1ʳ)

Similarly, the two is described as the letter *yāʾ* turned (*yāʾ mardūdan*), the three as the letter *hamza* turned, the four as the letter *ʿayn* linked to a straight line (*ʿayn mustaqīma fī l-khaṭṭ*), the five as its "opposite" (*ʿaksu-hu*) meaning a *ʿayn* that is not linked, the six as the letter *hā*ʾ, the seven as being the letter *ḥā*ʾ and nine the letter *wāw* (respectively: 2, 3,ـــ, ٥, ح ,ه, and و). The other numerals are not given any

[19] On an overview and quantitative analysis of the texts found in the region, see Djebbar and Moyon (2011) and Moyon (2021, 2022). For a selective overview of some manuscripts see Rebstock (1990).

correspondence to the Arabic alphabet but the 8 is likened to the shoulder blades (*katf*). This type of description is found in another anonymous fragment of text from the Saharan West, itself a commentary of a *naẓm.* It is written by an unknown Mauritanian author and is a commentary on the work of a sixteenth-century Moroccan author al-Simlālī (Rebstock 1990, 432–433). It describes nearly all the numerals as:

> The *alif*. The *yā*ʾ turned. "*ḥajj*" [written in Arabic it mimics the shapes of a 3: حج]. The ـ [Written as is in the text without description]. The *ʿayn* standing. The *hā*ʾ. The lean part of the belly (*mukhṭaf*). The scapula (*katf*). The *wāw*.[20]

This elaboration of al-Walātī on the text hints at the practice of learning the numerals, the *ḥurūf al-ghubār,* that are at the heart of Qalaṣādī's text. With the popularity of the *Kashf* in the region, it is no surprise that when numerals appear in West African texts dealing with arithmetic, the numerals used until the end of the nineteenth century are generally the occidental graph described in the *Kashf*: 1, 2, 3, ـ, ع, 6, 9 ,8 ,ح.[21]

The way those numerals are then spatially ordered on the paper or the board is crucial to solving problems in the text and to representing the correct final result. As such they are associated with the semantic fields related to drawing and arranging spatially:

- *Inzil*, *n-z-l* Pose.
- *Waḍaʿa*, *w-ḍ-ʿ* To place.
- *Jaʿala*, *j-ʿ-l* To put.
- *Ṣūra*, *ṣ-w-r* Image.
- *Shakl*, *sh-k-l* Shape, mainly in the introduction on numerals.

For example, when it comes to adding 48 and 97, the text reads:

If someone tells you: "add forty-eight to ninety-seven", pose (*inzil*) it like so:

——
48
97

Both the fact that the 48 is positioned above the 97 and the fact that there is a line yet above are crucial features and figure in all manuscript witnesses of the text.

> Then add the seven to the eight, making fifteen. Place five (*fa-ḍaʿ al-khamsa*) on the head of the line and pose the image of the one (*anzil bi-ṣūrat al-wāḥid*) under the nine and add it to it making ten. Then add it to the four making fourteen. Place four on the head of the line

[20] The original text reads:

ألف. وياء برد. و حج . وـ. وعين واقف. وهاء. ومخطف. وكتف. وواو.

("Rabbānī al-mubtadī", 2^r).

[21] This seems to change at some point in the twentieth century, presumably under the influence of printed texts circulating. A late manuscript like the *Nubdha fi ʿilm al- ḥisāb* of Aḥmad Bābir al-Arawānī (d. 1997) uses the oriental numerals: ٩ ,٨ ,٧ ,٦ ,٥ ,٤ ,٣ ,٢ ,١.

> (*fa-ḍaʿ al-arbaʿa ʿalā raʾs al-khaṭṭ*) and the ten after it. So it makes the result a hundred and forty-five like so: 145.

If we reproduce the procedure described, it goes like this:[22]

1	4	5
	4	8
	9	7
	1	

Hence, the text explicitly invites the reader to draw numerals as a way of calculating. I have found no example of calculations being fully put on paper.[23] The procedure also sometimes involves moving numbers, more precisely, moving them back (*qahqaru*). In opposition, the texts of the library dealing with inheritance do not use numerals. This does not necessarily mean that numerals were absent from the minds of practitioners, but it shows that numerals were not meant to be written on paper in pedagogical mathematical material. While the procedure to use numerals is described in the arithmetic manuals, they themselves only show the first step of the operation and the results using them, writing the rest using number-words.

This use of numerals translates into a different way to operate on fractions in treatises of arithmetic. To take an example from the *Kashf*:

> If someone says to you "add five sixths and three fourths of sixths to three sevenths and a fifth of a seventh" Then pose it like so: $\frac{53}{64}+\frac{31}{75}$.[24]

With modern writing, this would be written:

$$\left(\frac{5}{6}+\frac{\frac{3}{4}}{6}\right)+\left(\frac{3}{7}+\frac{\frac{1}{5}}{7}\right)$$

From this, Qalaṣādī first calculates the numerators of the two sums of fractions. He finds $5 \times 4 + 3 = 23$ and $3 \times 5 + 1 = 16$, respectively.

Anachronistically, this would yield:

$$\left(\frac{5\times 4+3}{6\times 4}\right)+\left(\frac{3\times 5+1}{7\times 5}\right)$$

[22] Mohamed Souissi in al-Qalaṣādī (1988b), reproduces those operations for all the examples presented in the *Kashf*.

[23] The phenomenon is not unique to the region or mathematic culture. See Chemla (2022, 27–31).

[24] al-Qalaṣādī (1988a, 71). On this way to write the "linked fraction" [*kasr muttaṣil*], see Djebbar (1992, 233).

He then calculates the numerator of the sum: $23 \times (7 \times 5) = 805$ and $16 \times (6 \times 4) = 384$ so the sum is 1189.

This yields:

$$\frac{1189}{7 \times 6 \times 5 \times 4}$$

Then, rather than multiplying the denominator to obtain $4 \times 5 \times 6 \times 7 = 840$, he divides successively the numerator by 4, 5, 6 and 7.

> Divide this result by the factors and this is by what is under the line: the seven first, after it the six and the five and the four.
>
> [...]
>
> So the result is one as an integer [*wāḥidan ṣaḥīḥan*] and two sevenths and five sixths of a seventh and a fifth of a sixth of a seventh and a fourth of a fifth of a sixth of a seventh like so: $1\frac{2521}{7654}$.

So the result is presented as $1 + \frac{2}{7} + \frac{5}{\frac{6}{7}} + \frac{\frac{4}{5}}{\frac{6}{7}}$.

As we see, in sharp contrast to *farāʾiḍ* texts dealing with inheritance shares, both manipulating fractions as mathematical objects with a numerator and a denominator and presenting results as fractions are features of arithmetic manuals.

10.4.2 Support for Calculations

This absence of numerals in manuals on *farāʾiḍ* in the nineteenth-century Saharan West is notable because of the comparison with North African texts. While the content in prose uses number-words and is generally very similar to what is found in the

	Share from the the *aṣl*	Share after *taṣḥīḥ*
Mother	4	12
Wife	3	9
3 Daughters	16	48
e	24	72

Fig. 10.2 The process of taṣḥīḥ

Sahara, in North African texts, results of the *taṣḥīḥ* are expressed in tables such as given in Fig. 10.2 that generally feature numerals:

Though tables feature in North African manuscripts on *farāʾiḍ*, the texts copied or written in the Sahara that I have been able to access have been devoid of them, possibly to save space on paper.

That does not preclude the possibility of another support being used for calculation. Outside of these tables, number-words are used in lieu of numerals. In the region, texts using Indian numerals are rare. However, their absence in texts does not necessarily indicate that knowledge of these numerals was rare. The name *ḥisāb ḥurūf al-ghubār* translates into "reckoning of the letters of dust". As such, those numerals are intrinsically associated to the *takht* dust board, a flat surface with a layer of dust on which operations could be conducted and then erased to let the writer start a new one. By their nature, those dust boards do not leave any exploitable traces for historians. However, very early on, authors suggested an adaptation of the methods to ink and paper. In the tenth century, Uqlidīsī wrote, right after his introduction of the methods of calculation using numerals on the *takht*:

> In this book we state all that is done by Hindi (schemes), not with *takht* or erasure, but with inkpot and paper. This is because many a man hates to expose the *takht* between his hands when he finds the need to use this art of calculation, for fear of the misinterpretation of the attendants or whoever may see it. It belittles him, for it is seen between the hands of the misbehaved who earn their living by astrology in the streets. Moreover, he who calculates on it finds it so difficult to reconsider what he has calculated to the extent that in most cases he repeats it, (not to mention) the exposure of the content to the blowing wind which changes the figures, apart from making the fingers dirty, over and above other things which distort orderliness. (al-Uqlīdisī 1978, 247)

Hence, the support of calculation gradually shifted to paper first in the eastern part of the Islamicate and then in Andalusia. According to Abdeljaouad, by the thirteenth century, textbook authors advocated the sole use of paper, pen, and ink for all calculations (Abdeljaouad 2006, 662).

I have found no example of calculations being put and carried out on paper in the manuscripts I study. In a region where paper was expensive before French colonization, the preferred medium for calculators to practice might have been the *alwāḥ*, wooden tablets used for the learning of the Arabic alphabet and the Qurʾān. Yacine Diallo mentions that at the beginning of colonization, they were used in Qurʾānic schools where students were also introduced to arithmetic, indicating that numeracy was taught on them alongside literacy (Diallo 1934, 154–160). We will now turn to cases that indubitably require calculations on the part of the practitioner.

10.4.3 The Treatment of Wills

Lastly, let us examine one of the places where the field of inheritance-share division and the field of arithmetic intersect: calculations involving wills. As we have seen in the introduction, part of the second book of al-Khwārizmī deals with this topic

(Rashed 2007, 236–239). He provides a typology of will problems and solutions using algebra. In the *Mukhtaṣar* and its commentaries in the library, the question is treated as follows. Suppose someone has a will that leaves a portion of their property to a person who is not among the standard heirs. First, one should calculate the common denominator of the shares of the standard heirs as above. Next, one should find the fraction of the estate that is left after the person named in the will obtains their share. Finally, the numerator of this fraction and the common denominator of the shares are compared so as to find their least common multiple, which now becomes the estate to be divided between the standard heirs (*al-bāqī baʿd al-waṣī min mukhraji-hā*).

For example, if someone has two sons, each inherits 1 from an estate of 2. If $\frac{1}{3}$ is willed to a third party, then $1-\frac{1}{3}=\frac{2}{3}$. So the common denominator of the shares of the estate, 2, and the numerator of the remainder, 2, are equal, and the estate to be divided between the sons remains 2. The problem is solved if everyone takes 1 from an estate of 3.

As should be expected by now, the solutions and examples are divided according to the result of the test of primality. So after equality comes harmony: if someone has four sons and willed $\frac{1}{3}$ of his fortune, each son inherits 1 from an estate of 4 and the remainder is, again 2. Then, having $4 = 2 \times 2$ so we multiply the denominator of the will by 2 and we get that each son takes 1 and the recipient of the will 2 from an estate of 6.

Finally, if the quantities are coprime, as with three sons and a third of the fortune willed, we have to multiply the estate of the case by the denominator of the rest. So each son gets 2 and the beneficiary of the will 3 from an estate of 9.

Then, Sīdiyyā looks at what happens when multiple people benefit from wills (*taʿdīd al-wārith*) via the example of someone who put in his will that one sixth of his fortune would go to someone and one seventh to someone else. We are instructed to compare the primality of 6 and 7. They disagree, so we multiply them together to get 42. We subtract from it the denominators of the share promised: $42 - 7 - 6 = 29$. Sīdiyyā then distinguishes two cases: either this 29 is coprime with the estate or it is not.

Sīdiyyā and his predecessors follow Khalīl's method and examples without deviation. In this chapter where the *Mukhtaṣar* itself uses a term associated with fractions, denominator (*mukhraj*), and is closest to operating on them, they place themselves in the uncharacteristic position of taking cases that are as simple as possible: the computation of the estate is trivialized by making all heirs sons and by the will being, more often than not, a third.

No mention is made of an algebraic solution using the method developed by al-Khwārizmī, who gives two examples that require more calculations. The value of his algebraic method is that the same type of calculation solves any problem, whereas that of the *faraḍiyyīn* method changes depending on the results of the test of primality. This is manifest in the number of examples given: al-Khwārizmī gives one for a will to one person and one for a will to two persons, compared to the five given in the commentary of Sīdiyyā.

10.5 Conclusion

Two concurrent ways to conceptualize and operate with fractions coexist in the texts of the library of Sīdiyyā. We have seen how the methods of dealing with fractions differ between the science of inheritance shares and arithmetic: they conceptualize and write fractions differently. One ostensibly has to do with a practical activity of inheritance calculation while the second is linked to the use of numerals.

It is then notable that both traditions know, recognize and name each other in my sources and beyond. For example, al-Walātī, in his *Nuzhat al-albāb*, does make an explicit link between inheritance and the chapter on the allotment:

> And it is one of the most important chapters needed to divide (*al-ḥāja fī qisma*) the allotments of associates or heirs or debtors (*ghuramāʾ*) [25]

And even more specifically to *farāʾiḍ* when he introduces the table of multiplication (*ḍaraba li-āḥād baʿḍa-hā fī baʿḍ*):

> The tables of multiplication are necessary for you to master, O inheritance specialist [...]
> Meaning: it is imperative for you, O master of *farāʾiḍ*[26]

So far I have not found any data allowing me to comment on this insularity. In an excellent article, Laabib Ezzaim interviewed *farāʾiḍ* practitioners in present day Morrocco (Laabib 2018). The multiplicity of their answers regarding why they used the *taṣḥīḥ* and not calculation of fractions might indicate a lack of consensus among the authors. Another hypothesis I propose is that the *taṣḥīḥ* was meant to be as accessible as possible since practically speaking, somebody had to be responsible for calculating inheritance shares. In this sense, it is possible that the need for practitioners trumped a desire for mathematical exhaustivity.

References

Manuscripts

al-Intishāʾī, Sīdiyyā b. al-Mukhtār b. al-Hayba al-Abyayrī. *Sharḥ bāb al-tarika min mukhtaṣar Khalīl b. Isḥāq*. University of Urbana Champaign, 83/9. Fond Stewart. https://waamd.lib.berkeley.edu/titles/924.

Qalaṣādī, ʿAli b. Muḥammad. *Kashf al-asrār*. University of Urbana Champaign. 54/10-54/11. Fond Stewart. https://waamd.lib.berkeley.edu/titles/384.

[25] The original text reads:
وهو من أهم الأبواب لقسيم الحاجة اليها في قسمة محاصات الشركاء والورثة والغرماء.
("Nuzhat al-albāb", 11^{v}).

[26] The original text reads:
ضرب الاحاد بعضها في بعض لازم ان تتقن يا ذا الفرض [...] اي لازم لك يا صاحب الفرائض.
("Nuzhat al-albāb", 6^{v}).

al-Qāsim, Muḥammad al-ʿArabī b. Sayyid Muḥammad b. ʿAbd al-Raḥmān. *Rabbānī al-mubtadī wa ṣāʾigh al-ḥujja al-muntahī*. Tīšīt, Maktabat Ahl Aḥmad aṡ-Ṣaġīr. MF Mau 0438. OMAR. https://dl.ub.uni-freiburg.de/omar/mfmau0438.
al-Walātī, Anbūya. *Nuzhat al-albāb*. University of Urbana Champaign, 59/4-59/5. Fond Stewart. https://waamd.lib.berkeley.edu/titles/461.

Printed Sources

al-Jundī, Khalīl b. Isḥāq, Muḥammad b. Muḥammad al-Ḥaṭṭāb, and Muḥammad b. Yūsuf Mawwāq. 2007. *Mawāhib al-jalīl li-sharḥ Mukhtaṣar Khalīl, wa-bi-asfali-hi al-Tāj wa-l-iklīl li-Mukhtaṣar Khalīl*. Al-Ṭabʻa 2. 8 vols. Beirut: Dār al-Kutub al-ʻIlmiyya.
al-Khirshī, Muḥammad, and ʿAlī al-ʿAdawī. 1900. *al-Khirshī ʿalā mukhtaṣar Khalīl wa-bi-hāmishi-hi ḥāshiyat al-ʿAdawī*, vol. 8. al-Maṭbaʿa al-ʿamīriyya al-kubrā.
al-Qalaṣādī, Muḥammad b. ʿAlī. 1988a. *Kashf al-asrār ʿan ʿilm ḥurūf al-ghubār*, edited by Mohamed Souissi. Carthage: Beit al-hikma.
Abdeljaouad, Mahdi. 2006. Issues in the History of Mathematics Teaching in Arab Countries. *Paedagogica Historica* 42 (4–5): 629–664.
Berggren, John Len. 2016. *Episodes in the Mathematics of Medieval Islam*. 2nd ed. New York, NY: Springer.
Brigaglia, Andrea, and Mauro Nobili. 2017. *The Arts and Crafts of Literacy: Islamic Manuscript Cultures in Sub-Saharan Africa. Studies in Manuscript Cultures*. Vol. 12. Berlin Boston: Walter de Gruyter.
Chemla, Karine. 2022. Cultures of Computation and Quantification in the Ancient World: An Introduction. In *Cultures of Computation and Quantification in the Ancient World: Numbers, Measurements, and Operations in Documents from Mesopotamia, China and South Asia. Why the Sciences of the Ancient World Matter 6*, ed. Agathe Keller, Christine Proust, and Karine Chemla, 1–140. Cham, Switzerland: Springer.
Coulson, Noel J. 1971. *Succession in the Muslim family*. Cambridge, Royaume-Uni de Grande-Bretagne et d'Irlande du Nord: University Press.
Diallo, Yacine. 1934. L'enseignement de l'arabe au Fouta-Djalon. *Bulletin de l'enseignement de l'Afrique occidentale française* 87:153–161.
Djebbar, Ahmed. 1992. Le traitement des fractions dans la tradition mathématique arabe du Maghreb. In *Histoire de fractions, fractions d'histoire. Science Networks Historical Studies*, ed. Paul Benoit, Karine Chemla, and Jim Ritter, vol. 10, 224–246. Basel; Boston: Birkhäuser Verlag.
Djebbar, Ahmed, and Marc Moyon. 2011. *Les sciences arabes en Afrique: mathématiques et astronomie, IXe-XIXe siècles*. Manuscrits du désert. Brinon-sur-Sauldre: Grandvaux.
Fadel, Mohammad. 1996. The Social Logic of Taqlīd and the Rise of the Mukhataṣar. *Islamic Law and Society* 3 (2): 193–233.
Hall, Bruce. 2013. Saharan Commerce and Islamic Law: The Question of Usury (Ribā) in the Nawāzil Literature of Mali and Mauritania, 1700-1929. *African Economic History* 41:1–18.
Hiskett, Mervyn. 1966a. Materials Relating to the Cowry Currency of the Western Sudan I: A Late Nineteenth-Century Schedule of Inheritance from Kano. Bulletin of the School of Oriental and African Studies, University of London 29 (1): 122–142.
Hiskett, Mervyn. 1966b. Materials Relating to the Cowry Currency of the Western Sudan II: Reflections on the Provenance and Diffusion of the Cowry in the Sahara and the Sudan. Bulletin of the School of Oriental and African Studies, University of London 29 (2): 339–366.
Hunwick, John Owen. 1999. Islamic Financial Institutions: Theoretical Structures and Aspects of Their Application in Sub-Saharan Africa. In *Credit, Currencies, and Culture: African Financial Institutions in Historical Perspective*. Uppsala, Sweden: Nordiska Afrikainstitutet.

ibn Anas, Mālik. 2004. Al-Muwatta': synthèse pratique de l'enseignement islamique traité de jurisprudence islamique, rite malikite. Trans. Muḥammad Dyāḫū. Beyrouth Paris: Albouraq diff. Librairie de l'Orient.

Laabib, Ezzaim. 2006. *Les techniques mathématiques dans la résolution des problèmes de partages successoraux dans le Maghreb médiéval à travers le Mukhtaṣar d'al-Hūfī (m. 1192): sources et prolongements*." Thèse de doctorat d'histoire des mathématiques. Université de Rabat.

Laabib, Ezzaim. 2018. *Une pratique mathématique qui vit en marge des mathématiques scolaires: la résolution des problèmes des héritages par des professionnels au Maroc*. Gennevilliers.

Layish, Aharon. 1983. The Mālikī Family 'Waqf' According to Wills and 'Waqfiyyāt. *Bulletin of the School of Oriental and African Studies, University of London* 46 (1): 1–32.

Lydon, Ghislaine. 2012. *On Trans-Saharan Trails: Islamic Law, Trade Networks, and Cross-Cultural Exchange in Nineteenth-Century Western Africa*. 1 paperback ed. Cambridge: Cambridge Univ. Press.

Marcus-Sells, Ariela. 2022. *Sorcery or Science? Contesting Knowledge and Practice in West African Sufi Texts. The Magic in History Series*. University Park, Pennsylvania: The Pennsylvania State University Press.

Meriwether, Margaret Lee. 1999. *The Kin Who Count: Family and Society in Ottoman Aleppo, 1770-1840*. 1st ed. Austin, Tex: University of Texas Press.

Moyon, Marc. 2021. Mathématiques et Astronomie dans les «manuscrits du désert»: Première approche. In *Revisiter l'histoire des sciences, des savoirs, des techniques et des arts au Moyen Âge, Latrach Edition*, ed. Khaled Kchir, 159–182. Laboratoire du Monde Arabo-Islamique Médiéval.

Moyon, Marc. 2011. Practical Geometries in Islamic Countries: The Example of the Division of Plane Figures. In *History and Epistemology in Mathematics Education*, ed. Kronfellner Manfred, Barbin Évelyne, and Tzanakis Costa, 527–538. Verlag Holzhausen GmbH. https://hal.archives-ouvertes.fr/hal-00933041.

Moyon, Marc. 2022. Scientific Practices in Sub-Saharan Africa. In *Routledge Handbook on the Sciences in Islamicate Societies: Practices from the 2nd/8th to the 13th/19th Centuries*. Routledge & CRC Press.

Pettigrew, Erin. 2007. Colonizing the Mahadra: Language, Identity, and Power in Mauritania Under French Control. *Ufahamu: A Journal of African Studies* 33 (2–3): 62–89.

Powers, David S. 1993. The Islamic Inheritance System: A Socio-Historical Approach. *Arab Law Quarterly* 8 (1): 13–29.

al-Qalaṣādī, Muḥammad b. ʿ Alī. 1988b. *Kashf al al-asrār ʿan ʿilm ḥurūf al-ghubār*. Trans. Mohamed Souissi. Carthage: Beit al-hikma.

Rashed, Roshdi. 2007. *Al-Khwārizmī: le commencement de l'algèbre. Collection Sciences dans l'histoire*. Paris: A. Blanchard.

Rebstock, Ulrich. 1990. Arabic Mathematical Manuscripts in Mauretania. *Bulletin of the School of Oriental and African Studies* 53 (03): 429–441.

Rebstock, Ulrich. 2009. Arithmetik (Ḥisāb) Und Erbteilungslehre (ʿIlm al-Farāʾiḍ): Symbiose Einer Islamischen Wissenschaftsdisziplin. *Geschichte Der Arabisch-Islamischen Wissenschaften* 18:269–295.

Rebstock, Ulrich. 2007. 'Homo Ludens' at Work Mauritanians' Skills in Determining Inheritance Shares. In *Studies in Islamic Law: A Festschrift for Colin Imber*, 233–243. Oxford: Oxford University Press.

Souissi, Mohamed. 1997. Ḳalaṣādī. In *The Encyclopaedia of Islam*, ed. Emeri J. van Donzel, vol. 4, 3rd ed., 476–477. Leiden: Brill.

Souissi, Mohamed. 1993. Les mathématiques appliquées aux partages successoraux en droit musulman. *Les Cahiers de Tunisie* 162–163:23–32.

Stewart, Charles C. 1973. *Islam and Social Order in Mauritania: A Case Study from the Nineteenth Century*. Oxford: Oxford University Press.
Stewart, Charles C. 1991. The Haroun Ould Sidia Collection of Arabic Manuscripts. *History in Africa* 18:349–358.
Stewart, Charles C., and Bruce S. Hall. 2011. The Historic 'Core Curriculum' and the Book Market in Islamic West Africa. In *The Trans-Saharan Book Trade: Manuscript Culture, Arabic Literacy, and Intellectual History in Muslim Africa. Library of the Written Word, The Manuscript World*, ed. Graziano Krätli and Ghislaine Lydon, 109–174. Leiden; Boston: Brill.
al-Uqlīdisī, Aḥmad ibn Ibrāhīm. 1978. *The Arithmetic of Al-Uqlīdisī: The Story of Hindu-Arabic Arithmetic as Told in Kitāb al-Fuṣūl Fī al-Ḥisāb al Hindī, by Abū al-Ḥasan, Aḥmed Ibn Ibrāhīm al-Uqlīdisī, Written in Damascus in the Year 341 (A.D. 952/3)*. Trans. Ahmad Salim Saidan. Dordrecht, Holland; Boston: D. Reidel Pub. Co.
Warscheid, Ismail. 2017. *Droit musulman et societe au Sahara premoderne: la justice islamique dans les oasis du Grand Touat (Algerie) aux XVIIe-XIXe siecles, Studies in the history and society of the maghrib*. Vol. 9. Leiden; Boston, Brill.
Warscheid, Ismail. 2018. Le Livre du désert. *Annales. Histoire, Sciences Sociales* 73 (2): 359–384.
Webb, Abu-Isa, Aaron Wannamaker, and Hisham Sharif, eds. 2016. *The Clear Quran: A Thematic English Translation of the Message of the Final Revelation*. Trans. Mustafa Khattab. Lombard, IL: Book of Signs Foundation.

GPSR Compliance

The European Union's (EU) General Product Safety Regulation (GPSR) is a set of rules that requires consumer products to be safe and our obligations to ensure this.

If you have any concerns about our products, you can contact us on ProductSafety@springernature.com

In case Publisher is established outside the EU, the EU authorized representative is:

Springer Nature Customer Service Center GmbH
Europaplatz 3
69115 Heidelberg, Germany

Batch number: 10683289

Printed by Printforce, the Netherlands